Advances in Xerogels: From Design to Applications

Advances in Xerogels: From Design to Applications

Editors

Francesco Caridi
Giuseppe Paladini
Andrea Fiorati

MDPI • Basel • Beijing • Wuhan • Barcelona • Belgrade • Manchester • Tokyo • Cluj • Tianjin

Editors

Francesco Caridi
Università degli Studi
di Messina
Messina
Italy

Giuseppe Paladini
University of Catania
Catania
Italy

Andrea Fiorati
Politecnico di Milano
Milan
Italy

Editorial Office
MDPI
St. Alban-Anlage 66
4052 Basel, Switzerland

This is a reprint of articles from the Special Issue published online in the open access journal *Gels* (ISSN 2310-2861) (available at: https://www.mdpi.com/journal/gels/special_issues/advances_xerogels).

For citation purposes, cite each article independently as indicated on the article page online and as indicated below:

LastName, A.A.; LastName, B.B.; LastName, C.C. Article Title. *Journal Name* **Year**, *Volume Number*, Page Range.

ISBN 978-3-0365-8162-0 (Hbk)
ISBN 978-3-0365-8163-7 (PDF)

Contents

About the Editors

Francesco Caridi

Francesco Caridi is a Researcher in Applied Physics (to Cultural Heritage, Environment, Biology and Medicine) at the Department of Mathematical and Computer Sciences, Physical and Earth Sciences of the University of Messina. He graduated in Physics cum laude in 2003 at the University of Messina, and received his PhD in Physics in 2007 from the same University. He obtained National Scientific Qualifications as Full Professor of Experimental Physics in 2020 and as Full Professor of Applied Physics in 2022. He is co-author of 148 publications in ISI-rated journals and more than 100 invited talks/communications at national and international conferences and events in the field of physics applied to cultural and environmental heritage. Citations: 1405, H-index: 27 (Scopus database). His research interests include, but are not limited to, a) application of invasive and non-invasive micro- and spectroscopic techniques for the analysis of materials largely employed in the field of cultural heritage, as well as their degradation forms; b) characterization of materials of particular historical artistic interest in terms of natural radioactivity content, in order to assess the radiological risk for humans; c) systematic implementation of operative strategies aimed at the evaluation of the radon exhalation rate in building materials; d) gamma spectrometry, alpha spectrometry, liquid scintillation counting (LSC), total alpha/beta counting, emanometry, gas-filled detectors.

Giuseppe Paladini

Giuseppe Paladini is a Researcher in Applied Physics (to Cultural Heritage, Environment, Biology and Medicine) at the Department of Physics and Astronomy "Ettore Majorana" of the University of Catania, in the context of the "SiciliAn MicronanOTecH Research And Innovation CEnter—SAMOTHRACE" innovation ecosystem, funded by the European Union under the National Recovery and Resilience Plan (PNRR). He graduated in Physics cum laude in 2016 from the University of Messina, obtained his PhD in Physics in 2019, and his National Scientific Qualification as Associate Professor of Applied Physics in 2023. He is co-author of 55 publications in ISI-rated journals and 35 invited talks/communications at national and international conferences and events. Citations: 268, H-index: 10 (Scopus database). His research interests include, but are not limited to, a) dynamical and structural properties of supramolecular systems for environmental and pharmaceutical applications; b) application of neutron and light spectroscopic techniques at nano-, micro- and macro-scale; c) characterization of environmental matrices in terms of specific activity of natural and anthropogenic radionuclides.

Andrea Fiorati

Andrea Fiorati is a Researcher in Materials Science and Technology at the Department of Chemistry, Materials and Chemical Engineering "Giulio Natta" of the Politecnico di Milano. He Graduated in Medicinal Chemistry and Pharmaceutical Technology in 2019 from the University of Milan, and received a PhD in Industrial Chemistry and Chemical Engineering in 2016. He is a co-author of 37 publications in ISI-rated journals, four book chapters, and 23 invited talks/ communications at National and International Conferences and Events. Citations: 573, H-index: 15 (Scopus database). His research interests include, but are not limited to, a) Chemical–physical approaches for materials characterization; b) nanocellulose-based materials: from environmental to biomedical applications; c) leveraging nano-dimensioned particles to improve the performance of polymeric materials; and d) optimization of mass transfer phenomena through polymeric materials for food packaging.

Preface to "Advances in Xerogels: From Design to Applications"

The field of material science and engineering has witnessed a surge of interest in the development of novel xerogels, solid materials derived from gels, for a wide range of applications. Despite potential shrinkage, xerogels possess remarkable properties, including high porosity and a large surface area, owing to the interconnected network of pores formed during drying. Consequently, xerogels find valuable applications as catalysts, adsorbents, sensors, membranes, and drug delivery systems.

The ability to conveniently control the structures and morphologies of xerogels during synthesis has garnered significant attention within the scientific community. In light of these considerations, this Special Issue of Gels compiles high-quality papers that showcase the latest advancements and breakthroughs in xerogel science, covering various aspects from design to application. Both theoretical and experimental cutting-edge studies exploring the structural, chemical, rheological, and dynamical properties of xerogels and their derivatives are included in this collection.

Francesco Caridi, Giuseppe Paladini, and Andrea Fiorati
Editors

gels

MDPI

Editorial

Editorial on the Special Issue: "Advances in Xerogels: From Design to Applications"

Andrea Fiorati [1,*], Francesco Caridi [2] and Giuseppe Paladini [3]

1 Department of Chemistry, Materials, and Chemical Engineering "G. Natta" INSTM Local Unit, Politecnico di Milano, 20131 Milan, Italy
2 Department of Mathematical and Computer Sciences, Physical Sciences and Earth Sciences, University of Messina, 98166 Messina, Italy; fcaridi@unime.it
3 Department of Physics and Astronomy "Ettore Majorana", University of Catania, 95123 Catania, Italy; giuseppe.paladini@dfa.unict.it
* Correspondence: andrea.fiorati@polimi.it; Tel.: +39-2399-3173

Xerogels are solid materials derived from gels which consist of interconnected particles or polymers dispersed in a liquid. This gel structure is subjected to a drying process (e.g., slow evaporation or freeze-drying), resulting in the removal of the liquid phase, leaving behind a solid material. During this drying process, the liquid is extracted from the gel while attempting to preserve its original shape and structure as much as possible. Although some shrinkage may occur, xerogels exhibit unique properties such as high porosity and a large surface area due to the interconnected network of pores formed during drying. These characteristics make xerogels valuable in a wide range of applications, including catalysts, adsorbents, sensors, membranes, and drug delivery systems. The field of materials science and engineering has seen a surge of interest in the development of innovative xerogels, leading to a focus on their design for various applications. With the ability to conveniently control their structures and morphologies during synthesis, xerogels have become a subject of great significance within the scientific community.

Taking these factors into account, the objective of this *Gels* Special Issue is to assemble high-quality papers that showcase recent progress and discoveries in xerogel science, encompassing design and practical utilization. We sought original contributions exploring traditional and non-traditional approaches to synthesizing and characterizing xerogels. The aim is to deepen our understanding of the fundamental and applied aspects of various organic and inorganic xerogel-like materials.

The recent advancements concerning the role of biopolymer-based xerogels in biomedical applications were thoroughly discussed by Khalil and colleagues. Through their systematic literature review, the authors provide insights into the biological properties of xerogels that make them suitable materials for various biomedical applications, including drug delivery, wound healing and dressing, tissue scaffolding, and biosensing [1].

Furthermore, xerogels are widely utilized for the adsorption of ions and metals. Putz et al. [2] reported the synthesis of ordered mesoporous silica materials capable of acting as sorbents for environmental remediation, specifically targeting Cu(II) and Pb(II). Xerogels' metal adsorption properties can be effectively harnessed to create efficient heterogeneous catalysts. For instance, Riva et al. [3] successfully catalyzed the Suzuki–Miyaura coupling reaction using Pd-loaded cellulose nanosponge as a heterogeneous catalyst. Additionally, Mahy et al. [4] described the preparation of mono- and bimetallic catalysts based on Fe, Ni, and Pd supported on silica through a sol–gel co-gelation process, resulting in porous materials with surface areas ranging from 100 to 400 m^2/g.

Although silica xerogels are widely recognized as the most commonly used type, it is known that metal oxides such as Al_2O_3, SiO_2, TiO_2, and CeO_2, which possess large specific surfaces and thermal stability, can serve as catalyst supports. Within our special issue, we feature the research of Romanczuk-Ruszuk and colleagues, who synthesized

Citation: Fiorati, A.; Caridi, F.; Paladini, G. Editorial on the Special Issue: "Advances in Xerogels: From Design to Applications". *Gels* **2023**, *9*, 446. https://doi.org/10.3390/gels9060446

Received: 25 May 2023
Accepted: 26 May 2023
Published: 27 May 2023

binary xerogel systems of Sr/Al using the sol–gel technique. These systems incorporated a metallic strontium precursor and were evaluated as supports for platinum catalysts. The researchers successfully achieved highly dispersed and stable strontium carbonate phases, leading to a remarkable dispersion (42–50%) of platinum nanoparticles [5].

The versatility of sol–gel approaches can be exploited to trap ions (as reported before) or molecules and macromolecules such as proteins to confer the final materials' specific properties. For the sake of an example, Morosanova and colleagues reported the synthesis of a promising biosensor development by incorporating enzymes (horseradish peroxidase and mushroom tyrosinase) and crude banana extract in a silane-based structure to create optical biosensors for hydrogen peroxide in the range of 0.2–3.5 mM [6].

The classic sol–gel process which involves hydrolysis reactions of (semi)metal alkoxides is a well-established method for creating organic–inorganic hybrid materials. However, alternative approaches utilizing non-aqueous or non-hydrolytic systems have also been investigated. In this context, Krupinski and colleagues presented a novel approach utilizing element chlorides and silylated precursors to produce "bridged" or "linearly expanded" silicas and silsesquioxanes through a non-hydrolytic sol–gel method. This study demonstrates that the non-hydrolytic approach to hybrid materials can be extended to other silylated precursors, as long as the reactivity of the corresponding chlorine compound is adequate [7].

The responsibility of porous materials toward external stimuli is crucial for the obtainment of smart innovative materials, and the chemistry of N-oxide moieties offers a wide range of possibilities. In this context, Trofimov and co-workers report the synthesis of 4-dialkylamino-2,5-dihydroimidazol-1-oxyls with moieties at position 2 and at the exocyclic nitrogen able to act as pH-Sensitive spin labels [8], while Jayabhavan and colleagues study the stimuli-responsivity supramolecular gels based on pyridyl-N-oxide amides [9].

Due to their versatility, xerogels and porous materials are becoming of great interest to scientists who focus their attention on food science and food packaging. Dried porous materials based on plant proteins have gained significant attention as potential sustainable food ingredients. However, plant proteins exhibit weaker gelling properties compared to animal proteins. To enhance plant protein gelling, optimization of gelation conditions involving protein concentration, pH, and ionic strength is necessary. De Berardinis and colleagues conducted a systematic study to investigate the impact of these factors on the gelation behavior of soy and pea protein isolates. The findings were used to create a map that identifies the gelation conditions for modulating the rheological properties of soy and pea protein hydrogels, with potential applications in the production of xerogels, cryogels, and aerogels [10]. Innovation in intelligent food packaging materials holds promise for enhancing food safety, quality, and control. Researchers are exploring the combination of biodegradable semi-synthetic polymers with natural polymers and additives to improve material functionality. In line with this, Vlad-Bubulac and colleagues developed composite films by casting a solution containing specific mass ratios of poly(vinyl alcohol) and chitosan as the polymeric matrix, supplemented with TiO_2 nanoparticles and a polyphosphonate as reinforcing additives. The favorable outcomes regarding precursor homogeneity, film quality, antimicrobial activity, and cytocompatibility demonstrate the potential suitability of these films for food packaging applications [11].

Xerogels offer unique properties for diverse applications, including catalysts, adsorbents, sensors, and drug delivery systems. Biopolymer-based xerogels show promise in biomedical applications, while metal oxides expand catalyst support options. Non-hydrolytic sol–gel approaches and additives enable hybrid materials. Intelligent food packaging and improved plant protein gelling are active areas of research. Composite films combining biodegradable polymers, natural polymers, and additives hold potential for food packaging. Exciting advancements in xerogel science drive innovation stimulating the interest of the scientific community.

Conflicts of Interest: The authors declare no conflict of interest.

References

1. Abdul Khalil, H.P.S.; Yahya, E.B.; Tajarudin, H.A.; Balakrishnan, V.; Nasution, H. Insights into the Role of Biopolymer-Based Xerogels in Biomedical Applications. *Gels* **2022**, *8*, 334. [CrossRef] [PubMed]
2. Putz, A.-M.; Ivankov, O.I.; Kuklin, A.I.; Ryukhtin, V.; Ianăşi, C.; Ciopec, M.; Negrea, A.; Trif, L.; Horváth, Z.E.; Almásy, L. Ordered Mesoporous Silica Prepared in Different Solvent Conditions: Application for Cu(II) and Pb(II) Adsorption. *Gels* **2022**, *8*, 443. [CrossRef] [PubMed]
3. Riva, L.; Nicastro, G.; Liu, M.; Battocchio, C.; Punta, C.; Sacchetti, A. Pd-Loaded Cellulose NanoSponge as a Heterogeneous Catalyst for Suzuki–Miyaura Coupling Reactions. *Gels* **2022**, *8*, 789. [CrossRef] [PubMed]
4. Mahy, J.G.; Delbeuck, T.; Tran, K.Y.; Heinrichs, B.; Lambert, S.D. Green Chemistry for the Transformation of Chlorinated Wastes: Catalytic Hydrodechlorination on Pd-Ni and Pd-Fe Bimetallic Catalysts Supported on SiO_2. *Gels* **2023**, *9*, 275. [CrossRef] [PubMed]
5. Romanczuk-Ruszuk, E.; Sztorch, B.; Oksiuta, Z.; Przekop, R.E. Metallic Strontium as a Precursor of the $Al_2O_3/SrCO_3$ Xerogels Obtained by the One-Pot Sol–Gel Method. *Gels* **2022**, *8*, 473. [CrossRef] [PubMed]
6. Morosanova, M.A.; Morosanova, E.I. Sol-Gel Films Doped with Enzymes and Banana Crude Extract as Sensing Materials for Spectrophotometric Determination. *Gels* **2023**, *9*, 240. [CrossRef] [PubMed]
7. Krupinski, K.; Wagler, J.; Brendler, E.; Kroke, E. A Non-Hydrolytic Sol–Gel Route to Organic-Inorganic Hybrid Polymers: Linearly Expanded Silica and Silsesquioxanes. *Gels* **2023**, *9*, 291. [CrossRef] [PubMed]
8. Trofimov, D.G.; Glazachev, Y.I.; Gorodetsky, A.A.; Komarov, D.A.; Rybalova, T.V.; Kirilyuk, I.A. 4-Dialkylamino-2,5-Dihydroimidazol-1-Oxyls with Functional Groups at the Position 2 and at the Exocyclic Nitrogen: The PH-Sensitive Spin Labels. *Gels* **2022**, *8*, 11. [CrossRef] [PubMed]
9. Jayabhavan, S.S.; Kristinsson, B.; Ghosh, D.; Breton, C.; Damodaran, K.K. Stimuli-Responsive Properties of Supramolecular Gels Based on Pyridyl-N-Oxide Amides. *Gels* **2023**, *9*, 89. [CrossRef] [PubMed]
10. De Berardinis, L.; Plazzotta, S.; Manzocco, L. Optimising Soy and Pea Protein Gelation to Obtain Hydrogels Intended as Precursors of Food-Grade Dried Porous Materials. *Gels* **2023**, *9*, 62. [CrossRef] [PubMed]
11. Vlad-Bubulac, T.; Hamciuc, C.; Rîmbu, C.M.; Aflori, M.; Butnaru, M.; Enache, A.A.; Serbezeanu, D. Fabrication of Poly(Vinyl Alcohol)/Chitosan Composite Films Strengthened with Titanium Dioxide and Polyphosphonate Additives for Packaging Applications. *Gels* **2022**, *8*, 474. [CrossRef] [PubMed]

gels

MDPI

Article

A Non-Hydrolytic Sol–Gel Route to Organic-Inorganic Hybrid Polymers: Linearly Expanded Silica and Silsesquioxanes

Katrin Krupinski [1], Jörg Wagler [1,2], Erica Brendler [3] and Edwin Kroke [1,2,*]

[1] Institute of Inorganic Chemistry, Department of Chemistry and Physics, Technische Universität Bergakademie Freiberg (TUBAF), Leipziger Strasse 29, 09596 Freiberg, Saxony, Germany; katrinlippe@gmx.net (K.K.); joerg.wagler@chemie.tu-freiberg.de (J.W.)

[2] Center of Efficient High Temperature Processes and Material Conversion (ZeHS), Technische Universität Bergakademie Freiberg (TUBAF), Winklerstr. 5, 09599 Freiberg, Saxony, Germany

[3] Institute of Analytical Chemistry, Department of Chemistry and Physics, Technische Universität Bergakademie Freiberg (TUBAF), Leipziger Strasse 29, 09596 Freiberg, Saxony, Germany; erica.brendler@chemie.tu-freiberg.de

* Correspondence: edwin.kroke@chemie.tu-freiberg.de; Tel.: +49-3731-39-3174; Fax: +49-3731-39-4058

Abstract: Condensation reactions of chlorosilanes ($SiCl_4$ and CH_3SiCl_3) and bis(trimethylsilyl)ethers of rigid, quasi-linear diols $(CH_3)_3SiO$–*AR*–$OSi(CH_3)_3$ (*AR* = 4,4′-biphenylene (**1**) and 2,6-naphthylene (**2**)), with release of $(CH_3)_3SiCl$ as a volatile byproduct, afforded novel hybrid materials that feature Si–O–C bridges. The precursors **1** and **2** were characterized using FTIR and multinuclear (^1H, ^{13}C, ^{29}Si) NMR spectroscopy as well as single-crystal X-ray diffraction analysis in case of **2**. Pyridine-catalyzed and non-catalyzed transformations were performed in THF at room temperature and at 60 °C. In most cases, soluble oligomers were obtained. The progress of these transsilylations was monitored in solution with ^{29}Si NMR spectroscopy. Pyridine-catalyzed reactions with CH_3SiCl_3 proceeded until complete substitution of all chlorine atoms; however, no gelation or precipitation was found. In case of pyridine-catalyzed reactions of **1** and **2** with $SiCl_4$, a Sol–Gel transition was observed. Ageing and syneresis yielded xerogels **1A** and **2A**, which exhibited large linear shrinkage of 57–59% and consequently low BET surface area of 10 $m^2 \cdot g^{-1}$. The xerogels were analyzed using powder-XRD, solid state ^{29}Si NMR and FTIR spectroscopy, SEM/EDX, elemental analysis, and thermal gravimetric analysis. The $SiCl_4$-derived amorphous xerogels consist of hydrolytically sensitive three-dimensional networks of SiO_4-units linked by the arylene groups. The non-hydrolytic approach to hybrid materials may be applied to other silylated precursors, if the reactivity of the corresponding chlorine compound is sufficient.

Keywords: inorganic/organic networks; non-aqueous gels; arylene bridged; chlorosilanes; pyridine; ^{29}Si solid state NMR

Citation: Krupinski, K.; Wagler, J.; Brendler, E.; Kroke, E. A Non-Hydrolytic Sol–Gel Route to Organic-Inorganic Hybrid Polymers: Linearly Expanded Silica and Silsesquioxanes. *Gels* **2023**, *9*, 291. https://doi.org/10.3390/gels9040291

Academic Editors: Francesco Caridi, Giuseppe Paladini and Andrea Fiorati

Received: 10 March 2023
Revised: 27 March 2023
Accepted: 30 March 2023
Published: 2 April 2023

1. Introduction

Organic-inorganic hybrid materials usually have properties intermediate between organic polymers and inorganic substances, such as oxide glasses or ceramics [1–3]. A well-known approach to organic-inorganic hybrid materials uses the classic Sol–Gel process based on hydrolysis reactions of (semi)metal alkoxides [4–7]. Non-aqueous or non-hydrolytic [8–11] and non-oxide [12,13] Sol–Gel systems have also been reported. The so-called non-hydrolytic routes frequently refer to oxidic products, which may be used for similar applications as oxide products derived from hydrolytic routes; these include catalyst supports [14], optical materials [15], or aerogels [16–18], among many others [4–9].

Non-hydrolytic Sol–Gel systems are frequently based on metal halides and their reactions with oxygen containing organic or organometallic compounds such as alkoxides or acetoxides. They yield polymeric networks that are comprised of M–O–M′ backbone units. Thus, these non-aqueous systems are basically alternatives to the classic aqueous

Sol–Gel routes as they afford oxides or metal oxide based organic-inorganic hybrid materials. The M–O–M′ bridges are generated via a nucleophilic attack of the oxygen atoms at the metal atom of the halide. Therefore, the common principle of most non-hydrolytic Sol–Gel processes relies on a scission of carbon-oxygen-bonds instead of hydrolysis involving a nucleophilic attack of OH-groups (Scheme 1a). The (semi)metals M and M′ may be identical. For example, the reaction of $SiCl_4$ with tetra-*n*-butoxysilane (Scheme 1b) has been used to generate silica gels upon formation of the liquid side product *n*-butylchloride [19]. The underlying reaction may be considered as an ether cleavage. This route can also be applied for other alkyl groups and may be simplified by using alcohols like methanol, ethanol, or isopropanol in order to form the alkoxide in situ via reactions with an element halide. In analogy to silica gels [20], other single component oxide gels (M = M′), such as alumina [21], titania [22] gels, or tungsten oxide [23], were prepared. Different mixed oxide ceramics such as TiO_2/SiO_2 [24], TiO_2/Al_2O_3 [25], SiO_2/Al_2O_3, or SiO_2/ZrO_2 [24,26] were synthesized using the same approach. This route has also been applied to prepare multinary solids such as YAG (yttrium aluminum garnet) powders [27], cordierites [28], or β-SiAlON:Eu powders [29]. In many cases, precipitation of the products, rather than Sol–Gel transition and monolith formation, has been reported. A number of related approaches, such as nonhydrolytic solvothermal synthesis of oxide nanoparticles under anhydrous conditions, e.g., [30], have been reported in recent years. In addition, numerous reports on the controlled aggregation and gelation of (non-hydrolytically) formed oxide colloids, i.e., nano-particle dispersions or sols, have been published [31].

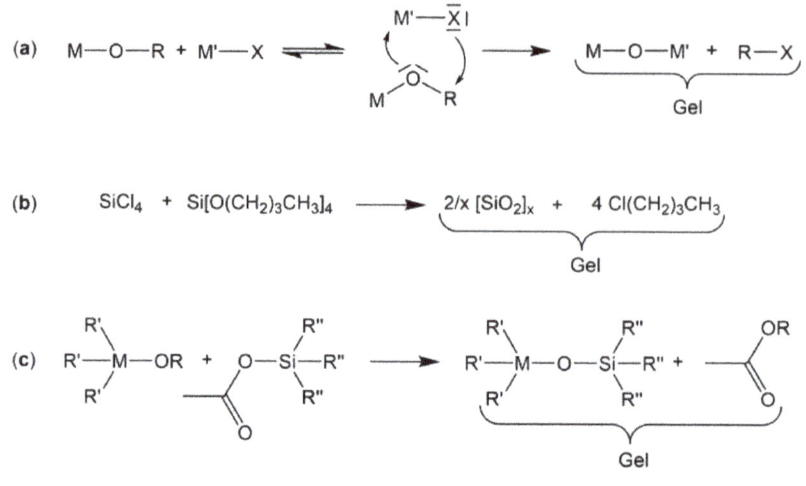

Scheme 1. Examples of non-hydrolytic oxide-based Sol–Gel strategies: (**a**) Generic reaction of metal alkoxide and metal halide moiety; (**b**) Reaction of silicon tetrachloride and tetra-*n*-butoxysilane; (**c**) Generic reaction of a metal alkoxide and an acetoxysilane.

Pronounced Lewis acidic metal halides, such as $TiCl_4$, react with organic ethers, e.g., diisopropyl ether or THF (tetrahydrofuran), to give mesoporous TiO_2-carbon nanocomposites [32]. In this process, which involves no additional solvent, the ether acts as both an oxygen donor and source of oxide and as the sole carbon source.

Acetoxysilanes and metal alkoxides (the latter in analogy to metal halides) react with formations of carboxylic acid esters and of non-aqueous oxide gels (Scheme 1c). The alkoxide may be a compound of the main group elements (silicon [33] and aluminum [21,33]) or of a transition metal (zirconium [26,33] or titanium [33,34]).

Further non-hydrolytic Sol–Gel systems are based on reactive organometallic compounds or alkoxides and further organic oxygen-containing compounds, e.g., acetone and zinc alkoxides react with formation of gels under enolization of the ketone and elimination of the corresponding alcohol [35]. This reaction behavior is observed only for non-acidic alkoxides.

Only few non-oxide Sol–Gel systems are known, i.e., Sol–Gel routes to sulfides, carbides, or nitrides, which lead to monolithic xerogels [12,13]. Some are related to the synthesis described here. In analogy to the hydrolysis and condensation reactions of the classic oxide Sol–Gel systems, compounds of the heavier chalcogenides may be considered as precursors [12,36].

Many papers on Sol–Gel synthesis routes to sulfides were published, but only a few Sol–Gel transitions were reported. For ZnS it was described that treatment of Et_2Zn or $[Zn(SR)_2]_x$ with H_2S generates precipitates, while reactions of dibenzyl trisulfide $(BnS)_2S$ and Et_2Zn in pentane (Scheme 2a) yielded transparent sols, gels, and xerogels after removal of the solvent [37]. The successful preparation of colloidal CdS and CdSe [38] gels, as well as CdSe sols obtained from $Cd(ethoxyacetate)_2$ and $Se(Si(CH_3)_3)_2$ [39], represents another non-oxide Sol–Gel strategy. CdS/CdSe aerogels can also be obtained from nanocrstyalline starting materials [40]. In another example, $Ge(OEt)_4$ was treated with H_2S to form sulfide gels [41]. Interesting nanostructured sulfide materials have been obtained from molybdenum chloride and $S(Si(CH_3)_3)_2$ in chloroform [42]. A similar approach, that is also based on the formation of trimethylchlorosilane, has been used to generate polydimethylsiloxane (PDMS) analogous silicon-sulfur polymers [43].

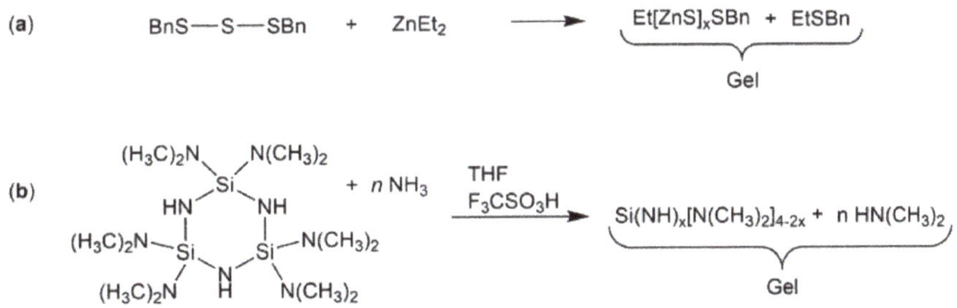

Scheme 2. Examples of non-hydrolytic non-oxide-based Sol–Gel strategies: (**a**) Reaction of dibenzyltrisulfide and diethylzinc; (**b**) Reaction of a dimethylamino substituted cyclotrisilazane and ammonia.

Even more work has been published on the synthesis of nitride materials via Sol–Gel routes. It may be assumed that simple replacement of water by ammonia or amines leads to nitride gels. However, it turns out that, in most cases, this approach is not successful, probably due to the much higher basicity of NH_3 and RNH_2 when compared to H_2O. The first example of a Sol–Gel process based on an ammonolysis reaction yielding Si/(C)/N gels starts from a dialkylamino substituted cyclic trisilazane (Scheme 2b); it can be described as a transamination reaction [44]. These so-called silicon diimide gels have subsequently been used for stationary phases in thin layer chromatography for organic acids [45].

Hexagonal boron nitride (h-BN) and ternary B/C/N ceramics have been synthesized using "silazanolysis" reactions of B-trichloroborazene and its alkyl substituted derivatives with hexa- and heptamethyl disilazane [46]. A three-dimensional network infiltrated with liquid $(CH_3)_3SiCl$ is formed that is analogous to the oxide polymer networks infiltrated by solvents and liquid reaction products. The dried xerogels showed low specific surface areas ($<35\ m^2{\cdot}g^{-1}$) when compared to oxide xerogels. Pyrolysis at 1200 °C produced hexagonal boron nitride.

Another approach to synthesize gel precursors for carbide and nitride materials is based on the pseudo-chalcogen concept [47]. In principle, the oxygen atoms of polysiloxanes are replaced by carbodiimide (–N=C=N–) groups in order to obtain (poly)silylcarbodiimide compounds [48]. Reactions of bis(trimethylsilyl)carbodiimide $(CH_3)_3Si–NCN–Si(CH_3)_3$ and dichlorosilanes (such as $(CH_3)_2SiCl_2$) give linear and cyclic oligomers [49], while reactions of this carbodiimide with trichlorosilanes ($RSiCl_3$ with R = alkyl, aryl or H) or with tetrachlorosilane may result in the formation of transparent gels (Scheme 3a) [50,51]. Several attempts to extend this Sol–Gel system through replacing the chlorosilanes with other element halides were not successful. For example, analogous reactions with titanium chlorides, such as $TiCl_4$ or $CpTiCl_3$, led to the formation of precipitates; the use of $RAlCl_2$ and gallium chlorides as starting materials gave soluble oligomers [52]. However, B-trichloroborazene ($B_3N_3H_3Cl_3$) reacts similarly to the chlorosilanes with bis(trimethylsilyl)carbodiimide and forms B/C/N gels (Scheme 3b) that can be transformed into B/C/N and B_4C ceramics [53].

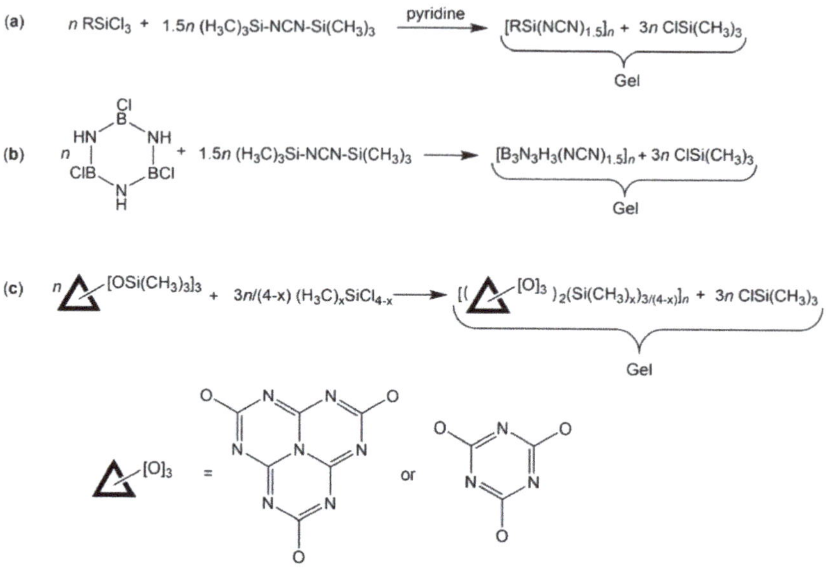

Scheme 3. Examples of non-hydrolytic Sol–Gel strategies with release of $(CH_3)_3SiCl$ as condensation product: (**a**) Generic reaction of trichlorosilanes and bis(trimethylsilyl)carbodiimide (R =, e.g., alkyl, aryl); (**b**) Reaction of B-trichloroborazene and bis(trimethylsilyl)carbodiimide; (c) Generic reaction of methylchlorosilanes and three-fold silylated cyameluric or cyanuric acid.

We modified the above-described Sol–Gel approach based on $(CH_3)_3SiCl$ formation by replacing bis(trimethylsilyl)carbodiimide with trimethylsilyl esters of cyanuric and cyameluric acid (Scheme 3c). Their reactions with $SiCl_4$ and CH_3SiCl_3 yielded networks consisting of three-fold bridged SiO_4 and CH_3SiO_3 units [54]. Similar three-fold bridged networks were formed upon acid catalyzed reactions of trihydroxybenzene and tetraethoxysilane (TEOS) [55]. The same approach was used with TEOS and dihydroxybenzenes as bifunctional precursors [56].

Here we present the extension of the concept of reaction of element chlorides and silylated precursors to obtain “bridged” or “linearly expanded” silicas and silsesquioxanes. Reactions of rigid, quasi-linear bis(trimethylsilyl)ethers of diols of the type $(CH_3)_3SiO$–*AR*–$OSi(CH_3)_3$ (*AR* = quasi-linear arylene spacer) and chlorosilanes ($SiCl_4$ and CH_3SiCl_3) afforded novel hybrid oligomers, polymers, and xerogels; the latter were investigated with X-ray diffraction (XRD), ^{29}Si nuclear magnetic resonance (NMR), and Fourier-transform infrared(FTIR) spectroscopy, as well as with scanning electron microcopy coupled with energy-dispersive X-ray spectroscopy (SEM/EDX), elemental analysis, thermal gravimetry with differential thermal analysis (TG/DTA), and TG-FTIR.

2. Results and Discussion

2.1. Selection, Synthesis and Characterisation of the Starting Materials

As mentioned in the introduction, the goal of the present work was to synthesize “linearly expanded silica and silsesquioxanes”. The latter term defines hybrid materials that differ from the well-known “organically bridged” silsesquioxanes $O_{1.5}Si$–[linker]–

$SiO_{1.5}$ [57–60] (Figure 1a), which feature various organic groups as linkers, and related systems such as polysilsesquioxanes $[R\text{–}SiO_{1.5}]_n$, with R being a terminal group (such as hydrocarbyl) [61,62]. These materials are based on the replacement of Si–O bonds by Si–C bonds. The present approach retains original Si–O motifs but alters the –O– bridge into an extended –O–[linker]–O– quasi-linear bidentate connector. Thus, these systems may be compared to metal organic frameworks (MOFs) [63], which may be considered "expanded" coordination centers (such as metal oxide centers) bridged by rigid bidentate ligands (such as terephthalate $[O_2C\text{–}C_6H_4\text{–}CO_2]^{2-}$) symbolized as L_yM–L″–[linker]–L″–ML_y (Figure 1b). This route to usually crystalline network structures has been extended to covalent organic frameworks (COFs) [64].

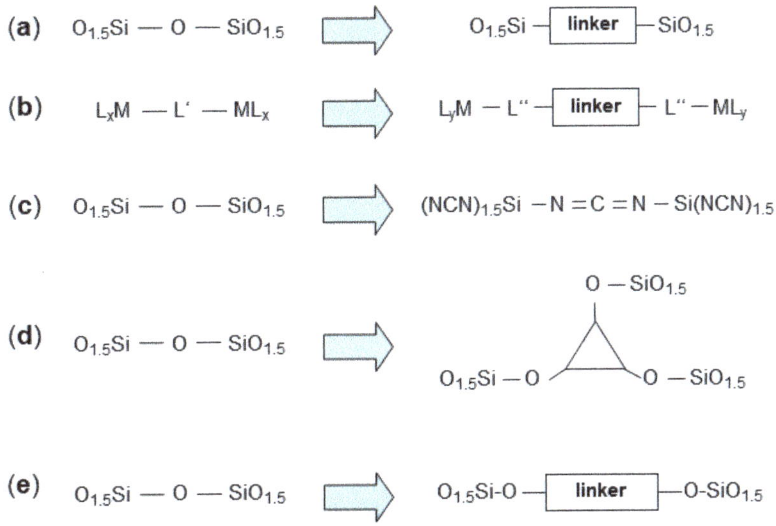

Figure 1. Approaches to "expanded oxides": (**a**) The most frequently prepared hybrid materials which may be described as organically bridged silsesquioxanes; (**b**) MOFs (metal organic frameworks) can be considered as "organically expanded" coordination centers; (**c**) Replacing the oxygen atoms in Si–O compounds by carbodiimide units results in non-oxides, which are "expanded" compared to the oxide analogues; (**d**) "Trigonally expanded silicas" as prepared earlier; (**e**) "Linearly expanded silicas" as presented here.

Silylcarbodiimides may also by viewed as "linearly expanded Si–O" compounds, since the oxygen atoms are replaced by NCN-units as alternative difunctional linkers of similar electronegativity [48–53]. This approach differs from the above oxides since Si–O bonds are replaced by Si–N bonds (Figure 1c). In the present work, we aimed to "expand" SiO_x-units with rigid spacers without changing the Si–O bonds, i.e., incorporating Si–O–C linkages. In a former study, we used three-fold functional units as organic bridges (Figure 1d) [54]. The obtained hybrid materials are highly cross-linked gels. In order to synthesize polymeric gels with a lower degree of cross-linking, we decided to use two-fold functionalized rigid bridges (Figure 1e).

All attempts to generate the target materials using reactions of commercially available diols, including 4,4′-dihydroxybiphenyl with alkoxysilanes such as TEOS or tetramethoxysilane (TMOS), were not successful. The poor solubility of these diols in polar and non-polar organic solvents may be one of the reasons. In order to start from homogeneous solutions, we decided to react silylated diols with chlorosilanes.

4,4′-Bis(trimethylsiloxy)biphenyl **1** was synthesized according to a previously reported protocol [65]; the silylated product was additionally purified by crystallization from *n*-hexane. 2,6-Bis(trimethylsiloxy)naphthalene **2** was prepared analogously by silylation of the diol with hexamethyldisilazane (HMDS) using KOtBu as a catalyst. Upon vacuum distillation, compound **2** solidified to afford a colorless crystalline solid. Both starting materials (**1** and **2**) showed sharp melting points and delivered satisfying elemental analyses. In the FTIR spectra, the expected absorption bands for C_{alkyl}–H, C_{aryl}–H, C=C, Si–O, and $Si(CH_3)_3$ groups were present. For both compounds, the ^{1}H, ^{13}C, and ^{29}Si NMR spectra exclusively exhibited the respective set of signals to be expected for the symmetric two-fold silylated products. No indications for monosilylated diols or for any other impurities were found.

In case of compound **2**, a single-crystal X-ray diffraction analysis was performed, providing complete structural information for this molecule (Figure 2). Selected bond lengths and angles are summarized in Table 1. All remaining structural data can be found in Table S1 (bond lengths), Table S2 (bond angles), and Table S3 (torsion angles).

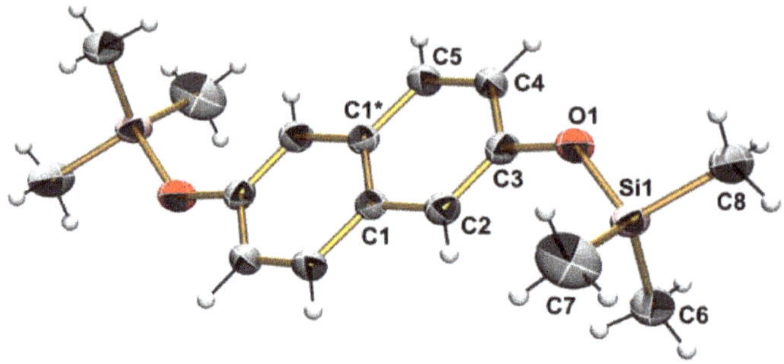

Figure 2. Molecular structure of **2** in the crystal (thermal displacement ellipsoids plotted at the 50% probability level). The bond C1–C1* of the molecule is located on a crystallographic center of inversion, the non-hydrogen atoms of the asymmetric unit are labeled, the asterisk * indicates a symmetry equivalent position. For clarity reasons hydrogen atoms (white balls) are not labeled.

Table 1. Selected bond lengths and angles for compound **2**. Symmetry equivalent atoms are marked with an asterisk *.

Bond Lengths [Å]		Bond Angles [°]	
C1–C2	1.424(2)	C1*–C1–C2	119.2(2)
C2–C3	1.364(2)	C1–C2–C3	120.5(1)
C3–C4	1.415(2)	C2–C3–C4	120.0(1)
C4–C5	1.364(2)	C3–C4–C5	120.6(1)
C5–C1*	1.413(2)	C4–C5–C1*	120.7(1)
C1–C1*	1.413(2)	C5*–C1–C1*	119.0(2)
		C5*–C1–C2	121.9(1)

The bonds between C4 and C5 (1.364 Å) and between C3 and C2 (1.364 Å) are significantly shorter than the remaining C–C bonds of the naphthalene group. Thus, four of the eleven bonds have an enhanced double bonding character. Nevertheless, all bond angles of the naphthalene unit are very close to 120° in accordance with a planar six-membered arene motif.

2.2. *Oligomer and Polymer Formation in Solution*

Solutions of the precursors (i.e., silylated diol **1** or **2**, chlorosilane CH_3SiCl_3, or $SiCl_4$—with and without addition of pyridine, in THF as the solvent) were allowed to react at room temperature and at 60 °C. Only the reactions with $SiCl_4$ and pyridine (**1aPy** and **2aPy**) afforded gels. All other reaction mixtures (**1a** = **1**/$SiCl_4$/THF; **2a** = **2**/$SiCl_4$/THF; **1b** = **1**/CH_3SiCl_3/THF; **2b** = **2**/CH_3SiCl_3/THF; **1bPy** = **1**/CH_3SiCl_3/pyridine/THF; **2bPy**= **2**/CH_3SiCl_3/pyridine/THF) were sealed in NMR tubes and investigated with ^{29}Si

NMR spectroscopy (using external D_2O). The ^{29}Si NMR spectrum of reaction mixture **1b** after one day at room temperature indicates a rather slow reaction; the spectrum is dominated by the signals of the starting materials, but the emergence of new signals at 31 ppm ($(CH_3)_3SiCl$) and −11 ppm (CH_3SiCl_2(O-aryl) confirm the transsilylation reaction even at room temperature and in absence of pyridine catalyst (cf. Figure S1 in the supporting information).

Upon further storage of reaction mixture **1b** (biphenyl precursor **1** and CH_3SiCl_3 in THF without pyridine) at room temperature, a new signal at −33 ppm (MeSiCl(O-aryl)$_2$ moieties) emerges in the ^{29}Si NMR spectrum; this is the sole product signal after 30 days (cf. Figure S2 in the supporting information). An analogous reaction mixture **2b** (using the naphthalene precursor **2** and CH_3SiCl_3 in THF without pyridine) exhibits similar behavior, but the emergence of the signal of CH_3SiCl(O-aryl)$_2$ groups (at −32 ppm) within one day and the disappearance of the signal of CH_3SiCl_2(O-aryl) groups (at −11 ppm) within 8 days indicates more rapid reaction of precursor **2** vs. **1** (cf. Figure S3 in the supporting information). Both systems have in common that transsilylation essentially stopped at the CH_3SiCl(O-aryl)$_2$ stage.

However, supplementing a reaction mixture such as **2b** with pyridine as a nucleophilic catalyst (**2bPy** = **2** plus CH_3SiCl_3 and pyridine in THF) enhances the already known transsilylation steps and gives rise to the emergence of a new ^{29}Si NMR signal at −52 ppm (CH_3Si(O-aryl)$_3$ moieties) within 9 hours; this is the exclusive high field signal after 6 days (cf. Figure S4 in the supporting information). Thus, with pyridine, a complete substitution of all three Cl atoms of CH_3SiCl_3 by O-atoms occurs, while the reaction seems to cease at the stage of a two-fold substitution without addition of pyridine. This rate-increasing effect of pyridine is observed in a similar way for the biphenyl derivative (**1bPy**).

For tetrachlorosilane, the reaction progress of non-gel-forming reaction mixtures can also be followed in solution by ^{29}Si NMR spectroscopy. The corresponding spectra are depicted in the supplementary material (Figures S5 and S6). They also indicate that the naphthalene derivative **2** reacts faster than the biphenyl compound **1**.

In the reaction mixtures of $SiCl_4$ with **1** and **2** plus pyridine (**1aPy** and **2aPy**), a complete chlorine substitution is observed as in the reactions of CH_3SiCl_3 with **1** and **2** plus pyridine (**1bPy** und **2bPy**), i.e., formation of "SiO_4" and "CH_3SiO_3" and acceleration of the reactions. In addition, gelation also occurred. This indicates that the degree of cross-linking in the case of starting material CH_3SiCl_3 is not sufficient to form a rigid network. The silsesquioxane units obviously lead to fewer cross-linked and THF soluble oligomeric structures.

2.3. Gel Synthesis and Characterisation

The reaction mixtures of both silylated precursors **1** and **2** with $SiCl_4$ and pyridine (**1aPy** and **2aPy**) formed gels within 24 hours. Over several days, strong syneresis was observed. After three weeks, the gels were dried in vacuum. The obtained xerogels **1A** and **2A** were brittle and glass-like; **1A** was white and **2A** was slightly brownish. Both xerogels were analyzed by elemental analysis, powder X-ray diffraction, and IR and solid state ^{29}Si NMR spectroscopy. The latter spectroscopic investigations were also used to characterize the hydrolytic sensitivity of xerogels **1A** and **2A**.

EDX studies (performed during SEM analyses, vide infra) of both xerogels showed the presence of the elements Si, O, and C, as expected. A determination of the chlorine content indicated minor contents of 1.15 wt% in case of **1A** and 0.89 wt% in case of **2A**. This corresponds well with the ^{29}Si NMR studies in solution (see Section 2.2), where a complete substitution of all three (in case of CH_3SiCl_3) and all four (in case of $SiCl_4$) chlorine atoms was observed for the pyridine catalyzed reactions.

The powder X-ray diffraction data indicated completely amorphous structures for xerogels **1A** and **2A**. In the vibrational spectra, all typical absorption bands for the characteristic C–H, C–C, C–O, Si–O, and Si–C valence and deformation vibrations were present (see materials and methods section).

In Figures 3 and 4, the solid state ^{29}Si NMR spectra of both xerogels **1A** and **2A** are shown. Despite the fact that stoichiometric amounts of the starting materials $SiCl_4$ and **1** or **2** were used, i.e., a molar ratio of 1:2, relatively strong signals for unreacted or partly reacted O–Si(CH_3)$_3$-groups in **1** (around 20 ppm), and some signals for potential side products with OSi(CH_3)$_3$-groups (signals in the range 10–16 ppm)), were found. The signals for the expected SiO_4-moieties in the xerogels are clearly detectable (around −100 ppm). As the spectrum essentially consists of these two groups of signals, we conclude that moieties of partial substitution (ClSi(O-aryl)$_3$, expected in the ^{29}Si NMR shift range (−80 to −85 ppm) are almost absent. In Figure 3, it can be seen that after storing the powdered xerogel **1A** for two days in ambient air, new upfield shifted signals emerge for both groups of signals; we attribute them to (aryl-O)$_3$Si–O–Si-groups. The spectrum of xerogel **1A** (in contrast to xerogel **2A**) already shows a sharp signal at 16.1 ppm before contact with air, which is attributed to adsorbed (CH_3)$_3$SiOH. After the contact with humidity from air, the unreacted Si(CH_3)$_3$)-groups in both xerogels were partially hydrolysed and bonded to the Si–O network of the xerogel. The backbone of xerogel **1A** is obviously more prone to hydrolysis when compared to that of xerogel **2A,** because after two days of storage with air contact, an upfield signal around −110 ppm is present in the spectrum of sample **1A** (Figure 3), but an analogous signal is less pronounced in the spectrum of **2A** upon the same treatment (Figure 4, bottom).

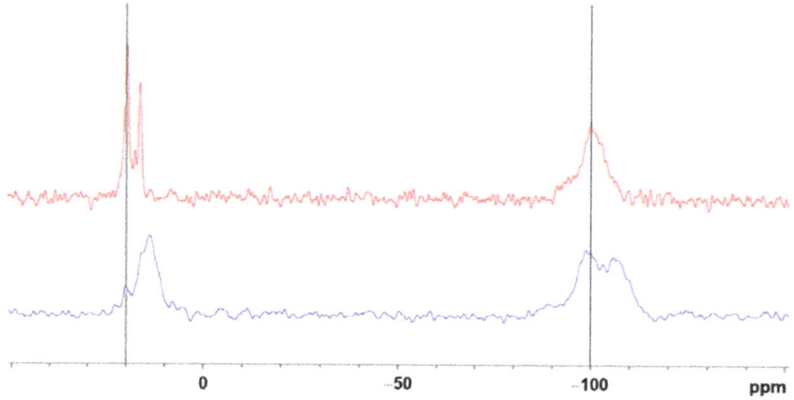

Figure 3. ^{29}Si cross-polarization magic angle spinning (CP MAS) NMR spectra of xerogel **1A** (obtained from **1**/$SiCl_4$/pyridine/THF), top (red trace): xerogel prepared under argon; bottom (blue trace): same sample of xerogel after two days in contact with ambient air. (The vertical lines at δ = +20 ppm and −100 ppm aid enhanced visualization of the chemical shift differences).

This higher sensitivity toward hydrolysis of **1A** may be caused by the fact that the biphenyl units provide more flexibility and more accessibility for water molecules due to the torsional freedom between both phenylene groups, while the naphthalene fragment is essentially rigid.

2.4. Morphological and Thermal Characterization of the Xerogels

The phenomenology of the Sol–Gel transition observed for the hybrid Sol–Gel system presented here is very similar to typical oxide Sol–Gel processes: a homogenous and transparent liquid reaction mixture solidifies—after a certain reaction time of typically hours to days—with formation of a shape retaining monolithic gel body that is usually translucent for visible light. In Figure 5, the appearance of gels of the system **1aPy** is depicted.

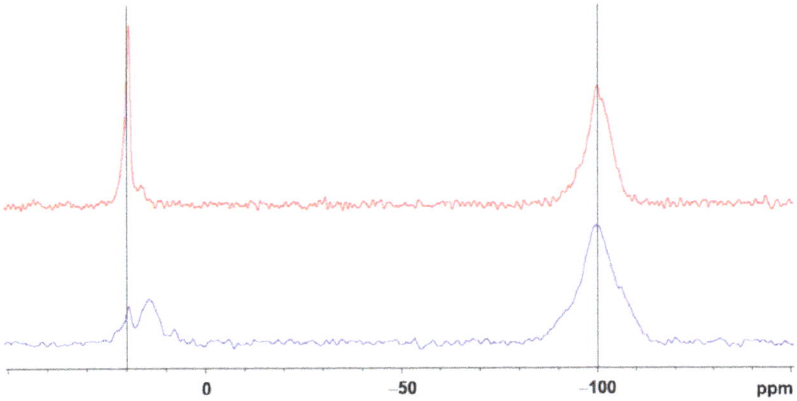

Figure 4. ^{29}Si CP MAS NMR spectra of xerogel **2A** (obtained from **2**/$SiCl_4$/pyridine/THF), top (red trace): xerogel prepared under argon; bottom (blue trace): same sample of xerogel after two days in contact with ambient air. (The vertical lines at δ = +20 ppm and −100 ppm aid enhanced visualization of the chemical shift differences).

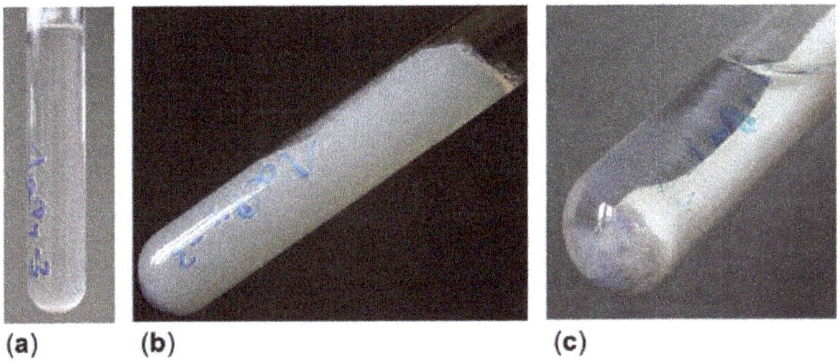

Figure 5. Reaction mixture **1aPy** and formation of gel **1A**: (**a**) at t = 0 h, (**b**) after one day gelation and partial syneresis already occurred, (**c**) aged for 21 days.

The morphology of the xerogels was investigated using scanning electron microscopy (SEM). The image in Figure 6 shows that xerogel **1A** is constituted of nanostructured primary particles with diameters of less than 50 nm. These primary particles are very strongly agglomerated, fused, or even coalesced.

Xerogel **2A** is characterized by a very similar microstructure, with primary particle sizes in the range below 50 nm, as can be seen in Figure 7. The degree of agglomeration is very similar to that of xerogel **1A**. The particles are almost fully coalesced to a dense homogeneous body, with only very little inter-particle spaces and corresponding porosity.

Nitrogen adsorption measurements, analyzed according to the Brunauer–Emmett–Teller (BET) model, indicated that xerogel **2A** has a surface area of approximately 10 $m^2 \cdot g^{-1}$. This relatively low value corresponds very well with the SEM images, which show that most of the porosity and surface area (which was most likely present in the wet gels) strongly decreases during ageing and drying of the gels. This is also supported by the large linear shrinkage of 57% for xerogel **1A** and 59% for xerogel **2A**.

The thermal behavior of the xerogels was investigated under argon atmosphere applying a heating rate of 5 $K \cdot min^{-1}$ up to 800 °C. Further TG-FTIR studies were performed in the temperature range up to 750 °C, again with a heating rate of 5 $K \cdot min^{-1}$ under argon. The results of the measurements are shown in Figures 8 and 9.

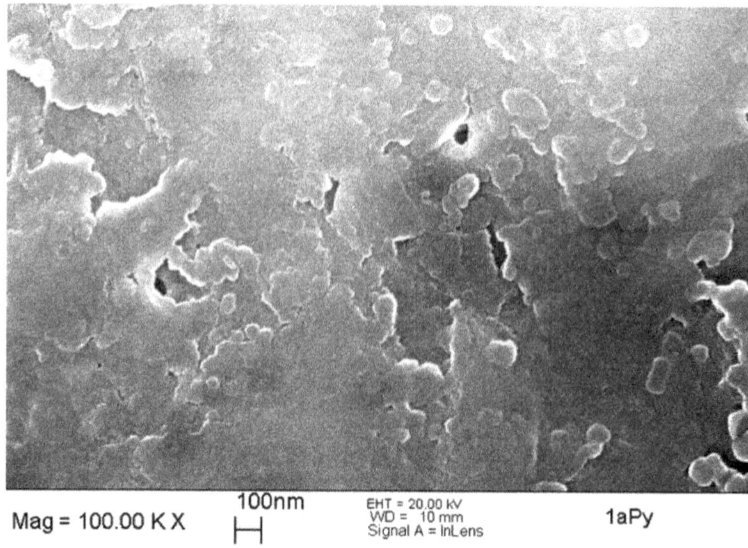

Figure 6. SEM image of xerogel **1A** (obtained from **1**/$SiCl_4$/pyridine/THF).

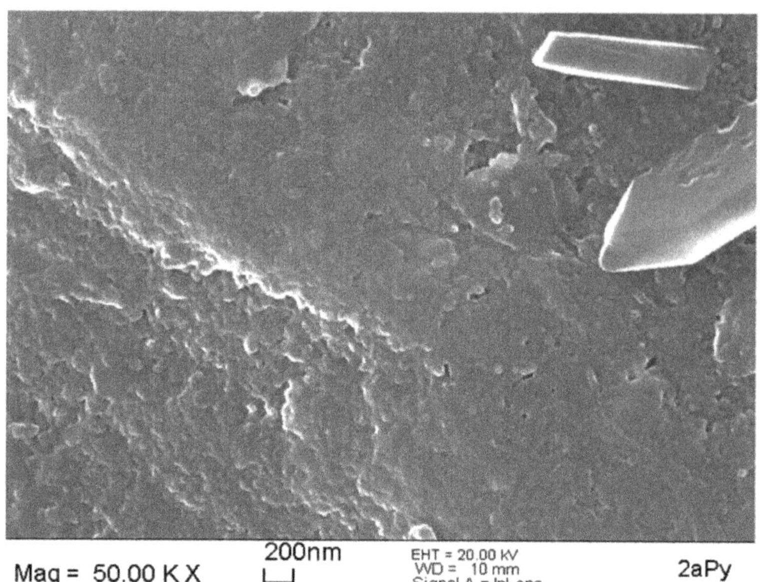

Figure 7. SEM image of xerogel **2A** (obtained from **2**/$SiCl_4$/pyridine/THF). The composition and identity of the larger crystalline-appearing particle in the upper right corner of the image could not be resolved.

The TG/DTG curves of xerogels **1A** and **2A** are similar, showing three peaks in the comparable temperature ranges (**1A**: ca. 220 °C, 520 °C, and 600 °C; **2A**: ca. 150 °C, 520 °C, and 600 °C). For both xerogels, the mass loss below 250 °C can be attributed at least in part to the desorption of residual THF and pyridine. This was supported by FTIR spectra of the evolved gases. Desorption of volatile species containing –O–Si(CH_3)$_3$ groups and also condensation reactions of –O–Si$(CH_3)_3$ terminal groups also occurs. Hexamethyldisiloxane and trimethylsilanol were clearly detectable in the FTIR spectra. At temperatures around 500 °C and 600 °C, decomposition reactions of aromatic C–H and remaining trimethylsilyl units led to the formation of methane for both xerogels.

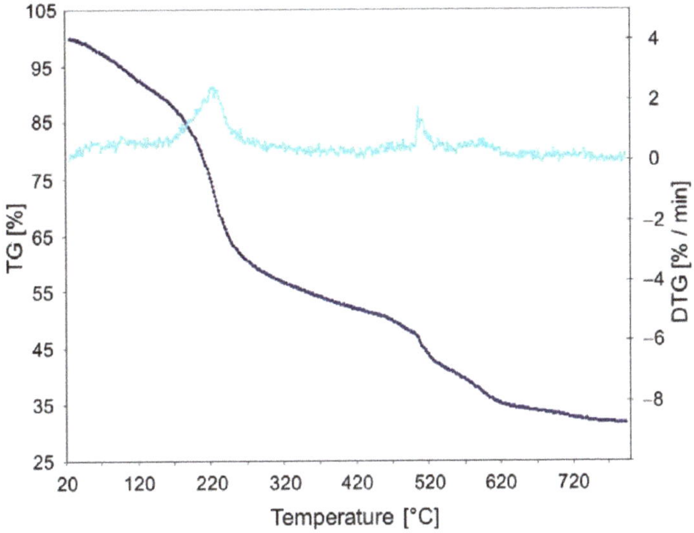

Figure 8. TG/DTG measurement of xerogel **1A** (obtained from **1**/$SiCl_4$/pyridine/THF; TG trace dark blue, DTG trace light blue).

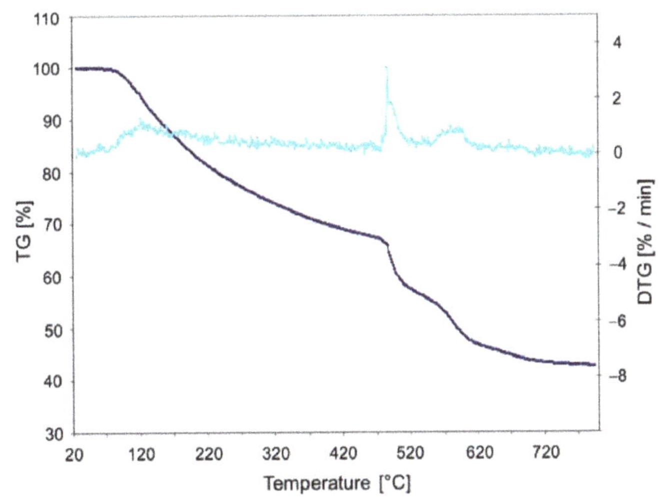

Figure 9. TG/DTG measurement of xerogel **2A** (obtained from **2**/$SiCl_4$/pyridine/THF; TG trace dark blue, DTG trace light blue).

In the case of xerogel **1A**, the total mass loss in the temperature range from 20 °C to 800 °C is 68.2%; xerogel **2A** displays a lower mass loss of 57.3% in the same temperature range. Although the pyrolytic residues were not analyzed, it is expected that Si/C/O/(H) materials are formed at 800 °C. Thus, a slightly higher thermal stability and a higher ceramic yield is found for xerogel **2A**; this may be attributed to a higher degree of "graphitization" and, in turn, a higher thermal stability of the naphthalene moieties when compared to the biphenylene units.

3. Conclusions

Here we reported on a non-hydrolytic Sol–Gel system for "organically bridged" or "linearly expanded" silica and silsesquisiloxane hybrid materials. The synthesis is based on reactions of $SiCl_4$ and CH_3SiCl_3 with rigid aromatic diols, which are two-fold trimethysilylated, namely compounds **1** and **2**. The driving force for these reactions is the formation of $(CH_3)_3SiCl$.

The reactions proceed smoothly at room temperature and are accelerated by the addition of catalytic amounts of pyridine. Most importantly, pyridine addition also causes a complete substitution of all three or four chlorine atoms in the stoichiometric reaction mixtures with CH_3SiCl_3 and $SiCl_4$, respectively, even at room temperature.

A Sol–Gel transition is observed only in case of reactions of $SiCl_4$ in the presence of pyridine. The gels exhibit strong shrinkage, which leads to rather dense, nearly non-porous xerogels **1A** and **2A**, both of which are hydrolytically sensitive.

The route to organic-inorganic hybrid polymers and gels presented here may be applied to many related systems. We expect that other chlorosilanes with at least three chlorine atoms as well as other element halides with similar reactivity may be combined in suitable inert solvents with trimethylsilylated di-, tri-, or polyols to form $(CH_3)_3SiCl$ or $(CH_3)_3SiX$ and more-or-less branched and cross-linked network structures. The products may be calcined to form amorphous organic-inorganic hybrids with tailorable elemental composition and properties.

The hydrolytic sensitivity of as prepared xerogels may not necessarily be a disadvantage. This property might be exploitable for the development of environmentally degradable materials or for a targeted release of incorporated components triggered by contact with water.

4. Materials and Methods

General Techniques and Chemicals. All manipulations were performed under an inert (Ar or N_2) atmosphere using standard Schlenk techniques or in a glovebox. The solvents, tetrahydrofuran (THF), and *n*-hexane were dried by distillation from a sodium/ benzophenone mixture. The remaining starting materials, i.e., 4,4′-dihydroxybiphenyl, 2,6-dihydroxynaphthalene, hexamethyldisilazane (HMDS), potassium tert-butanolate (KOtBu), as well as the chlorosilanes $SiCl_4$ and CH_3SiCl_3, were used as received without further purification.

Synthesis of the starting materials. 4,4′-Bis(trimethylsiloxy)biphenyl (**1**): The compound was synthesized according to a procedure reported in the literature [65]. However, the brownish product was impure and had to be purified. This was performed by extracting 45.0 g (136 mmol) of the raw product with 50 mL boiling n-hexane. In the extract solution, white crystals of the product formed at room temperature. After filtration and drying under vacuum, 40.9 g (124 mmol) analytically pure **1** were obtained. Mp 64 °C. NMR: ^{1}H (400 MHz, $CDCl_3$) δ [ppm]: 7.42 (m, 4H, $^3J_{HH}$ = 8.4 Hz, H2,2′), 6.88 (m, 4H, $^3J_{HH}$ = 8.4 Hz, H3,3′), 0.31 (s, 18H, CH_3); ^{13}C (100 MHz, $CDCl_3$) δ [ppm]: 154.3 (C4, 4′), 134.2 (C1,1′), 127.7 (C2, 2′), 120.2 C3,3′), 0.2 (CH_3); ^{29}Si (79.5 MHz, $CDCl_3$) δ [ppm]: 19.7 (s). IR (KBr pellet), ν [cm^{-1}]: 3035–2960 (w) [aromatic C–H valence vibration]; 2900 (w) [C–H valence vibration of CH_3-groups]; 1500 (m), 1600 (m) [aromatic C=C valence vibration]; 1263–1248 (s) [C–O valence vibration]; 1260–1250 (s), 843–817 (s), 760–752 (m) [$Si(CH_3)_3$ valence vibration]; 1174 (w), 1102 (w), 936–919 (s) [Si–O–C(aryl) valence vibration]. Elemental analysis calcd (%) for $C_{18}H_{26}O_2Si_2$ (M = 330.56 g/mol): C 65.40, H 7.93, found: C 65.35, H 8.09.

2,6-Bis(trimethylsiloxy)naphthalene (2): In a round bottomed flask, 5.0 g (31 mmol) 2,6-dihydroxynaphthalene was mixed with 18 mL (86 mmol) of hexamethyldisilazane (HMDS) and a catalytic amount (~50 mg) of potassium tert-butanolate. The reaction mixture was heated under reflux with vigorous stirring for 1 h. The end of the reaction was indicated when the ammonia formation has ceased. Isolation of the product was accomplished using fractioned distillation. After removal of excessive HMDS, pure **2** was collected at 130–134 °C/1.3–2.6 × 10^{-2} Pa as a colorless liquid. Upon cooling to room temperature, the distillate formed a white crystalline solid. Yield: 9.2 g (30 mmol, 97%). Mp 109 °C. NMR: ^{1}H (400 MHz, $CDCl_3$) δ [ppm]: 7.58 (d, 2H, $^{3}J_{HH}$ = 8.8 Hz, $H^{4,8}$), 7.15 (d, 2H, $^{4}J_{HH}$ = 1.8 Hz, $H^{1,5}$), 7.03 (dd, 2H, $^{3}J_{HH}$ = 8.8 Hz, $^{4}J_{HH}$ = 1.8 Hz, $H^{3,7}$), 0.3 (s, 18H, CH_3); ^{13}C (100 MHz, $CDCl_3$) δ [ppm]: 151.5 (C2), 130.3 (Cq), 128.0 (C4), 122.3 (C3), 114.9 (C1), 0.3 (CH_3); ^{29}Si (79.5 MHz, $CDCl_3$) δ [ppm]: 19.9 (s). IR (KBr pellet), ν [cm^{-1}]: 3050 (w), 2955 (w) [aromatic C–H valence vibration]; 2900 (w) [C–H valence vibration of CH_3-groups]; 1600 (m), 1500 (m) [aromatic C=C valence vibration]; 1390–1375 (m) [CH_3-deformation vibration]; 1270–1250 (s), 1121 (m), 1160 (m) [C–O valence vibration]; 1261–1244 (s), 874–847 (vs), 811 (w), 761 (m) [$Si(CH_3)_3$ valence vibration]; 966–958 (m) [Si–O–C(aryl) valence vibration]. Elemental analysis calcd (%) for $C_{16}H_{24}O_2Si_2$ (M = 304.53 g/mol): C 63.11, H 7.94, found: C 62.85, H 8.18. A single-crystal X-ray structure analysis of **2** was reported earlier [66] and provided further detailed information on this compound.

Polymer and Gel Synthesis. The polymer synthesis was performed analogous to the preparation of s-triazine- and s-heptazine-based hybrid materials, as described in [54]. In a typical experiment, 1 mmol of **1** was dissolved in 5 mL of dry THF and 0.5 mmol of $SiCl_4$ was added at room temperature in a Schlenk tube. For comparison, the same experiment was conducted with 0.1 mmol of pyridine added as a catalyst. The liquid homogeneous reaction mixtures were stored at room temperature or heated to 60 °C. For the synthesis of less cross-linked polymers, 0.33 mmol of methyltrichlorosilane and 0.5 mmol of **1** were dissolved in 4 ml of dry THF, in absence and presence of 0.05 mmol of pyridine, and stored at room temperature or heated to 60 °C. Syntheses of polymers derived from the naphthalene derivative **2** were conducted analogously. Further details are summarized in Tables 2 and 3.

Table 2. Overview about the reactions of precursor **1** with $SiCl_4$ and CH_3SiCl_3 in THF.

Reaction Mixture	n(1) [mmol]	THF [mL]	n($SiCl_4$) [mmol]	n(CH_3SiCl_3) [mmol]	$n_{Pyridine}$ [mmol]	Observation
1a	1	5	0.5	-	-	no gelation after three weeks at 60 °C
1aPy	1	5	0.5	-	0.1	gelation after 1 day; strong syneresis in following weeks, linear shrinkage of about 57% for xerogel 1A
1b	0.5	4	-	0.33	-	no gelation after three weeks at 60 °C
1bPy	0.5	4	-	0.33	0.05	

Table 3. Overview about the reactions of precursor **2** with $SiCl_4$ and CH_3SiCl_3 in THF.

Reaction Mixture	n(2) [mmol]	THF [mL]	n($SiCl_4$) [mmol]	n(CH_3SiCl_3) [mmol]	$n_{Pyridine}$ [mmol]	Observations
2a	1	5	0.5	-	-	no gelation after three weeks at 60 °C
2aPy	1	5	0.5	-	0.1	gelation after 1 day; strong syneresis in following weeks, linear shrinkage of 59% for xerogel 2A
2b	0.5	4	-	0.33	-	no gelation after three weeks at 60 °C
2bPy	0.5	4	-	0.33	0.05	

Gelation was observed only in the cases of the pyridine catalyzed experiments with $SiCl_4$. After an aging period of 21 days at 60 °C, the gels were carefully dried in vacuum. The obtained xerogels **1A** and **2A** consisted of glass-like bulk xerogel bodies. Yield and linear shrinkage were determined followed by characterization with powder XRD, gas adsorption (BET), elemental analysis, simultaneous thermal analysis (TG/DTA), and IR and ^{29}Si CP/MAS NMR spectroscopy:

1A: Yield = 173 mg of a white, solid xerogel. NMR (CP/MAS at 5 kHz) δ [ppm]: ^{29}Si (79.5 MHz): −100.2 (broad, SiO_4-units), further sharp signals at 19.4, 17.4 and 16.1 (O–Si(CH_3)$_3$ groups). IR (KBr-pellet), ν [cm^{-1}]: 2960(vw) [aromatic C–H valence vibration]; 1600 (w), 1500 (w) [aromatic C=C valence vibration]; 1250 (m) [C–O valence vibration]; 1251 (m), 845 (w), 821 (w) [Si(CH_3)$_3$ valence vibration]; 1169 (w), 1107 (w), 936 (w), 913 (w) [Si–O–Ar valence vibration]. Elemental analysis calcd (%) for [$C_{24}H_{16}O_4Si$]$_n$ (M = 396.45 g/mol): Cl 0.0., Si 7.08, found: Cl 1.15, Si 8.79.

2A: Yield = 153 mg of a white and very rigid xerogel. NMR (CP/MAS at 5 kHz) δ [ppm]: ^{29}Si (79.5 MHz): −100.1 (broad, SiO_4-units), 19.0 (sharp, O- Si(CH_3)$_3$ groups). IR (KBr-pellet), ν [cm^{-1}]: 3065 (vw), 2958 (vw) [aromatic C–H valence vibration]; 1600 (m), 1505 (w) [aromatic C=C valence vibration]; 1375 (m) [CH_3-deformation vibration]; 1230 (m), 1119 (w), 1150 (w) [C–O valence vibration]; 870 (w), 802 (w), 750 (vw) [Si(CH_3)$_3$ valence vibration]; 983 (m) [Si–O–Ar valence vibration]. Elemental analysis calcd (%) for [$C_{20}H_{12}O_4Si$]$_n$ (M = 344.38 g/mol): Cl 0.0., Si 8.2, found: Cl 0.89, Si 10.8.

Methods of Characterization. All manipulations of gels, xerogels, and pyrolysis products were performed under inert gas using glove box and/or Schlenk techniques.

IR Spectra were recorded using the standard KBr pellet method. For the starting materials, a Nicolet 510 FTIR spectrometer was used. The IR investigation of the xerogels was performed on a Varian FTIR spectrometer.

Solution ^{1}H, ^{13}C and ^{29}Si NMR spectra were recorded on a BRUKER DPX 400 instrument at room temperature (TMS as internal standard). The ^{29}Si NMR spectra were measured using inverse gated decoupling (IGATED).

Solid State ^{29}Si NMR data of the dried gels **1A** and **2A** were acquired on a BRUKER AvanceTM WB 400 MHz spectrometer with a resonance frequency of 79.51 MHz using magic-angle spinning (MAS) at 5 kHz in 7 mm ZrO_2 rotors and cross polarization (CP) with 5 ms contact time. The chemical shift is referenced externally to octakis(trimethylsiloxy)octasilsesquioxane Q_8M_8 (the most upfield signal of its Q^4 groups at δ_{iso} = −109 ppm) and reported relative to tetramethylsilane (TMS).

Powder X-ray Diffraction (XRD): The patterns of the xerogels were obtained with a Siemens D-5000 diffractometer with Cu K_α radiation (λ = 1.5418 Å).

Single-crystal X-ray diffraction analysis of **2**: When the freshly vacuum-distilled 2,6-bis(trimethylsiloxy)naphthalene had solidified, a tiny piece of the solid was found to be suitable for X-ray structure analysis. X-ray diffraction data were collected on a BRUKER-NONIUS X8 APEXII-CCD diffractometer with Mo-K_α-radiation (λ = 0.71073 Å). The structure was solved with direct methods and refined with full-matrix least-squares methods. All non-hydrogen atoms were refined anisotropically. Hydrogen atoms were placed in idealized positions and refined isotropically (riding model). Structure solution and refinement of F^2 against all reflections were performed with SHELXS-97 and SHELXL-97 (G.M. Sheldrick, Universität Göttingen, Göttingen, Germany, 1986–1997).

2,6-bis(trimethylsiloxy)naphthalene: $C_{16}H_{24}O_2Si_2$, M_r = 304.53, colourless piece, 0.45 × 0.40 × 0.06 mm^3, trigonal space group R-3, $a = b$ = 21.0302(4), c = 10.5261(4) Å, $\alpha = \beta = 90°$, $\gamma = 120°$, V = 4031.67(19) Å^3, Z = 9, $\rho_{calcd.}$ = 1.129 g·cm^{-3}, $2\theta_{max}$ = 60°, F(000) = 1476, μ = 0.197 mm^{-1}, no absorption correction, T = 203(2) K, 24,421 recorded reflections ($-29 \leq h \leq 29$, $-29 \leq k \leq 27$, $-14 \leq l \leq 14$), 2608 unique and 2026 observed reflections with $F_o > 2\,\sigma(F_o)$, 91 parameters, $R1$ = 0.0401, $wR2$ = 0.1083 [$I > 2\sigma(I)$], $R1$ = 0.0578, $wR2$ = 0.1179 (all data), residual electron density (highest peak and deepest hole) 0.410 and −0.341 eÅ^{-3}.

Crystallographic data (excluding structure factors) for the structure reported in this paper have been deposited earlier with the Cambridge Crystallographic Centre as private communication CCDC 622212. These data can be obtained free of charge from the Cambridge Crystallographic Data Centre via https://www.ccdc.cam.ac.uk/structures/ (accessed on 27 February 2023).

High Resolution Scanning Electron Microscopy (HR-SEM) was performed with an LEO1530 microscope equipped with an energy dispersive X-ray (EDX) system.

Nitrogen Adsorption Measurements (Micromeritics ASAP 2000) were carried out after desorption at 150 °C/~0.5 mbar for 12 h of samples (~0.5 g) which were ground in a mortar under inert conditions.

Thermal Analysis was first performed using a Seiko Instruments TG/DTA 22 under argon atmosphere applying a heating rate of $5K{\cdot}min^{-1}$. Further TG-FTIR studies were performed on a Setaram Sensys TG-DSC coupled with a Varian FTIR instrument in the temperature range up to 750 °C with a heating rate of $5\ K{\cdot}min^{-1}$ under argon. Alumina crucibles were used.

Elemental Analyses of the starting materials **1** and **2** were carried out on a "Foss Heraeus CHN-O-Rapid" analyzer while the determination of chlorine and silicon contents in **1A** and **2A** was performed by the "Mikroanalytisches Labor Pascher" (Remagen-Bandorf, Germany).

Supplementary Materials: The following supporting information can be downloaded at: https://www.mdpi.com/article/10.3390/gels9040291/s1, Figure S1: ^{29}Si solution NMR spectrum of 1b (1 with CH_3SiCl_3) after 1 day at room temperature; Figure S2: ^{29}Si solution NMR spectra of reaction mixture 1b (1/CH_3SiCl_3/THF), top: after 1 day; center: after 8 days; bottom: after 30 days at room temperature; Figure S3: ^{29}Si solution NMR spectra of reaction mixture 2b (2/CH_3SiCl_3/THF), top: after 1 day; center: after 8 days; bottom: after 30 days at room temperature; Figure S4: ^{29}Si solution NMR spectra of reaction mixture 2bPy (2/CH_3SiCl_3/pyridine/THF), top: after 9 h; center: after 6 days; bottom: after 27 days at room temperature; Figure S5: ^{29}Si solution NMR spectra of reaction mixtures 1a (1/$SiCl_4$/pyridine/THF) [top: after 9 h (with broad absorption band for the glas of the NMR tube); bottom: after 7 days]; Figure S6: ^{29}Si solution NMR spectra of reaction mixtures 2a (2/$SiCl_4$/pyridine/THF) [top: after 18 h (with broad absorption band for the glas of the NMR tube); bottom: after 7 days]; Figure S7: Molecular structure of 2 in the crystal (thermal displacement ellipsoids plotted at the 50% probability level, H-atoms are omitted for clarity). The bond C1–C1* of the molecule is located on a crystallographic center of inversion, the atoms of the asymmetric unit are labeled, the asterisk * indicates a symmetry equivalent position. Figure S8: Xerogel **1A** after drying at 60 °C in vacuum for several hours. Table S1: Bond lengths [Å] of compound 2 (in its crystal structure); Table S2: Bond angles [deg.] of compound 2 (in its crystal structure); Table S3: Torsion angles [deg.] of compound 2 (in its crystal structure).

Author Contributions: Conceptualization, E.K. and J.W.; methodology, K.K., J.W., E.B. and E.K.; validation, E.K. and J.W.; investigation, K.K. (synthesis, analysis), J.W. (X-ray structures) and E.B. (solid state NMR studies); resources, E.K.; data curation, J.W.; writing, E.K.; writing—review and editing, J.W. and E.B.; visualization, K.K. and J.W.; supervision, E.K. and J.W.; All authors have read and agreed to the published version of the manuscript.

Funding: This work was supported by TU Bergakademie Freiberg (Saxony, Germany).

Institutional Review Board Statement: Not applicable.

Informed Consent Statement: Not applicable.

Data Availability Statement: Not applicable.

Acknowledgments: The authors gratefully acknowledge financial support by Deutsche Forschungsgemeinschaft (DFG, Bonn). Beate Kutzner, Conny Wiltzsch, and Florian Hoffmann are acknowledged for performing the solution NMR measurements. Marcus Schwarz is acknowledged for performing the REM/EDX measurements.

Conflicts of Interest: The authors declare no conflict of interest. The funders had no role in the design of the study; in the collection, analyses, or interpretation of data; in the writing of the manuscript; or in the decision to publish the results.

References

1. Judeinstein, P.; Sanchez, C. Hybrid organic–inorganic materials: A land of multidisciplinarity. *J. Mater. Chem.* **1996**, *6*, 511–525. [CrossRef]
2. Kickelbick, G. *Hybrid Materials: Synthesis, Characterization, and Applications*; Wiley-VCH: Weinheim, Germany, 2006.

3. Faustini, M.; Nicole, L.; Ruiz-Hitzky, E.; Sanchez, C. History of Organic–Inorganic Hybrid Materials: Prehistory, Art, Science, and Advanced Applications. *Adv. Funct. Mater.* **2018**, *28*, 1704158. [CrossRef]
4. Brinker, C.J.; Scherer, G.W. *Sol–Gel Science*; Academic Press: San Diego, CA, USA, 1990.
5. Hench, L.L.; West, J.K. The Sol–Gel Process. *Chem. Rev.* **1990**, *90*, 33–72. [CrossRef]
6. Corriu, R.J.P.; Leclerq, D. Recent developments of molecular chemistry for Sol–Gel processes. *Angew. Chem. Int. Ed.* **1996**, *35*, 1420–1436. [CrossRef]
7. Sakka, S. (Ed.) *Sol–Gel Science and Technology*; 4 volume set; Springer: New York, NY, USA, 2003.
8. Vioux, A. Nonhydrolytic Sol−Gel Routes to Oxides. *Chem. Mater.* **1997**, *9*, 2292–2299. [CrossRef]
9. Hay, J.N.; Raval, H.M. Synthesis of Organic−Inorganic Hybrids via the Non-hydrolytic Sol−Gel Process. *Chem. Mater.* **2001**, *13*, 3396–3403. [CrossRef]
10. Debecker, D.P.; Mutin, P.H. Non-hydrolytic sol–gel routes to heterogeneous catalysts. *Chem. Soc. Rev.* **2012**, *41*, 3624–3650. [CrossRef]
11. Smeets, V.; Styskalik, A.; Debecker, D.P. Non-hydrolytic sol–gel as a versatile route for the preparation of hybrid heterogeneous catalysts. *J. Sol-Gel Sci. Technol.* **2021**, *97*, 505–522. [CrossRef]
12. Kroke, E. Novel Sol–Gel Routes to Non-Oxide Ceramics. In *9th CIMTEC—World Ceramics Congress, Ceramics: Getting into the 2000's—Part C*; Vincenzini, P., Ed.; Techna: Faenza, Italy, 1999; pp. 123–134.
13. Hector, A.L. Synthesis and processing of silicon nitride and related materials using preceramic polymer and non-oxide Sol–Gel approaches. *Coord. Chem. Rev.* **2016**, *323*, 120–137. [CrossRef]
14. Rex, A.; dos Santos, J.H.Z. The use of Sol–Gel processes in the development of supported catalysts. *J. Sol–Gel Sci. Technol.* **2023**, *105*, 30–49. [CrossRef]
15. Almeida, R.M.; Martucci, A.; Santos, L.; Rojas-Hernandez, R.E. (Eds.) *Sol–Gel Derived Optical and Photonic Materials*; Elsevier Ltd.: London, UK, 2021.
16. Aegerter, M.A.; Leventis, N.; Koebel, M.M. (Eds.) *Aerogels Handbook*; Springer: New York, NY, USA, 2011.
17. Hu, P.; Liu, L.; Zhao, M.; Wang, J.; Ma, X.; Wang, J. Review on design, synthesis, and use of high temperature resistant aerogels exceeding 800 °C. *ES Mater. Manuf.* **2022**, *15*, 14–33.
18. Eychmueller, A. Nanoparticle-Based Aerogels and Their Prospective Future Applications. *J. Phys. Chem. C* **2022**, *126*, 19011–19023. [CrossRef]
19. Gerrard, W.; Jones, J.V. Stability of isomeric butoxysilanes with respect to silicon tetrachloride and hydrogen chloride. *J. Chem. Soc.* **1952**, 1690–1693. [CrossRef]
20. Corriu, R.J.P.; Leclercq, D.; Lefèvre, P.; Mutin, P.H.; Vioux, A. Preparation of monolithic gels from silicon halides by a non-hydrolytic Sol–Gel process. *J. Non-Cryst. Solids* **1992**, *146*, 301–303. [CrossRef]
21. Acosta, S.; Corriu, R.J.P.; Leclerq, D.; Mutin, P.H.; Vioux, A. Novel non-hydrolytic Sol–Gel route to metal oxides. *J. Sol–Gel Sci. Technol.* **1994**, *2*, 25–28. [CrossRef]
22. Arnal, P.; Corriu, R.J.P.; Leclerq, D.; Mutin, P.H.; Vioux, A. A Solution Chemistry Study of Nonhydrolytic Sol−Gel Routes to Titania. *Chem. Mater.* **1997**, *9*, 694–698. [CrossRef]
23. Oakton, E.; Siddiqi, G.; Fedorov, A.; Coperet, C. Tungsten oxide by non-hydrolytic sol–gel: Effect of molecular precursor on morphology, phase and photocatalytic performance. *New J. Chem.* **2016**, *40*, 217–222. [CrossRef]
24. Andriamananarivelo, M.; Corriu, R.J.P.; Leclerq, D.; Mutin, P.H.; Vioux, A. Mixed oxides SiO_2–ZrO_2 and SiO_2–TiO_2 by a non-hydrolytic sol–gel route. *J. Mater. Chem.* **1996**, *6*, 1665–1671. [CrossRef]
25. Andriamananarivelo, M.; Corriu, R.J.P.; Leclerq, D.; Mutin, P.H.; Vioux, A. Nonhydrolytic Sol−Gel Process: Aluminum Titanate Gels. *Chem. Mater.* **1997**, *9*, 1098–1102. [CrossRef]
26. Jansen, M.; Guenther, E. Oxide gels and ceramics prepared by a nonhydrolytic sol- gel process. *Chem. Mater.* **1995**, *7*, 2110–2114. [CrossRef]
27. Nassar, E.J.; dos Santos Pereira, P.F.; de Oliveira Nassor, E.C.; Ávila, L.R.; Ciuffi, K.J.; Calefi, P.S. Nonhydrolytic Sol–Gel synthesis and characterization of YAG. *J. Mater. Sci.* **2007**, *42*, 2244–2249. [CrossRef]
28. Janković-Častvan, I.; Lazarević, S.; Jordović, B.; Petrović, R.; Tanasković, D.; Janaćković, D. Electrical properties of cordierite obtained by non-hydrolytic sol–gel method. *J. Eur. Ceram. Soc.* **2007**, *27*, 3659–3661. [CrossRef]
29. Gao, Y.; Hamana, D.; Iwasaki, R.; Iihama, J.; Honda, S.; Kumari, M.; Hayakawa, T.; Bernard, S.; Thomas, P.; Iwamoto, Y. Chemical route for synthesis of β-SiAlON:Eu^{2+} phosphors combining polymer-derived ceramics route with non-hydrolytic Sol–Gel chemistry. *J. Sol–Gel Sci. Technol.* **2022**, *104*, 711–723. [CrossRef]
30. Koshevaya, E.; Mikhaylov, V.; Sitnikov, P.; Krivoshapkina, E.; Krivoshapkin, P. Electrosurface properties and acid-base equilibria of Ta2O5 and Ta2O5:Eu nanoparticles in NaCl solutions. *Surf. Interfaces* **2022**, *29*, 101713. [CrossRef]
31. Shrestha, S.; Wang, B.; Dutta, P. Nanoparticle processing: Understanding and controlling aggregation. *Adv. Colloid Interface Sci.* **2020**, *279*, 102162. [CrossRef] [PubMed]
32. Escamilla-Perez, A.M.; Louvain, N.; Boury, B.; Brun, N.; Mutin, P.H. Ethers as Oxygen Donor and Carbon Source in Non-hydrolytic Sol–Gel: One-Pot, Atom-Economic Synthesis of Mesoporous TiO_2-Carbon Nanocomposites. *Chem. Eur. J.* **2018**, *24*, 4982–4990. [CrossRef]
33. Corriu, R.J.P.; Leclerq, D.; Lefevre, P.; Mutin, P.H.; Vioux, A. Preparation of monolithic binary oxide gels by a nonhydrolytic Sol–Gel process. *Chem. Mater.* **1992**, *4*, 961–963. [CrossRef]

34. Iwasaki, M.; Yasumori, A.; Shibata, S.; Yamane, M. Preparation of high homogeneity BaO-TiO_2-SiO_2 gel. *J. Sol–Gel Sci. Technol.* **1994**, *2*, 387–391. [CrossRef]
35. Goel, S.C.; Chiang, M.Y.; Gibbons, P.C.; Buhro, W.F. New Chemistry for the Sol–Gel Process: Acetone as a New Condensation Reagent. *Mater. Res. Soc. Symp. Proc.* **1992**, *271*, 3–13. [CrossRef]
36. Mehrotra, R.C. Present status and future potential of the Sol–Gel process. In *Chemistry, Spectroscopy and Applications of Sol–Gel Glasses*; Reisfeld, R., Jorgensen, C.K., Eds.; Springer: Berlin, Germany, 1992; p. 20.
37. Guiton, T.A.; Czekaj, C.L.; Pantano, C.G. Organometallic Sol/Gel Chemistry of Metal Sulfides. *J. Non-Cryst. Solids* **1990**, *121*, 7–15. [CrossRef]
38. Gacion, T.; Malier, L.; Boilot, J.-P. New Transparent Chalcogenide Materials Using a Sol–Gel Process. *Chem. Mater.* **1997**, *9*, 1502–1504.
39. Ptatschek, V.; Schreder, B.; Herz, K.; Hilbert, U.; Ossau, W.; Schottner, G.; Rahäuser, O.; Bischof, T.; Lermann, G.; Materny, A.; et al. Sol–Gel Synthesis and Spectroscopic Properties of Thick Nanocrystalline CdSe Films. *J. Phys. Chem. B* **1997**, *101*, 8898–8906. [CrossRef]
40. Rusch, P.; Luebkemann, F.; Borg, H.; Eckert, J.G.; Dorfs, D.; Bigall, N.C. Influencing the coupling between network building blocks in CdSe/CdS dot/rod aerogels by partial cation exchange. *J. Chem. Phys.* **2022**, *156*, 234701. [CrossRef]
41. Seddon, A.B.; Hodgson, S.N.B.; Scott, M.G. Sol–Gel approach to preparing germanium disulphide. *J. Mater. Sci.* **1991**, *26*, 2599–2602. [CrossRef]
42. Leidich, S.; Buechele, D.; Lauenstein, R.; Kluenker, M.; Lind, C. "Non-hydrolytic" sol–gel synthesis of molybdenum sulfides. *J. Solid State Chem.* **2016**, *242*, 175–181. [CrossRef]
43. Gerwig, M.; Schwarz, M.; Brendler, E.; Kroke, E. $[SiS_xH_y]_n$—Perhydridopolysilathianes—Cross-linked thio-analogs of polysiloxanes. *Eur. J. Inorg. Chem.* **2016**, *2016*, 5028–5035. [CrossRef]
44. Rovai, R.; Lehmann, C.W.; Bradley, J.S. Non-Oxide Sol–Gel Chemistry: Preparation from Tris(dialkylamino)silazanes of a Carbon-Free, Porous, Silicon Diimide Gel. *Angew. Chem. Int. Ed.* **1999**, *38*, 2036–2038. [CrossRef]
45. Kitney, S.P.; Sajedin, S.M.; Rocher, V.; Cheng, F.; Kelly, S.M. Silicon diimide gel as an efficient stationary phase in thin layer chromatography for acid-sensitive organic compounds. *Chem. Commun.* **2017**, *53*, 11080–11082. [CrossRef]
46. Paine, R.T. Processing of boron-nitrogen preceramic polymers. *J. Inorg. Organomet. Polym.* **1992**, *2*, 183–195. [CrossRef]
47. Jäger, L.; Köhler, H. Pseudochalkogene. *Sulfur Reports* **1992**, *12*, 159–203. [CrossRef]
48. Riedel, R.; Kroke, E.; Greiner, A.; Gabriel, A.O.; Ruwisch, L.; Nicolich, J.; Kroll, P. Inorganic Solid State Chemistry with Main Group Element Carbodiimides. *Chem. Mater.* **1998**, *10*, 2964–2979. [CrossRef]
49. Lippe, K.; Wagler, J.; Kroke, E.; Herkenhoff, S.; Ischenko, V.; Woltersdorf, J. Cyclic Silylcarbodiimides as Precursors for Porous Si/C/N Materials: Formation, Structures, and Stabilities. *Chem. Mater.* **2009**, *21*, 3941–3949. [CrossRef]
50. Cheng, H.; Li, Y.; Kroke, E.; Herkenhoff, S. In situ synthesis of Si_2N_2O/Si_3N_4 composite ceramics using polysilyloxycarbodiimide precursors. *J. Eur. Ceram. Soc.* **2013**, *33*, 2181–2189. [CrossRef]
51. Voelger, K.W.; Hauser, R.; Kroke, E.; Riedel, R.; Ikuhara, Y.H.; Iwamoto, Y. Synthesis and characterization of novel non-oxide Sol–Gel derived mesoporous amorphous Si-CN membranes. *J. Ceram. Soc. Japan* **2006**, *114*, 567–570. [CrossRef]
52. Krupinski, K.; Brendler, E.; Gericke, R.; Wagler, J.; Kroke, E. A new aspect of the "pseudo water" concept of bis(trimethylsilyl)carbodiimide—"pseudohydrates" of aluminum. *Z. Naturforsch.* **2018**, *B73*, 911–918. [CrossRef]
53. Kroke, E.; Völger, K.W.; Klonczynski, A.; Riedel, R. A Sol–Gel Route to B_4C. *Angew. Chem. Int. Ed.* **2001**, *40*, 1698–1700. [CrossRef]
54. El-Gamel, N.E.A.; Schwarz, M.; Brendler, E.; Kroke, E. s-Triazine and tri-s-triazine based organic-inorganic hybrid gels prepared from chlorosilanes by exchange reactions. *Chem. Commun.* **2006**, 4741–4743. [CrossRef]
55. Mahadik, D.B.; Wang, Q.; Meti, P.; Lee, K.-Y.; Gong, Y.-D.; Park, H.-H. Synthesis of multi-functional porous superhydrophobic trioxybenzene cross-linked silica aerogels with improved textural properties. *Ceram. Internat.* **2020**, *46*, 17969–17977. [CrossRef]
56. Wang, Q.; Mahadik, D.B.; Meti, P.; Gong, Y.-D.; Lee, K.-Y.; Park, H.-H. Dioxybenzene-bridged hydrophobic silica aerogels with enhanced textural and mechanical properties. *Micropor. Mesopor. Mater.* **2020**, *294*, 109863. [CrossRef]
57. Small, J.H.; Shea, K.J.; Loy, D.A. Arylene- and alkylene-bridged polysilsesquioxanes. *J. Non-Cryst. Solids* **1993**, *160*, 234–246. [CrossRef]
58. Shea, K.J.; Loy, D.A. Bridged Polysilsesquioxanes: Molecular Engineering of Hybrid Organic–Inorganic Materials. *MRS Bulletin* **2001**, *26*, 368–375. [CrossRef]
59. Pfeifer, S.; Schwarzer, A.; Schmidt, D.; Brendler, E.; Veith, M.; Kroke, E. Precursors for pyromellit-bridged silica sol–gel hybrid materials. *New J. Chem.* **2013**, *37*, 169–180.
60. Pfeifer, S.; Brendler, E.; Veith, M.; Kroke, E. Hybrid-Coatings Derived from Pyromellitic Acid Bridged Alkoxy-silylalkyl Precursors. *J. Sol–Gel Sci. Technol.* **2014**, *70*, 191–202. [CrossRef]
61. Haas, K.-H. Hybrid Inorganic–Organic Polymers Based on Organically Modified Si-Alkoxides. *Adv. Eng. Mater.* **2000**, *2*, 571–582. [CrossRef]
62. Croissant, J.G.; Cattoen, X.; Durand, J.-O.; Wong Chi Man, M.; Khashab, N.M. Organosilica hybrid nanomaterials with a high organic content: Syntheses and applications of silsesquioxanes. *Nanoscale* **2016**, *8*, 19945–19972. [CrossRef] [PubMed]
63. Furukawa, H.; Cordova, K.E.; O'Keeffe, M.; Yaghi, O.M. The Chemistry and Applications of Metal-Organic Frameworks. *Science* **2013**, *341*, 974. [CrossRef] [PubMed]

64. Ding, S.-Y.; Wang, W. Covalent organic frameworks (COFs). From design to applications. *Chem. Soc. Rev.* **2013**, *42*, 548–568. [CrossRef] [PubMed]
65. Ismail, R.M. Silylierung mit Hexamethyldisilazan. *Z. Naturforsch.* **1963**, *18b*, 582. [CrossRef]
66. Lippe, K.; Kroke, E.; Wagler, J. *CSD Private Communication*, 2016; CCDC 622212.

gels

MDPI

Article

Green Chemistry for the Transformation of Chlorinated Wastes: Catalytic Hydrodechlorination on Pd-Ni and Pd-Fe Bimetallic Catalysts Supported on SiO_2

Julien G. Mahy [1,2,*], Thierry Delbeuck [1], Kim Yên Tran [1], Benoît Heinrichs [1] and Stéphanie D. Lambert [1]

[1] Department of Chemical Engineering—Nanomaterials, Catalysis & Electrochemistry, University of Liège, B6a, Quartier Agora, Allée du six Août 11, 4000 Liège, Belgium

[2] Institut National de la Recherche Scientifique (INRS), Centre-Eau Terre Environnement, Université du Québec, 490, Rue de la Couronne, Québec, QC G1K 9A9, Canada

* Correspondence: julien.mahy@uliege.be; Tel.: +32-3663563

Abstract: Monometallic catalysts based on Fe, Ni and Pd, as well as bimetallic catalysts based on Fe-Pd and based on Ni-Pd supported on silica, were synthesized using a sol–gel cogelation process. These catalysts were tested in chlorobenzene hydrodechlorination at low conversion to consider a differential reactor. In all samples, the cogelation method allowed very small metallic nanoparticles of 2–3 nm to be dispersed inside the silica matrix. Nevertheless, the presence of some large particles of pure Pd was noted. The catalysts had specific surface areas between 100 and 400 m^2/g. In view of the catalytic results obtained, the Pd-Ni catalysts are less active than the monometallic Pd catalyst (<6% of conversion) except for catalysts with a low proportion of Ni (9% of conversion) and for reaction temperatures above 240 °C. In this series of catalysts, increasing the Ni content increases the activity but leads to an amplification of the catalyst deactivation phenomenon compared to Pd alone. On the other hand, Pd-Fe catalysts are more active with a double conversion value compared to a Pd monometallic catalyst (13% vs. 6%). The difference in the results obtained for each of the catalysts in the Pd-Fe series could be explained by the greater presence of the Fe-Pd alloy in the catalyst. Fe would have a cooperative effect when associated with Pd. Although Fe is inactive alone for chlorobenzene hydrodechlorination, when Fe is coupled to another metal from the group VIIIb, such as Pd, it allows the phenomenon of Pd poisoning by HCl to be reduced.

Keywords: noble metal; environmental catalysis; sol–gel process; porous materials; green chemistry

Citation: Mahy, J.G.; Delbeuck, T.; Tran, K.Y.; Heinrichs, B.; Lambert, S.D. Green Chemistry for the Transformation of Chlorinated Wastes: Catalytic Hydrodechlorination on Pd-Ni and Pd-Fe Bimetallic Catalysts Supported on SiO_2. *Gels* **2023**, *9*, 275. https://doi.org/10.3390/gels9040275

Academic Editors: Francesco Caridi, Giuseppe Paladini and Andrea Fiorati

Received: 3 March 2023
Revised: 21 March 2023
Accepted: 23 March 2023
Published: 25 March 2023

1. Introduction

Over time, the presence of chlorinated organic compounds in industry increased, both in the chemical and petrochemical sectors and in the agricultural and electronic sectors. However, the discovery of the carcinogenic, mutagenic and toxic effects of these compounds led to their production being reduced and then banned. These products are nevertheless present in residual stocks and in certain chemical industry process fumes [1,2]. It is therefore necessary to develop treatment methods which allow their emissions, as well as the quantities of these compounds which have already been produced, to be eliminated.

Thermal or catalytic incineration [2,3] is the most widespread process for the treatment of these compounds because it is already in place for other wastes and is therefore well-known and simple to implement. However, the incineration of organochlorine compounds leads to the formation of other harmful compounds such as dioxins. The development of other methods is therefore necessary.

Hydrodechlorination (HDC) seems to be an interesting alternative. This process, the principle of which can be summarized by Equation (1) [4], has the advantages of producing recoverable hydrocarbons, of recovering chlorine in the form of hydrochloric acid and of not producing dioxins. This process could therefore prove to be attractive from an economic and environmental point of view [2,5].

$$C_xH_yCl_z + z\,H_2 \rightarrow C_xH_{y+z} + z\,HCl \quad (1)$$

There are a wide variety of chlorinated organic compounds [6–8]. This work focuses on the study of the hydrodechlorination of chlorinated aromatic compounds because these are the most widespread and the most persistent. The study of the HDC reaction for this family of compounds is made through the use of a model molecule, monochlorobenzene [9–11], C_6H_5Cl, which is the simplest molecule in this family of compounds. The equation corresponding to the reaction carried out is:

$$C_6H_5Cl + H_2 \rightarrow C_6H_6 + HCl. \quad (2)$$

This reaction must be controlled well to avoid the formation of cyclohexane. Otherwise, the need for a complex separation would arise. In addition, benzene is of much greater economic interest. High selectivity of the monochlorobenzene hydrodechlorination reaction is therefore a very important criterion in this study.

The use of a catalyst is necessary for this reaction to happen. Palladium has been identified as the most active metal for the HDC reaction of chlorobenzene [8,11–13]. However, palladium presents a major problem: the palladium-based catalyst is deactivated by adsorption of the HCl produced during the HDC reaction [14]. Nickel is the metal most often cited after palladium for the HDC of chlorinated compounds [2,11,15,16]. Nickel is less active than palladium but has the advantage of being less sensitive than the latter to deactivation [17]. Some authors have observed an increase in activity and/or stability, i.e., resistance to deactivation, when nickel and palladium are alloyed with other metals such as iron [5]. It is known that certain bimetallic catalysts exhibit superior performance to monometallic catalysts [18–20]. Many authors [12,21,22] demonstrated that the selectivity of non-destructive catalytic reactions is often higher for a bimetallic catalyst than a monometallic catalyst. Additionally, bimetallic catalysts are often more resistant to poisoning than monometallic catalysts [11,14]. Among the supports used, carbon [23,24], alumina [11,14] and silica [10,17] are frequently encountered. Only the latter two allow oxidative regeneration of spent catalysts.

There are several existing methods of synthesizing bimetallic catalysts [12]. Among those, the sol–gel method of cogelation allows the synthesis of the support and the dispersion of the precursors of the active metal sites in the porosity of this support to be combined in a single step to produce Ni-Cu/SiO_2 [22], Cu/SiO_2 [25], Pd/SiO_2 [26–28] and Fe/SiO_2 [28–30]. The sol–gel method is a synthetic process carried out at a low temperature and at a low pressure which leads to the formation of engineered and controlled ceramic nanopowders. This approach could be successfully employed to fabricate oxide, non-oxide and composite nanopowders with high purity [31]. Previous studies which characterized catalysts prepared using this method have demonstrated the method's remarkable properties [25,26,28,32]: very high dispersion of the active metallic phase, formation of nanoalloys, localization of metallic nanoparticles in silica cages allowing sintering of the metal to be avoided while keeping it perfectly accessible, hierarchical porous structure allowing easy distribution of reactants and products and the possible regeneration of the catalyst.

The objective of this work is to evaluate the extent to which the sol–gel method of cogelation for the synthesis of catalysts for the HDC of monochlorobenzene is of interest. The novelty resides in the production of Pd-Ni/SiO_2 and Pd-Fe/SiO_2 catalysts in one step by the cogelation process, something that is not reported in the literature. Different Pd-Ni and Pd-Fe bimetallic catalysts, supported on silica and synthesized by cogelation, were studied and their performances were evaluated. Then, the performances of the catalysts were related to their physico-chemical properties. A series of characterizations were carried out, including X-ray diffraction, transmission electron microscopy, measurement of nitrogen adsorption–desorption isotherms and ICP. Bimetallic catalysts Pd-Ni/SiO_2 and Pd-Fe/SiO_2

and monometallic catalysts Fe/SiO_2, Ni/SiO_2 and Pd/SiO_2 were be synthesized. For each bimetallic catalyst formulation, a series of catalysts with different metal contents was studied, the Pd content remained constant and the value of the second metal varied. Thus, the influence of the presence of the second metal can be studied.

2. Results and Discussion

The amounts of all reagents used for material synthesis are detailed in Table 1.

Table 1. Operating synthesis variables for the catalysts Pd-Fe/SiO_2 et Pd-Ni/SiO_2.

	Pd Amount (wt.%) [a]	Fe Amount (wt.%) [a]	$n_{Pd(NO3)2}$ (mmol)	$n_{Fe(acac)3}$ (mmol)	n_{TEOS} (mmol)	n_{EDAS} (mmol)	$n_{water+NH3}$ (mmol)	n_{C2H5OH} (mmol)
Pd100	1.50	-	0.96014	-	109.71	1.92	555.76	1116.31
Pd33-Fe67	1.50	1.57	0.97731	1.95463	103.99	7.82	549.28	1118.11
Pd50-Fe50	1.50	0.79	0.96865	0.96865	106.88	4.84	552.55	1117.21
Pd67-Fe33	1.50	0.39	0.96438	0.48219	108.30	3.37	554.16	1116.76
Fe100	-	0.79	-	0.95349	108.80	2.86	554.16	1116.76
	Pd Amount (wt.%) [a]	**Ni Amount (wt.%) [a]**	**$n_{Pd(NO3)2}$ (mmol)**	**$n_{Ni(acac)3}$ (mmol)**	**n_{TEOS} (mmol)**	**n_{EDAS} (mmol)**	**$n_{water+NH3}$ (mmol)**	**n_{C2H5OH} (mmol)**
Pd33-Ni67	1.50	1.66	2.44925	4.89851	245.69	34.29	357.03	2799.78
Pd50-Ni50	1.50	0.83	0.96982	0.96982	104.05	7.76	549.35	1118.10
Pd67-Ni33	1.50	0.41	2.41239	1.2062	267.24	12.06	381.42	2793.00
Ni100	1.50	0.83	-	0.995462	106.02	5.73	551.58	1117.48

[a] = measured by ICP-AES.

2.1. Synthesis

For all catalysts, no weight loss could be observed during the different steps of synthesis, calcination and reduction, and the wt.% of metal was as expected in each material (Table 1). Therefore, all the reagents reacted as was previously observed with this kind of sol–gel synthesis [26,28].

The gel time for all samples is given in Table 2. The gel time is the time which elapsed between the introduction of the last reagent into the solution and gelation in the oven at 80 °C. Gelling is defined as the moment when the liquid no longer flows when the bottle is tilted at an angle of 45°.

Table 2. Physico-chemical properties of samples Pd-Fe/SiO_2 and Pd-Ni/SiO_2.

	Gel Time (min)	d_{XRD} (nm) [a]	d_{TEM} (nm)	S_{BET} (m^2/g)	[EDAS]/[TEOS]
Pd100	55	-	2.1	220	0.018
Pd33Fe67	22	25	1.5	395	0.075
Pd50Fe50	25	22	2.1	360	0.045
Pd67Fe33	30	19	2.6	300	0.031
Fe100	35	-	-	110	0.026
	Gel Time (min)	**d_{XRD} (nm) [a]**	**d_{TEM} (nm)**	**S_{BET} (m^2/g)**	**[EDAS]/[TEOS]**
Ni100	30	6	-	230	0.054
Pd33Ni67	18	13	2.6	320	0.140
Pd50Ni50	25	11	2.5	290	0.075
Pd67Ni33	35	4	2.5	250	0.045

[a] = measured with Scherrer equation [33].

As described in [28,34], EDAS reacts faster than TEOS for the hydrolysis and condensation reactions. The evolution of the gel time observed in Table 2 follows this trend.

Indeed, for example, in the Pd-Fe series the gel time increases from 22 min to 30 min when the EDAS/TEOS ratio decreases from 0.075 to 0.031. The impact of the EDAS on the catalyst texture will be discussed later when analyzing the specific surface area results (see Section 2.4).

2.2. Composition of the Catalysts

The ICP results (Table 1) showed that the catalysts were well doped with Pd, Ni and Fe but it does not give information about the nature of these species. XRD measurements were performed to determine the phases of the metallic species. Figures 1 and 2 represent all the XRD patterns of the samples.

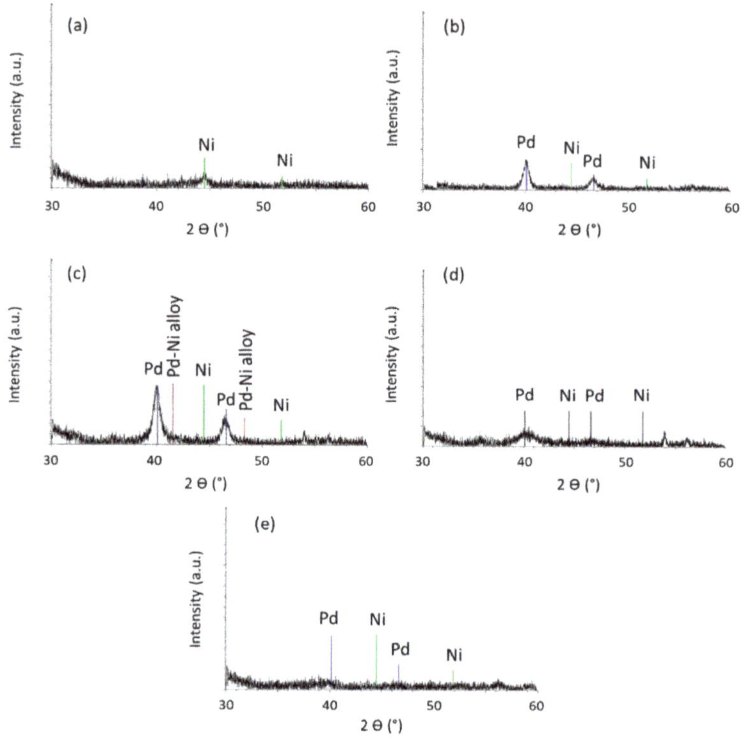

Figure 1. XRD patterns of (**a**) Ni100, (**b**) Pd33Ni67, (**c**) Pd50Ni50, (**d**) Pd67Ni33 and (**e**) Pd100 samples. Reference peaks corresponding to Pd (JCPD No. 46-1043 [24]), Ni (JCPD No. 04-0850 [16]) and Pd-Ni alloy (JCPD No. 65-9444 [35]) are denoted on the figures.

In Figure 1, the Pd-Ni/SiO_2 series is represented. In Figure 1a, which represents the Ni100 catalyst, two weak crystallographic reflections are observed coinciding with two of the characteristic lines of Ni at 44.41° and 51.90°. The low intensities of these peaks and their spreading could be explained by two factors: (i) the very small size of the metallic particles; (ii) the presence of amorphous Ni oxide or hydroxide which cannot be detected. In Figure 1b–d which illustrates the Pd-Ni/SiO_2 catalysts, two crystallographic reflections characteristic of Pd are observed, but none of Ni. The presence of a Pd-Ni alloy would be marked by the presence of a crystallographic reflection between the characteristic lines of Pd and Ni. Their absence at the location of the characteristic peaks of the Pd-Ni alloy and Ni does not mean these two compounds are absent. Indeed, seeing the crystallographic reflection obtained in Figure 1a for the Ni100 sample, one can easily conceive that, with such a weak signal, it is very difficult to detect this compound. Moreover, assuming that a

Pd-Ni alloy has been formed, the signal would be even weaker and would more than likely be confused with the background noise.

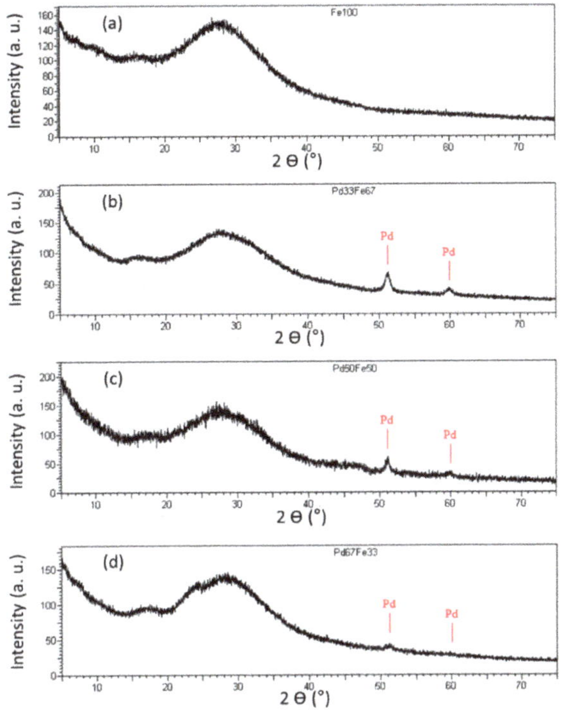

Figure 2. XRD patterns of (**a**) Fe100, (**b**) Pd33Fe67, (**c**) Pd50Fe50 and (**d**) Pd67Fe33 samples. Reference peaks corresponding to Pd are denoted on the figures.

As regards the Pd-Fe/SiO_2 series of catalysts, the results obtained are almost identical to the Pd-Ni/SiO_2 series (Figure 2), except that no compound is detected in the case of the Fe100 sample. As a reminder, the diffractograms in this series extend from $2\theta = 0°$ to $80°$. The crystallographic reflections present for angles of less than approximately 40° are due to the support, namely, the silica. The diffractogram corresponding to Fe100 is shown in Figure 2a.

Although a signal can be detected in the case of the Ni100 sample (Figure 1a), no signal is detected in the case of the Fe100 sample (Figure 2a). Two possibilities could explain this phenomenon: (i) the size of the iron particles is very small; (ii) the presence of amorphous iron oxide or hydroxide which cannot be detected. If, with Ni it could already be conceived that the presence of an alloy was possible, with Fe it can be conceived even more easily that a Pd-Fe alloy could be present. Indeed, since Fe was not even detected on the diffractogram for the Fe100 sample (Figure 2a), it is more than likely that if a Pd-Fe alloy existed in the catalysts synthesized for this study, it would not be detected either. Moreover, it seems obvious that if the Fe is not detected in the Fe100 catalyst, it will also not be detected on the other catalysts in the Pd-Fe series, as can be seen on the diffractograms (Figure 2b–d).

The size of the metallic particles (d_{XRD}) can be estimated directly from the diffractograms of the catalysts by applying Scherrer's formula based on the width of the peaks (Equation (3)). The particle sizes of the different catalysts are listed in Table 2 for all the samples. The size is in the range of 5–15 nm for the Pd-Ni/SiO_2 series and around 20–25 nm for the Pd-Fe/SiO_2 series.

2.3. Morphology

Micrographs of some Pd-Ni/SiO_2 and Pd-Fe/SiO_2 catalysts are presented in Figure 3. For most of the samples, similar morphology is obtained with bigger particles (corresponding to silica particles around 20–30 nm) with smaller darker particles inside (metallic particles around 1–3 nm). With the TEM micrographs, one can measure the distribution of the metal particles within the silica matrix and obtain their average metallic size (d_{TEM}), as denoted in Table 2. In these images, the metallic particles are generally fairly well distinguishable (zoom on Figure 3b). By measuring the different particles, a mean particle size can be calculated, d_{TEM}. For Ni100 (Figure 3a) and Fe100 samples, no metallic nanoparticle is observed inside the silica particle. The presence of oxide in these catalysts, despite the reduction, cannot be ruled out in view of the results obtained using X-ray diffraction. As the contrast is much worse on the TEM images for the oxides, it can be deduced that it would not be possible to distinguish the oxides, which could be one of the explanations for the lack of visible particles for these catalysts. For the other samples, the metallic particle sizes are between 1.5 and 2.6 nm.

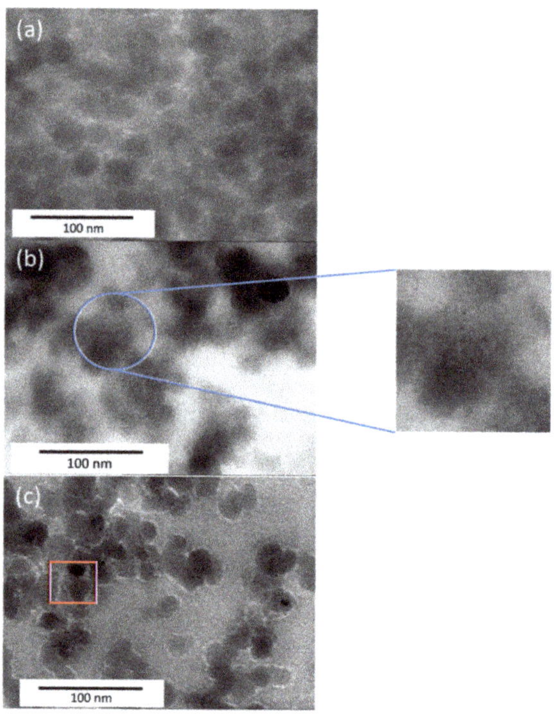

Figure 3. TEM micrographs of (**a**) Ni100, (**b**) Pd33Ni67 and (**c**) Pd50Fe50 samples.

It is observed that the average sizes obtained by this method are very different from the sizes determined by X-ray diffraction (Table 2). This can be explained by the presence of very large particles which are relatively rare on the images as can be seen in Figure 3c (red square).

2.4. Texture

The specific surface area for each sample has been calculated and is denoted in Table 2. Two nitrogen adsorption–desorption isotherms are represented in Figure 4 for the Fe100 and Pd50Ni50 samples, illustrating the two shapes observed for all samples. Indeed, for all samples except for the Fe100 catalyst, the isotherms look like Figure 4b with a sudden increase in adsorbed volume at low relative pressures followed by a plateau, characteristic of microporous adsorbents (type I [36]). In the case of the Fe100 sample (Figure 4a), at low relative pressures no sudden increase in absorbed volume is observed, which testifies to the absence of micropores. At relative pressures close to 1, the isotherms correspond in all cases to a type IV isotherm (macroporous adsorbents [36]).

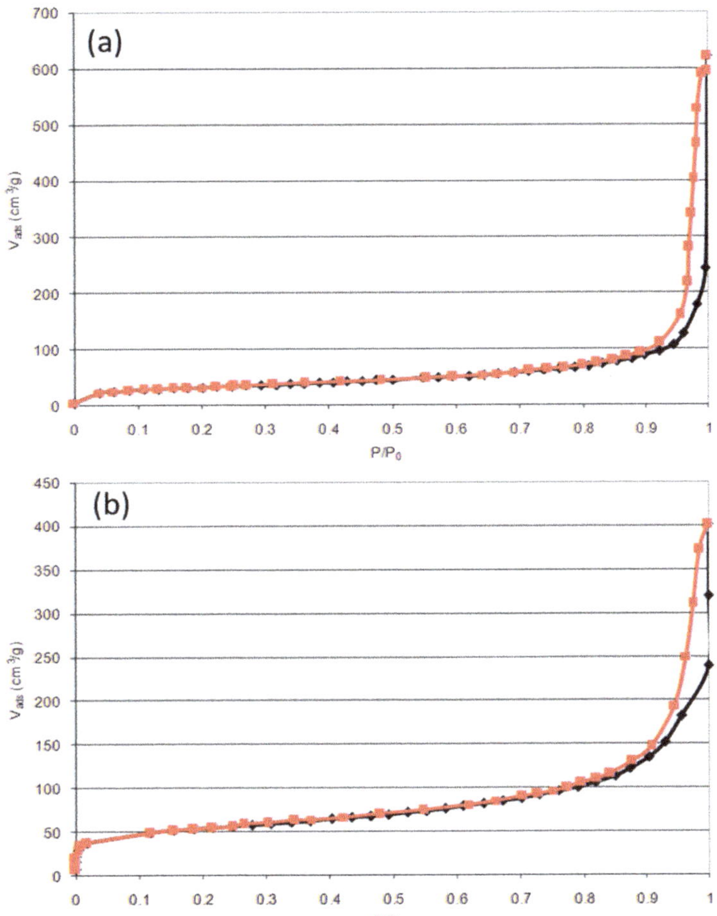

Figure 4. Nitrogen adsorption–desorption isotherms for (**a**) Fe100 and (**b**) Pd50Ni50 samples.

All the isotherms present a hysteresis characteristic of the phenomenon of capillary condensation in the mesopores.

For the S_{BET} values, it can be observed that in each series the specific surface area increases with the EDAS/TEOS ratio as previously explained in [26,28,29,34]. This is due to the role of the nucleating agent EDAS which reacts faster than TEOS (see gel time in Table 2 and Section 2.1) and is therefore the nucleating agent of the silica particles. As was previously described in [26,28,29,34], when the amount of EDAS increases it leads to a higher specific surface area. For example, in the Pd-Fe/SiO_2 series (Table 2), when the EDAS/TEOS ratio is equal to 0.031 for the Pd67Fe33 sample, the S_{BET} value is 300 m^2/g, while when the EDAS/TEOS ratio increases to 0.075 for the Pd33Fe67 sample, the S_{BET} is 395 m^2/g.

2.5. Catalytic Activity

It should be noted that the benzene selectivity is equal to 100% for each catalyst evaluated in this study.

2.5.1. Pd-Ni/SiO_2 Series

Figure 5 shows the conversion of chlorobenzene to benzene as a function of time for the different catalysts in the Pd-Ni/SiO_2 series. The mean conversion values on the plateau at 220 °C are denoted in Table 3 for all samples.

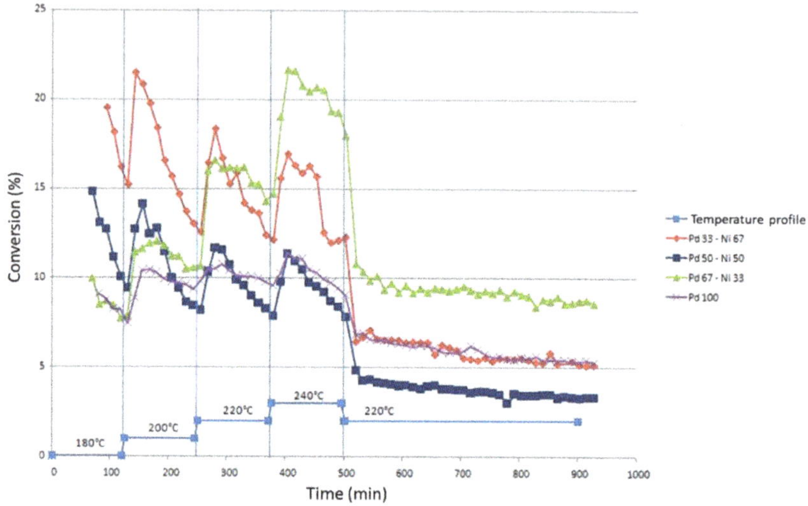

Figure 5. Comparison of the conversions of chlorobenzene to benzene as a function of time for the Pd-Ni/SiO_2 series.

Table 3. Mean chlorobenzene conversion at 220 °C.

	Chlorobenzene Conversion (%)	Standard Deviation (%)
Pd100	6.0	0.7
Pd33Fe67	5.3	1.1
Pd50Fe50	8.3	1.1
Pd67Fe33	12.5	1.0
Fe100	0.0	0.0
	Chlorobenzene Conversion (%)	**Standard Deviation (%)**
Ni100	0.0	0.0
Pd33Ni67	6.0	0.7
Pd50Ni50	3.7	0.6
Pd67Ni33	8.7	0.9

It must be remembered that the mass of the catalyst for the Pd-Ni/SiO_2 series is 0.2 g compared to the Pd-Fe series and the Pd monometallic sample where it is 0.1 g. This difference in mass is necessary for the Pd-Ni/SiO_2 series as otherwise the initial conversion is too high and leads to a rapid deactivation.

There are two different types of behavior in this same series: (i) a relatively constant activity on each level, that is to say a fairly low deactivation. This behavior is characteristic for Pd100 and Pd67Ni33 catalysts; (ii) a strong deactivation. The increase in temperature increases the activity of the catalyst. The deactivation is such that this increase does not make it possible to obtain a better conversion than the maximum conversion at the previous level. This behavior is characteristic of the Pd33Ni67 and Pd50Ni50 catalysts.

Note that the Ni100 catalyst is not shown in Figure 5. In fact, the catalytic experiment carried out on this catalyst revealed virtually zero activity throughout the test, despite the temperature variations (less than 1%).

Taking into account the mass difference introduced into the reactor between the Pd-Ni/SiO_2 series and the monometallic Pd catalyst, it is observed that the presence of Ni in the catalyst does not improve the conversion of chlorobenzene into benzene. Indeed, given that in the case of the Pd100 sample a mass of catalyst twice as small as the mass introduced into the reactor for the Pd-Ni/SiO_2 series samples was introduced into the reactor, one should obtain a significantly lower conversion of chlorobenzene to benzene than in the case of bimetallic catalysts. However, in Figure 5, it is observed that the conversion values of chlorobenzene into benzene for the Pd100 sample are similar to those obtained for the Pd50Ni50 sample, whereas the mass present in the reactor for the Pd100 catalyst is double that for the Pd50Ni50 sample.

When the nickel content increases in the bimetallic catalysts, the deactivation of the bimetallic catalyst takes place more and more rapidly. Therefore, the Pd67Ni33 catalyst is the most attractive of the Pd-Ni/SiO_2 catalysts tested. Seeing the increase in conversion with temperature, it could be thought that this catalyst would only become more interesting than the monometallic Pd catalyst at higher temperatures.

The differences in activity and behavior between the different Pd-Ni/SiO_2 catalysts could be the consequence of the formation of different alloys, the nature of which differs according to the quantity of Ni present in the bimetallic catalyst.

According to the phase diagram of the Pd-Ni alloy (Figure 6), at the temperatures of the hydrodechlorination reaction of chlorobenzene to benzene (between 453 K and 513 K), Pd and Ni form a solid solution, and this is the case in the entire range of concentration between pure Pd and pure Ni. According to this phase diagram, the Pd-Ni/SiO_2 catalysts studied in this work therefore form an alloy with an FCC structure (face centered cubic). However, since the metal particles present in the Pd-Ni/SiO_2 catalysts are very small in size (d_{TEM}, Table 2), the metals may not follow exactly the same phase diagram [37]. Nevertheless, in Figure 1, the main palladium lines appear. According to Scherrer's formula as applied to these peaks (Equation (3)), particle sizes between 4 and 13 nm are obtained (d_{XRD}, Table 2). It could therefore be that the large particles observed using TEM (Figure 3c) are particles of pure palladium. It is also hypothesized that the small particles which are not detectable by using X-ray diffraction consist of a mixture of the two metals, Pd and Ni.

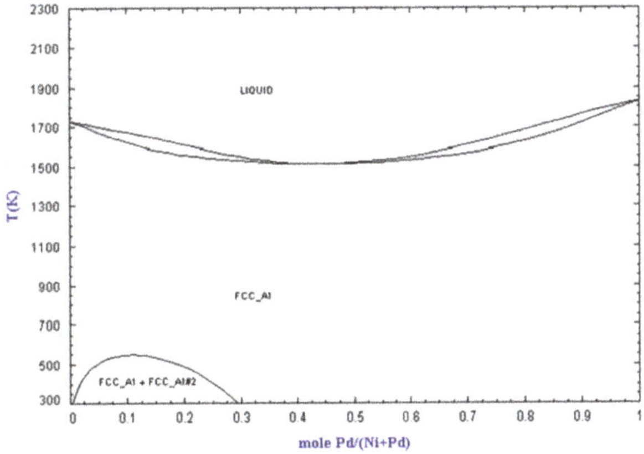

Figure 6. Phase diagram of the Pd-Ni alloy, reproduced from [38].

The Pd-Ni alloy is an alloy that has been studied very little in the literature on the hydrodechlorination reaction of chlorinated compounds. These catalysts are reported, as is the case with this study, to provide good activity, which, however, is lower than in the case of Pd-Fe catalysts. However, Pd-Ni catalysts have a selectivity close to or equal to 100% for the hydrodechlorination of chlorinated compounds [12,39].

An explanation for the better activity of catalysts containing a large amount of Ni has been given by Simagina et al. [39]. The cause may be a Pd segregation phenomenon on the surface of the bimetallic particles. At a high Ni concentration, the Pd is essentially found isolated in small aggregates on the surface of the Ni-rich particles and is therefore more active than in a case where it is found in larger aggregates. Indeed, according to Śrębowata et al. [40], there would be a segregation of Pd at the surface of the Pd-Ni particles.

The deactivation of catalysts containing Ni has already been observed in previous works [2,41]. The cause of this deactivation seems to be the adsorption of HCl. The adsorbed HCl modifies the cubic crystalline structure of Ni to form a hexagonal structure of the $NiCl_2$ type which induces an irreversible deactivation of the catalyst. The deactivation of Pd-based catalysts supported on alumina is usually caused by coking, which leads to a rapid decrease in catalytic activity [12,42]. Nevertheless, in this work it can be assumed that coking is a minor phenomenon on the silica support [12,19] and that catalyst deactivation is mainly due to Ni deactivation by chlorine atoms. This would explain the increased deactivation of the catalysts when the Ni content increases.

2.5.2. Pd-Fe/SiO_2 Series

Figure 7 shows the conversion of chlorobenzene to benzene for the different catalysts tested in the Pd-Fe/SiO_2 series. The mean conversion values on the plateau at 220 °C are denoted in Table 3 for all samples.

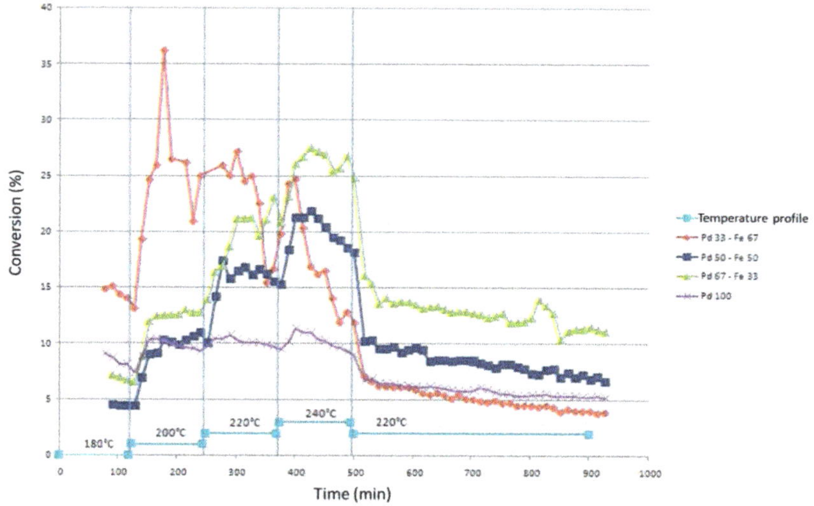

Figure 7. Comparison of the conversions of chlorobenzene to benzene as a function of time for the Pd-Fe/SiO_2 series.

Again, the same two types of behavior observed for the Pd-Ni/SiO_2 series are present, namely: (i) a relatively constant activity on each level, i.e., a fairly low deactivation. This behavior is characteristic of Pd100, Pd67Fe33 and Pd50Fe50 catalysts; (ii) a strong deactivation. The increase in temperature increases the activity of the catalyst. The deactivation is such that this increase does not make it possible to obtain a better conversion than the maximum conversion at the previous level. This behavior is characteristic of the Pd33Fe67 catalyst, other than between the first two levels at 180 °C and 200 °C.

It can be noted that the Fe100 catalyst is not shown in Figure 7. Indeed, the catalytic experiment carried out on this catalyst revealed zero activity throughout the test despite the temperature variations.

In Figure 7, it is observed that the presence of Fe in the presence of Pd leads to different results depending on the quantity of Fe present in the bimetallic catalyst: (i) When compared to the Pd100 catalyst, the presence of a small amount of iron (Pd67Fe33) improves the performance of the catalyst without affecting its deactivation; (ii) the presence of a quantity of Fe equivalent to the quantity of Pd in the catalyst Pd50-Fe50 reduces catalytic performance at a low temperature (180 °C and 200 °C). The conversion values of chlorobenzene into benzene are only higher than those of the Pd100 sample from the third stage, namely 220 °C. Deactivation remains low; (iii) the presence of a greater quantity of Fe than of Pd (Pd33Fe67) improves catalytic performance, compared to the Pd100 and Pd67Fe33 catalysts. On the other hand, the deactivation of the catalyst is much stronger than in the other Pd-Fe/SiO_2 catalysts.

According to the literature [43], the difference in the behavior of a catalyst with the presence of a second metal can be explained by the formation of alloys. The change in the electronic state of the catalyst would then be responsible for the modification of the catalytic properties [44].

On the Fe-Pd phase diagram [38] (Figure 8), by tracing a horizontal line at 400 °C (673 K), there are several possible alloys depending on the atomic percentage of Pd: a mixture of Fe + FePd, the FePd alloy or the $FePd_3$ alloy. It is hypothesized that the nature of the particles detected on micrographs (Figure 3) follows this phase diagram, although the metallic particles are very small in size [37].

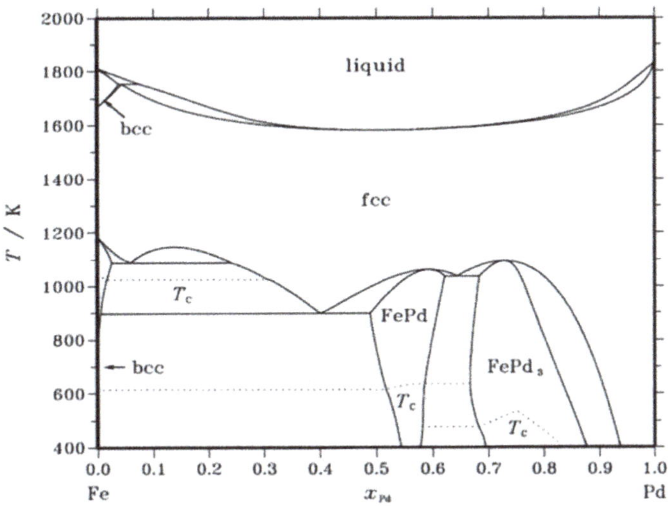

Figure 8. Phase diagram of the Pd-Fe alloy, reproduced from [38].

Pure palladium particles were detected by using XRD (Figure 2). According to Scherrer's formula (Equation (3)), the dimensions of these pure palladium particles are between 22 and 25 nm (Table 2). These particles would thus correspond to the large particles observed by TEM. As these are quite rare (Figure 3c), it is considered that the hydrodechlorination activity of Pd-Fe/SiO_2 catalysts mainly result from the small metal particles (Table 2).

For a percentage close to 33 atomic % in Pd, which would correspond to the catalyst Pd33Fe67, the palladium and the iron are organized in a Fe + FePd mixture. The same analysis can be carried out for the two other catalysts, Pd50Fe50 and Pd67Fe33, therefore corresponding, respectively, to 50% and 67% atomic Pd. In these two cases, the alloys present are in different proportions or in different natures. For the Pd50-Fe50 sample, the FePd alloy would be almost alone with a small proportion of monometallic Fe, while for the Pd67Fe33 sample a mixture of alloys with FePd and $FePd_3$ structure is present.

In view of these results (Figure 7), it seems that the FePd-type alloy only allows a better conversion of chlorobenzene into benzene at higher temperatures than for the Pd100 sample. The presence of iron in the unalloyed state, although not active if used alone in a catalyst, could greatly favor the conversion of chlorobenzene into benzene but nevertheless leads to accelerated deactivation of the catalyst. Finally, the $FePd_3$ structural alloy shows better activity than the FePd structural alloy while keeping a reasonable deactivation rate.

Zhang et al. [45] suggest an explanation for the roles played by the two metals. The Pd would act as a catalyst while the Fe would act as an electron donor (reducer). According to Kim et al. [46], the presence of Fe improves the activity of Pd-based catalysts. Tests revealed an alloy with an FePd structure. This alloy would be responsible for the improvement of the catalytic activity because of the cooperative effect of Fe during the hydrogen transfer step.

According to some authors [9,47,48], the particles of Pd-Fe catalysts would be enriched at the surface with Fe atoms and therefore the formation of $FeCl_3$ on the surface of the catalysts would be thermodynamically favored compared to the formation of $PdCl_2$. In this case, the poisoning of Pd sites by HCl is reduced by the elimination of chlorine in the form of $FeCl_3$. In this way, the stability of the bimetallic catalysts is higher than that of the monometallic catalyst based on Pd, as is the case here for the catalysts Pd67Fe33 and Pd50Fe50. Too high a quantity of Fe can nevertheless lead to a phenomenon of encapsulation of the Pd active sites by $FeCl_3$. This would lead to lower stability than in the case of Pd alone, as is the case with Pd33Fe67 [43].

Lingaiah et al. [47] also highlight the importance of particle sizes in the context of the HDC of chlorobenzene. Low dispersion would improve the activity of the catalysts tested, thus demonstrating that the hydrodechlorination reaction of chlorobenzene is sensitive to the surface structure of the catalysts. In addition, interest in low dispersion would be based on an improvement in catalytic stability. Large Pd particles would be less prone to quenching by HCl. These results show that the process of drying using microwaves (MW) is of particular interest. This technique would favor the increase of particle sizes compared to traditional thermal drying methods. In the case of this present study, the metal particles remain small in size, and the activity remains high.

Lingaiah et al. [47] have also shown that even catalysts with high dispersion can exhibit activity and stability comparable to catalysts with low dispersion. Their findings also show a higher activity of monometallic Pd catalysts compared to those of bimetallic catalysts. However, Babu et al. [5] highlight the possibility of forming particular Pd species which are more active than others. It could therefore be thought that the species formed in this work are much more active in the case of bimetallic alloys than for monometallic Pd.

2.6. Comparison with the Literature

In Table 4, the present study is compared to the literature by summarizing the most important parameters and results of each study. First, it can be observed that it is quite difficult to make a comparison because many conditions such as the temperature, the amount of the catalyst, or the H_2 fraction differ greatly from one study to another. From Table 4, it can be also seen that some studies do not give all the information needed for a proper comparison.

Second, it is observed that the catalysts developed in this paper allow a 100% selectivity at a reasonably low temperature (220 °C) with a dilute amount of H_2 (2.6 vol%). Conversion is lower, but the idea of this study was to stay at lower conversion in order to consider a differential reactor. The benzene selectivity is often missing in the other studies and, as expected, conversion increases when the temperature is higher.

Table 4. Comparison of the chlorobenzene (CB) catalytic activity with the literature.

References	Catalyst (Optimal Active Phase Content)	Catalytic Experiment Conditions	Optimal Catalytic Results
Present study	• $Pd-Ni/SiO_2$ (1.5 wt% of Pd and 0.41 wt% of Ni) • $Pd-Fe/SiO_2$ (1.5 wt% of Pd and 0.39 wt% of Fe)	• Temperature 220 °C • Catalytic mass of 0.1 g (for Fe series) or 0.2 g (for the Ni series) • vol% of H_2 with H_2:CB molar ratio of 1000:1	• 12.5% conversion at 220 °C with a benzene selectivity of 100% for the best of Ni series • 9% conversion at 220 °C with a benzene selectivity of 100% for the best of Fe series
[2]	• Ni/Al_2O_3 (6 wt% of Ni)	• Temperature range 150–350 °C • Catalytic mass of 0.1 g • H_2:CB molar ratio of 50:1	• 100% conversion at 300 °C with a benzene selectivity of 20%
[16]	• Ni/Al_2O_3 (90 wt% of Ni)	• Temperature range 350–650 °C • Catalytic mass of 0.1 g • 97 vol% of H_2	• 100% conversion at 350 °C with a benzene selectivity of 100% under pure H2 flow
[5]	• $Pd-Fe/Al_2O_3$ (0.5 wt% of both Pd and Fe)	• Temperature 140 °C • Catalytic mass of 1 g • H_2:CB molar ratio of 10:1	• Conversion not specified • Selectivity of benzene of 98%
[9]	• Pd-Fe/Carbon (2 wt% of both Pd and Fe)	• Temperature range 150–200 °C • Catalytic mass not specified • H_2:CB molar ratio of 3:1	• 100% conversion at 200 °C • Selectivity not specified
[10]	• Ni/MCM-41 (10 wt% of Ni)	• Temperature 100 °C • Catalytic mass not specified • H_2 partial pressure of 1.5 MPa	• 100% conversion at 200 °C • Selectivity not specified
[11]	• $Pd-Ni/Al_2O_3$ (0.005 wt% of both Pd and Ni)	• Temperature range 150–350 °C • Catalytic mass of 50 mg • H_2:CB molar ratio of 50:1	• 87% conversion at 200 °C • Selectivity not specified
[14]	• $Pd-Ni/Al_2O_3$ (0.5 wt% of both Pd and Ni)	• Temperature 140 °C • Catalytic mass of 0.8 g • H_2:CB molar ratio of 3:1	• 100% conversion at 140 °C • Selectivity not specified
[17]	• Ni_2P/SiO_2 (5 wt% of Ni)	• Temperature range 250–400 °C • Catalytic mass of 0.5 g • H_2 flow of 60 mL/min	• 100% conversion at 325 °C • Selectivity not specified
[23]	• Pd/Carbon (0.9 wt% of Pd)	• Temperature range 100–300 °C • Catalytic mass of 8 mg • H_2:CB molar ratio of 30:1	• 85% conversion at 300 °C • Selectivity not specified

3. Conclusions

Two series of bimetallic catalysts, $Pd-Ni/SiO_2$ and $Pd-Fe/SiO_2$, were synthesized by the sol–gel process for the catalysis of the hydrodechlorination of chlorobenzene into benzene. This synthesis process uses two silicon alkoxides, TEOS and EDAS. EDAS is a modified alkoxide with a diamine function allowing the metal to be highly dispersed in the silica matrix. In all cases, the sol–gel process allows very small metallic nanoparticles in the order of 2–3 nm to be dispersed inside the silica matrix. Nevertheless, the presence of some large particles of pure Pd in the analyzed catalysts can be noted. The catalysts have a specific surface between 100 and 400 m^2/g.

All bimetallic catalysts can catalyze the hydrodechlorination of chlorobenzene. In order to consider a differential reactor, the work was carried out at low conversion conditions. The Pd-Ni combination is nevertheless less active than the Pd-Fe combination. Monometallic catalysts Ni/SiO_2 and Fe/SiO_2 are inactive for the hydrodechlorination reaction of chlorobenzene, with no conversion.

Pd-Ni/SiO_2 catalysts are less active than monometallic Pd catalyst (mean conversion of 6%). For catalysts with a low proportion of Ni such as Pd67Ni33 (mean conversion of 9%), one can only foresee a possible improvement in activity compared to that of the monometallic Pd catalyst for reaction temperatures above 240 °C. In this series of catalysts, increasing the Ni content increases the activity but leads to an amplification of the catalyst deactivation phenomenon compared to Pd alone (mean conversion drops to 4% with the Pd50Ni50 sample). The increase in activity could be due to an overall effect, i.e., the Ni would dilute the Pd into small aggregates, and thus increase the surface area of active Pd. The deactivation would be due to the formation of $NiCl_2$.

The Pd-Fe catalysts are overall more active than the monometallic Pd catalyst. The difference in the results obtained for each of the catalysts in the series could be explained by the significant presence of FePd alloy in the catalyst. The combined action of Fe and Pd would be responsible for this increase in activity. Fe would have a cooperative effect when associated with Pd, although it is inactive alone. The presence of Fe could also reduce the phenomenon of Pd poisoning by HCl. The best Pd-Fe catalyst (Pd67Fe33) doubled the mean conversion to 13% compared to the Pd monometallic catalyst (mean conversion of 6%).

4. Materials and Methods

4.1. Catalyst Synthesis

Reagents: Fe(III) acetylacetonate (Fe(acac)$_3$, ($CH_3COCH=C(O)CH_3)_3Fe$, Sigma-Aldrich, 97%, St. Louis, MO, USA); Pd (II) acetylacetonate (Pd(acac)$_2$, CH3COCH=C(O)CH$_3$)$_2$Pd, Sigma-Aldrich, 99%); Ni(II) acetylacetonate (Ni(acac)$_2$, $CH_3COCH=C(O)CH_3)_2Ni$, Sigma-Aldrich, 97%); Ammonia solution (NH_4OH, Sigma-Aldrich, 25%); distilled water; tetraethoxysilane (TEOS, $Si(OC_2H_5)_4$, Sigma-Aldrich, 98%); N-[3-(triméthoxysilyl)propyl]ethylenediamine (EDAS, $(OCH_3)_3$-Si-$(CH_2)_3$-NH-$(CH_2)_2$-NH, Sigma-Aldrich, 97%). The amounts of all reagents used for material synthesis are detailed in Table 1.

Pd(acac)$_2$ and the salt of the second metal, (Ni(acac)$_2$ or Fe(acac)$_3$), were mixed with EDAS and half the volume of ethanol (or only one metallic salt if a monometallic catalyst was prepared). The mixture was then stirred at room temperature until the solution became clear. After the addition of TEOS, a 0.54 M aqueous ammoniacal solution diluted in the remaining half of the ethanol was added with vigorous stirring. The synthesis bottle was then hermetically sealed and placed in the oven at 80 °C for 3 days in order to undergo gelling and aging of the gel.

The wet gels were then dried under vacuum at 80 °C for 3 days, during which the pressure was slowly reduced to 1200 Pa in order to avoid the gel exploding. The gels were then dried at 150 °C at 1200 Pa for 12 h. The dried gels were then calcined in air (0.1 mmol/s) for 12 h at 400 °C with a ramp of 120 °C/h (air flow rate 0.02 mmol/s). The calcined gels used for the catalytic tests were finally reduced in situ under hydrogen (0.05 mmol/s) for 6 h at 400 °C and 1.25 bar. The catalysts used for the characterization were reduced under hydrogen (0.25 mmol/s) for 6 h at 400 °C and at atmospheric pressure. The synthesis process is represented in Figure 9.

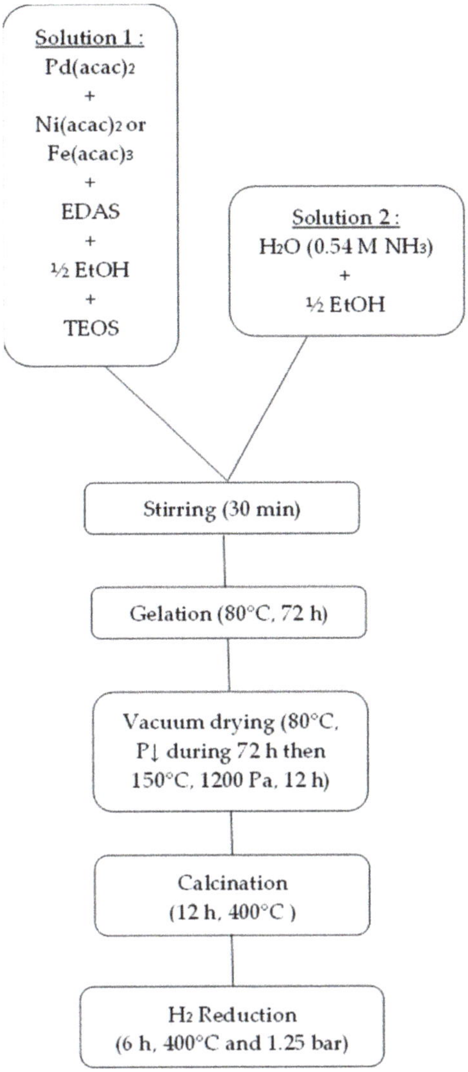

Figure 9. Schematic diagram of the catalyst preparation.

4.2. Characterizations

Nitrogen adsorption–desorption isotherms were obtained thanks to an ASAP 2420 multi-sampler adsorption–desorption volumetric device from Micromeritics.

The crystallographic properties were observed through the X-ray diffraction (XRD) patterns recorded with a Siemens D5000 powder diffractometer using Cu-K_α radiation for the Ni series and using Fe-K_α radiation for the Fe series. The Scherrer formula (Equation (3)) was used to determine the size of the metal crystallites (i.e., d_{XRD}) [33]:

$$d_{XRD} = 0.9\frac{\lambda}{(B\cos(\theta))} \quad (3)$$

where d_{XRD} is the crystallite size of metals (nm), B the peak full width at half maximum after correction of the instrumental broadening (rad), λ the wavelength (nm) and θ the Bragg angle (rad).

The sizes of the metallic nanoparticles were estimated by transmission electron microscopy (TEM) with a Philips CM100 device, with measurements being taken on approximately one hundred particles. Each sample was prepared before being analyzed using TEM. This preparation was necessary to disperse the samples and thus allow their analysis under the electron microscope. The preparation procedure was as follows:

(i) the catalysts to be observed were embedded in the EPON 812 resin;
(ii) embedded catalysts were placed under vacuum for a few hours to allow the resin to penetrate the pores of the catalyst;
(iii) the catalysts then remained for 3 days at 60 °C for the resin to polymerize;
(iv) very thin slices of polymerized resin containing the catalyst were cut using a crystal knife;
(v) this thin slice was then placed on a copper grid.

The grids had a diameter of 3.05 mm and included 400 mesh (37 μm thick).

The sections were then placed directly on the sample holder of the transmission electron microscope.

Real palladium, nickel and iron contents in the catalysts were measured using inductively coupled plasma atomic emission spectroscopy (ICP-AES) on a Spectroflam from Spectro Analytical Instruments. Samples were dissolved in concentrated acids (18 M H_2SO_4, 22 M HF, 14 M HNO_3). Palladium, nickel and iron contents were obtained by comparison with standard solutions in the same medium.

4.3. Catalytic Experiments

The installation used for the hydrodechlorination of chlorobenzene is represented in Figure 10. The reactor was removed and charged with 0.2 g of Pd-Ni/SiO_2 catalyst or 0.1 g of Pd-Fe/SiO_2 catalyst. It was then moved back to its original location. The system was purged with helium. After a purge time of 5 to 10 min, the reduction procedure could be started: a temperature ramp of 5 °C/min up to a plateau of 400 °C was maintained for 6 h, the 2 stages being carried out under a flow rate of 4 NL/h of hydrogen. The reduction temperature was determined by temperature programmed reduction (TPR) (see Supplementary Materials). The temperature of the oven was then brought back to 180 °C, the starting temperature of the actual test cycle. The hydrogen flow was reduced to 1 NL/h and a flow of 37 NL/h of helium was added.

The temperature program was made up of different levels of 180 °C, 200 °C, 220 °C, 240 °C and finally a return to 220 °C. Each level was maintained for a period of 2 h except for the last which was maintained for a minimum of 6 h. The temperature ramps were set at a rate of 5 °C/min. An illustration of this temperature program can be seen in Figure 11. The start of the program coincided with the pivoting of the valve, allowing continuous injection of chlorobenzene.

Throughout the experiment, the chromatograph analyzed the reactor effluents at regular intervals, i.e., approximately every 12 min. Each experiment was carried out three times to assess reproducibility. The effluent was analyzed by gas chromatography (ThermoFinnigan with FID) using a Porapak Q5 packed column.

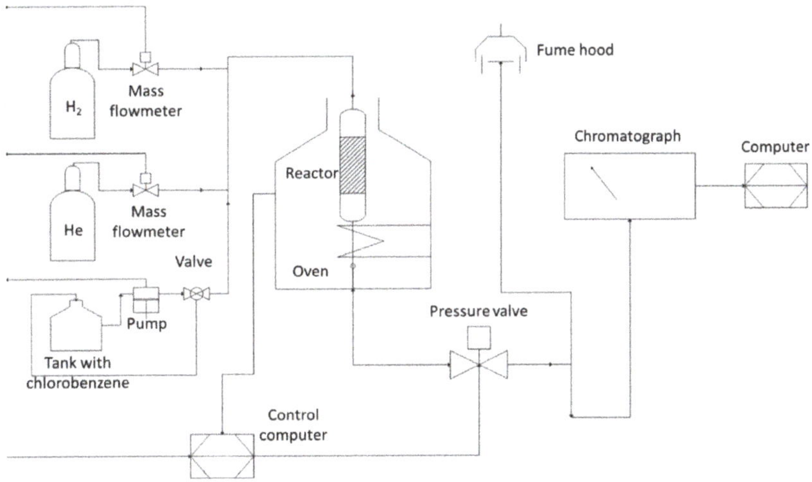

Figure 10. Installation Scheme.

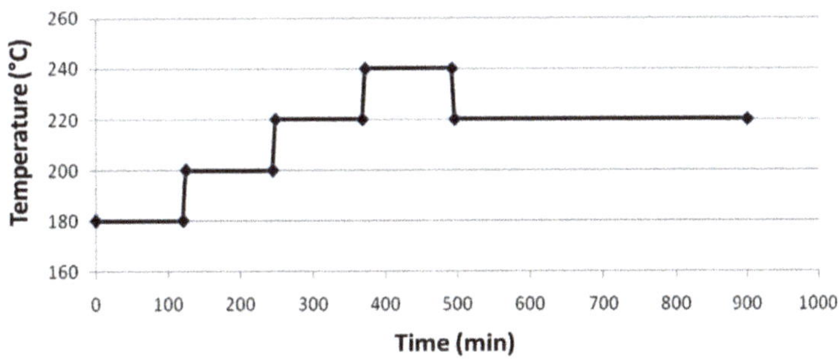

Figure 11. Temperature profile used for the catalytic experiments.

Supplementary Materials: The following supporting information can be downloaded at: https://www.mdpi.com/article/10.3390/gels9040275/s1, Figure S1. TPR of Pd50-Ni50 sample; Figure S2. TPR of Pd50-Fe50 sample.

Author Contributions: J.G.M.: Methodology, Investigation, Formal analysis, Writing—original draft, writing—review and editing. T.D., K.Y.T. and B.H.: Conceptualization, investigation, formal analysis, writing—review and editing. S.D.L.: Methodology, formal analysis, writing—review and editing, funding acquisition and project administration. All authors have read and agreed to the published version of the manuscript.

Funding: This research received no external funding.

Institutional Review Board Statement: Not applicable.

Informed Consent Statement: Not applicable.

Data Availability Statement: The raw/processed data required to reproduce these findings cannot be shared at this time as these data are part of an ongoing study.

Acknowledgments: Julien G. Mahy and Stéphanie D. Lambert thank the F.R.S.-FNRS for his Post-doctoral Researcher position and her Research Director position, respectively. J.G.M. is grateful to the Rotary for a District 2160 grant, to the University of Liège and the FNRS for financial support for a postdoctoral stay in INRS Centre Eau, Terre, Environnement in Québec, Canada.

Conflicts of Interest: The authors declare no conflict of interest.

References

1. McGrath, M. Catalysts and Systems for the Removal of Volatile Oganic Compounds. *Appl. Catal. B* **1995**, *5*, N25. [CrossRef]
2. Ryaboshapka, D.A.; Lokteva, E.S.; Golubina, E.V.; Kharlanov, A.N.; Maslakov, K.I.; Kamaev, A.O.; Shumyantsev, A.V.; Lipatova, I.A.; Shkol'nikov, E.I. Gas-Phase Hydrodechlorination of Chlorobenzene over Alumina-Supported Nickel Catalysts: Effect of Support Structure and Modification with Heteropoly Acid HSiW. *Kinet. Catal.* **2021**, *62*, 127–145. [CrossRef]
3. Ritter, E.R.; Bozzelli, J.W.; Dean, A.M. Kinetic Study on Thermal Decomposition of Chlorobenzene Diluted in H_2. *J. Phys. Chem.* **1990**, *94*, 2493–2504. [CrossRef]
4. Kovenklioglu, S.; Cao, Z.; Shah, D.; Farrauto, R.J.; Balko, E.N. Direct Catalytic Hydrodechlorination of Toxic Organics in Wastewater. *AIChE J.* **1992**, *38*, 1003–1012. [CrossRef]
5. Babu, N.S.; Lingaiah, N.; Kumar, J.V.; Prasad, P.S.S. Studies on Alumina Supported Pd-Fe Bimetallic Catalysts Prepared by Deposition-Precipitation Method for Hydrodechlorination of Chlorobenzene. *Appl. Catal. A Gen.* **2009**, *367*, 70–76. [CrossRef]
6. Xu, L.; Stangland, E.E.; Dumesic, J.A.; Mavrikakis, M. Hydrodechlorination of 1,2-Dichloroethane on Platinum Catalysts: Insights from Reaction Kinetics Experiments, Density Functional Theory, and Microkinetic Modeling. *ACS Catal.* **2021**, *11*, 7890–7905. [CrossRef]
7. Zawadzki, B.; Kowalewski, E.; Asztemborska, M.; Matus, K.; Casale, S.; Dzwigaj, S.; Śrębowata, A. Palladium Loaded BEA Zeolites as Efficient Catalysts for Aqueous-Phase Diclofenac Hydrodechlorination. *Catal. Commun.* **2020**, *145*, 106113. [CrossRef]
8. Deka, J.R.; Saikia, D.; Chen, P.H.; Chen, K.T.; Kao, H.M.; Yang, Y.C. N-Functionalized Mesoporous Carbon Supported Pd Nanoparticles as Highly Active Nanocatalyst for Suzuki-Miyaura Reaction, Reduction of 4-Nitrophenol and Hydrodechlorination of Chlorobenzene. *J. Ind. Eng. Chem.* **2021**, *104*, 529–543. [CrossRef]
9. Lingaiah, N.; Uddin, M.A.; Muto, A.; Sakata, Y. Hydrodechlorination of Chlorinated Hydrocarbons over Metal-Carbon Composite Catalysts Prepared by a Modified Carbothermal Reduction Method. *Chem. Commun.* **1999**, *17*, 1657–1658. [CrossRef]
10. Wu, N.; Zhang, W.; Li, B.; Han, C. Nickel Nanoparticles Highly Dispersed with an Ordered Distribution in MCM-41 Matrix as an Efficient Catalyst for Hydrodechlorination of Chlorobenzene. *Microporous Mesoporous Mater.* **2014**, *185*, 130–136. [CrossRef]
11. Golubina, E.V.; Rostovshchikova, T.N.; Lokteva, E.S.; Maslakov, K.I.; Nikolaev, S.A.; Egorova, T.B.; Gurevich, S.A.; Kozhevin, V.M.; Yavsin, D.A.; Yermakov, A.Y. Chlorobenzene Hydrodechlorination on Bimetallic Catalysts Prepared by Laser Electrodispersion of NiPd Alloy. *Pure Appl. Chem.* **2018**, *90*, 1685–1701. [CrossRef]
12. Flid, M.R.; Kartashov, L.M.; Treger, Y.A. Theoretical and Applied Aspects of Hydrodechlorination Processes—Catalysts and Technologies. *Catalysts* **2020**, *10*, 216. [CrossRef]
13. Ali, A.M.; Podila, S.; Daous, M.A.; Al-Zahrani, A.A.; Mahpudz, A. Highly Efficient Hydrotalcite Supported Palladium Catalyst for Hydrodechlorination of 1, 2, 4-Tri Chlorobenzene: Influence of Pd Loading. *J. Chem. Sci.* **2020**, *132*, 39. [CrossRef]
14. Seshu Babu, N.; Lingaiah, N.; Sai Prasad, P.S. Characterization and Reactivity of Al_2O_3 Supported Pd-Ni Bimetallic Catalysts for Hydrodechlorination of Chlorobenzene. *Appl. Catal. B* **2012**, *111–112*, 309–316. [CrossRef]
15. Hussain, A.A.; Nazir, S.; Irshad, R.; Tahir, K.; Raza, M.; Khan, Z.U.H.; Khan, A.U. Synthesis of Functionalized Mesoporous Ni-SBA-16 Decorated with MgO Nanoparticles for Cr (VI) Adsorption and an Effective Catalyst for Hydrodechlorination of Chlorobenzene. *Mater. Res. Bull.* **2021**, *133*, 111059. [CrossRef]
16. Mishakov, I.V.; Vedyagin, A.A.; Bauman, Y.I.; Potylitsyna, A.R.; Kadtsyna, A.S.; Chesnokov, V.V.; Nalivaiko, A.Y.; Gromov, A.A.; Buyanov, R.A. Two Scenarios of Dechlorination of the Chlorinated Hydrocarbons over Nickel-Alumina Catalyst. *Catalysts* **2020**, *10*, 1446. [CrossRef]
17. Cecilia, J.A.; Jiménez-Morales, I.; Infantes-Molina, A.; Rodríguez-Castellón, E.; Jiménez-López, A. Influence of the Silica Support on the Activity of Ni and Ni_2P Based Catalysts in the Hydrodechlorination of Chlorobenzene. Study of Factors Governing Catalyst Deactivation. *J. Mol. Catal. A Chem.* **2013**, *368–369*, 78–87. [CrossRef]
18. Liao, Y.; Wang, Y.; Zhang, Y. Preparation and Catalytic Hydrodechlorination Property of Nano Bimetallic Catalyst Pd–Ni/ΓAl$_2$ O$_3$–SiO$_2$. *Catalysts* **2022**, *12*, 370. [CrossRef]
19. Liao, Y.; Wang, W.; Zhang, Y.; Wang, Y.; Guo, M. Preparation, Characterization and Catalytic Hydrodechlorination Property for Bleached Shellac of Pd–Ni@SiO_2 Bimetallic Nano-Catalyst. *React. Kinet. Mech. Catal.* **2020**, *130*, 1043–1061. [CrossRef]
20. Xiong, J.; Tian, L.; Cheng, R. Promoted Catalytic Hydrodechlorination for Deep Degradation of Chlorophenols over Rh-La/SiO_2 Catalyst. *J. Hazard. Mater.* **2021**, *416*, 125913. [CrossRef]
21. Guczi, L. Bimetallic Nano-Particles: Featuring Structure and Reactivity. *Catal. Today* **2005**, *101*, 53–64. [CrossRef]
22. Pirard, S.L.; Mahy, J.G.; Pirard, J.-P.; Heinrichs, B.; Raskinet, L.; Lambert, S.D. Development by the Sol-Gel Process of Highly Dispersed Ni-Cu/SiO_2 Xerogel Catalysts for Selective 1,2-Dichloroethane Hydrodechlorination into Ethylene. *Microporous Mesoporous Mater.* **2015**, *209*, 197–207. [CrossRef]
23. Klokov, S.V.; Lokteva, E.S.; Golubina, E.V.; Maslakov, K.I.; Levanov, A.V.; Chernyak, S.A.; Likholobov, V.A. Effective Pd/C Catalyst for Chlorobenzene and Hexachlorobenzene Hydrodechlorination by Direct Pyrolysis of Sawdust Impregnated with Palladium Nitrate. *Catal. Commun.* **2016**, *77*, 37–41. [CrossRef]
24. Lei, M.; Tang, Y.; Wang, H.; Zhu, L.; Zhang, G.; Zhou, Y.; Tang, H. A Catalytic Strategy for Rapid Cleavage of C-Cl Bond under Mild Conditions: Effects of Active Hydrogen Induced by Pd Nanoparticles on the Complete Dechlorination of Chlorobenzenes. *Chem. Eng. J.* **2021**, *419*, 129510. [CrossRef]
25. Lambert, S.; Cellier, C.; Ferauche, F.; Gaigneaux, É.M.; Heinrichs, B. On the Structure-Sensitivity of 2-Butanol Dehydrogenation over Cu/SiO_2 Cogelled Xerogel Catalysts. *Catal. Commun.* **2007**, *8*, 2032–2036. [CrossRef]

26. Lambert, S.; Tran, K.Y.; Arrachart, G.; Noville, F.; Henrist, C.; Bied, C.; Moreau, J.J.E.; Wong Chi Man, M.; Heinrichs, B. Tailor-Made Morphologies for Pd/SiO_2 Catalysts through Sol-Gel Process with Various Silylated Ligands. *Microporous Mesoporous Mater.* **2008**, *115*, 609–617. [CrossRef]
27. Lambert, S.; Sacco, L.; Ferauche, F.; Heinrichs, B.; Noels, A.; Pirard, J.P. Synthesis of SiO_2 Xerogels and Pd/SiO_2 Cogelled Xerogel Catalysts from Silylated Acetylacetonate Ligand. *J. Non Cryst. Solids* **2004**, *343*, 109–120. [CrossRef]
28. Mahy, J.G.; Tasseroul, L.; Tromme, O.; Lavigne, B.; Lambert, S.D. Hydrodechlorination and Complete Degradation of Chlorinated Compounds with the Coupled Action of Pd/SiO_2 and Fe/SiO_2 Catalysts: Towards Industrial Catalyst Synthesis Conditions. *J. Env. Chem. Eng.* **2019**, *7*, 103014. [CrossRef]
29. Mahy, J.G.; Tasseroul, L.; Zubiaur, A.; Geens, J.; Brisbois, M.; Herlitschke, M.; Hermann, R.; Heinrichs, B.; Lambert, S.D. Highly Dispersed Iron Xerogel Catalysts for P-Nitrophenol Degradation by Photo-Fenton Effects. *Microporous Mesoporous Mater.* **2014**, *197*, 164–173. [CrossRef]
30. Mahy, J.G.; Tasseroul, L.; Herlitschke, M.; Hermann, R.P.; Lambert, S.D. Fe^{3+}/Iron Oxide/SiO_2 Xerogel Catalysts for p-Nitrophenol Degradation by Photo-Fenton Effects: Influence of Thermal Treatment on Catalysts Texture. *Mater. Today Proc.* **2016**, *3*, 464–469. [CrossRef]
31. Najafi, A.; Sharifi, F.; Mesgari-Abbasi, S.; Khalaj, G. Influence of PH and Temperature Parameters on the Sol-Gel Synthesis Process of Meso Porous ZrC Nanopowder. *Ceram. Int.* **2022**, *48*, 26725–26731. [CrossRef]
32. Lambert, S.; Ferauche, F.; Brasseur, A.; Pirard, J.P.; Heinrichs, B. Pd-Ag/SiO_2 and Pd-Cu/SiO_2 Cogelled Xerogel Catalysts for Selective Hydrodechlorination of 1,2-Dichloroethane into Ethylene. *Catal. Today* **2005**, *100*, 283–289. [CrossRef]
33. Sing, K.S.W.; Rouquerol, J. 3 Characterization of Solid Catalysts. *Handb. Heterog. Catal.* **1997**, *438*, 428–582. [CrossRef]
34. Lambert, S.; Alié, C.; Pirard, J.P.; Heinrichs, B. Study of Textural Properties and Nucleation Phenomenon in Pd/SiO_2, Ag/SiO_2 and Cu/SiO_2 Cogelled Xerogel Catalysts. *J. Non Cryst. Solids* **2004**, *342*, 70–81. [CrossRef]
35. She, Y.; Lu, Z.; Fan, W.; Jewell, S.; Leung, M.K.H. Facile Preparation of PdNi/RGO and Its Electrocatalytic Performance towards Formic Acid Oxidation. *J. Mater. Chem. A Mater.* **2014**, *2*, 3894–3898. [CrossRef]
36. Thommes, M.; Kaneko, K.; Neimark, A.V.; Olivier, J.P.; Rodriguez-Reinoso, F.; Rouquerol, J.; Sing, K.S.W. Physisorption of Gases, with Special Reference to the Evaluation of Surface Area and Pore Size Distribution (IUPAC Technical Report). *Pure Appl. Chem.* **2015**, *87*, 1051–1069. [CrossRef]
37. Samsonov, V.M.; Bazulev, A.N.; Sdobnyakov, N.Y. On Applicability of Gibbs Thermodynamics to Nanoparticles. *Cent. Eur. J. Phys.* **2003**, *3*, 474–484. [CrossRef]
38. Centre de Recherche en Calcul Thermodynamique de l'Ecole Polytechnique de Montréal Noble Metal Alloy Database. Available online: https://www.crct.polymtl.ca/fact/documentation/ (accessed on 4 January 2023).
39. Simagina, V.; Likholobov, V.; Bergeret, G.; Gimenez, M.T.; Renouprez, A. Catalytic Hydrodechlorination of Hexachlorobenzene on Carbon Supported Pd-Ni Bimetallic Catalysts. *Appl. Catal. B* **2003**, *40*, 293–304. [CrossRef]
40. Śrebowata, A.; Juszczyk, W.; Kaszkur, Z.; Karpiński, Z. Hydrodechlorination of 1,2-Dichloroethane on Active Carbon Supported Palladium-Nickel Catalysts. *Catal. Today* **2007**, *124*, 28–35. [CrossRef]
41. Choi, Y.H.; Lee, W.Y. Effect of Ni Loading and Calcination Temperature on Catalyst Performance and Catalyst Deactivation of Ni/SiO_2 in the Hydrodechlorination of 1,2-Dichloropropane into Propylene. *Catal. Lett.* **2000**, *67*, 155–161. [CrossRef]
42. Ordóñez, S.; Díez, F.V.; Sastre, H. Characterisation of the Deactivation of Platinum and Palladium Supported on Activated Carbon Used as Hydrodechlorination Catalysts. *Appl. Catal. B* **2001**, *31*, 113–122. [CrossRef]
43. Golubina, E.V.; Lokteva, E.S.; Lunin, V.V.; Telegina, N.S.; Stakheev, A.Y.; Tundo, P. The Role of Fe Addition on the Activity of Pd-Containing Catalysts in Multiphase Hydrodechlorination. *Appl. Catal. A Gen.* **2006**, *302*, 32–41. [CrossRef]
44. Coq, B.; Figueras, F. Bimetallic Palladium Catalysts: Influence of the Co-Metal on the Catalyst Performance. *J. Mol. Catal. A Chem.* **2001**, *173*, 117–134. [CrossRef]
45. Zhang, W.-X.; Wang, C.-B.; Lien, H.-L. Treatment of Chlorinated Organic Contaminants with Nanoscale Bimetallic Particles. *Catal. Today* **1998**, *40*, 387–395. [CrossRef]
46. Kim, J.K.; Lee, J.K.; Kang, K.H.; Lee, J.W.; Song, I.K. Catalytic Decomposition of Phenethyl Phenyl Ether to Aromatics over Pd-Fe Bimetallic Catalysts Supported on Ordered Mesoporous Carbon. *J. Mol. Catal. A Chem.* **2015**, *410*, 184–192. [CrossRef]
47. Lingaiah, N.; Seshu Babu, N.; Gopinath, R.; Siva Sankara Reddy, P.; Sai Prasad, P.S. Hydrodechlorination of Chlorobenzene over Supported Metal Catalysts. *Catal. Surv. Asia* **2006**, *10*, 29–39. [CrossRef]
48. Kulkarni, P.P.; Kovalchuk, V.I.; D'itri, J.L. Oligomerization Pathways of Dichlorodifluoromethane Hydrodechlorination Catalyzed by Activated Carbon Supported Pt-Cu, Pt-Ag, Pt-Fe, and Pt-Co. *Appl. Catal. B* **2002**, *36*, 299–309. [CrossRef]

gels

MDPI

Article

Sol-Gel Films Doped with Enzymes and Banana Crude Extract as Sensing Materials for Spectrophotometric Determination

Maria A. Morosanova and Elena I. Morosanova *

Analytical Chemistry Division, Chemistry Department, Lomonosov Moscow State University, 119234 Moscow, Russia
* Correspondence: emorosanova@gmail.com

Abstract: Chromogenic enzymatic reactions are very convenient for the determination of various biochemically active compounds. Sol-gel films are a promising platform for biosensor development. The creation of sol-gel films with immobilized enzymes deserves attention as an effective way to create optical biosensors. In the present work, the conditions are selected to obtain sol-gel films doped with horseradish peroxidase (HRP), mushroom tyrosinase (MT) and crude banana extract (BE), inside the polystyrene spectrophotometric cuvettes. Two procedures are proposed: the use of tetraethoxysilane-phenyltriethoxysilane (TEOS-PhTEOS) mixture as precursor, as well as the use of silicon polyethylene glycol (SPG).In both types of films, the enzymatic activity of HRP, MT, and BE is preserved. Based on the kinetics study of enzymatic reactions catalyzed by sol-gel films doped with HRP, MT, and BE, we found that encapsulation in the TEOS-PhTEOS films affects the enzymatic activity to a lesser extent compared to encapsulation in SPG films. Immobilization affects BE significantly less than MT and HRP. The Michaelis constant for BE encapsulated in TEOS-PhTEOS films almost does not differ from the Michaelis constant for a non-immobilized BE. The proposed sol-gel films allow determining hydrogen peroxide in the range of 0.2–3.5 mM (HRP containing film in the presence of TMB), and caffeic acid in the ranges of 0.5–10.0 mM and 2.0–10.0 mM (MT- and BE-containing films, respectively). BE-containing films have been used to determine the total polyphenol content of coffee in caffeic acid equivalents; the results of the analysis are in good agreement with the results obtained using an independent method of determination. These films are highly stable and can be stored without the loss of activity for 2 months at +4 °C and 2 weeks at +25 °C.

Keywords: sol-gel films; optical sensors; tetraethoxysilane; phenyltriethoxysilane; silicon polyethylene glycol; peroxidase; tyrosinase; crude banana extract; determination of total polyphenol content

Citation: Morosanova, M.A.; Morosanova, E.I. Sol-Gel Films Doped with Enzymes and Banana Crude Extract as Sensing Materials for Spectrophotometric Determination. *Gels* **2023**, *9*, 240. https://doi.org/10.3390/gels9030240

Academic Editors: Francesco Caridi, Giuseppe Paladini and Andrea Fiorati

Received: 28 February 2023
Revised: 16 March 2023
Accepted: 16 March 2023
Published: 18 March 2023

1. Introduction

Sol-gel materials are widely used in analytical practice. The immobilization of proteins in sol-gel matrices has been attempted in a number of works, given the good properties of the resulting materials: preservation of the biocatalytic properties of the encapsulated proteins and excellent optical properties. Since the first reported case of a successful immobilization of active alkaline phosphatase via the sol-gel method [1], sol-gel silicate has become a desired immobilization matrix for the design of active biocomposite materials. Sol-gel materials as a matrix for immobilized proteins or biomolecules can have many applications, such as stationary phase [2], drug delivery materials [3], and coatings [4,5]. Often sol-gel materials with immobilized enzymes are created for analytical enzymatic applications [1,6–15].

Silica sol-gel films present a major advantage comparing to other sol-gel materials for enzyme immobilization: the closeness of the immobilized enzymes to the solid-solution interface. This increases the accessibility of the enzymes to substrates from the aqueous phase, making silica sol-gel film a promising platform for biosensor development. Sol-gel films are particularly often employed for both optical [6–12], and electrochemical biosensors [6,13–15].

Chromogenic enzymatic reactions are very convenient for the determination of various biochemically active compounds. In our opinion, the use of sol-gel films to create optical biosensors deserves attention as an effective way to simplify analysis for its further implementation in field conditions.

To obtain sol-gel materials doped with biomolecules, the latter are often encapsulated in matrices of Ormosils—materials obtained by the hydrolysis of tetraethoxysilane in the presence of organic silicon alkoxides, primarily 3-aminopropyltriethoxysilane [16,17]. However, the alcohol formed during the hydrolysis and condensation of the alkoxide precursors could negatively affect the entrapped enzymes' activity [17]. This is an important problem that must be solved before employing the sol-gel process as a universal method of protein or other biomolecules encapsulation. In the literature, special schemes are described for the synthesis of sol-gel materials doped with enzymes to preserve the enzymatic activity of immobilized enzymes [18–23]. The approaches described can be divided into two groups: the use of tetraethoxysilane and its derivatives as precursors while using various ways to minimize the contact of the enzyme with ethanol released upon hydrolysis of the precursors and the use of silicate and glycerates as precursors, the hydrolysis of which does not release ethanol.

The approaches in the first group include the modification of the procedure while using standard alkoxide precursors [9,10,18]. To minimize the contact of the enzyme with ethanol, a two-stage synthesis scheme is proposed: the first stage is the hydrolysis of tetraethoxysilane or its mixtures with organic silicon alkoxides and the preparation of the sol; the second stage is the addition of the enzyme to the sol-gel solution and the formation of gels. Sometimes even the removal of the alcohol by rotavaporization method is performed before the second stage [18]. The technology of "kinetic doping" is also proposed where the nascent sol-gel film is submerged into enzyme-containing buffer solution which provides alcohol dilution [9,10].

The second group implies the use of different precursors: sodium silicate [19,20] or glycerol-derived silicates [21–23], which provide an alcohol-free sol. Both these routes possess some limitations in their application: the glycerol-derived silicate precursors need to be synthesized and sodium silicate precursors produce high sodium concentration levels in the sol.

Using crude extracts as enzyme sources in biocomposite sol-gel materials seems a promising approach. It was shown for crude extracts with polyphenol oxidase activity that, other conditions being equal, the enzymatic activity of sol-gel materials doped with the extracts is significantly higher than the activity of sol-gel materials doped with commercial tyrosinase [11,12]. Based on our experience of studying crude extracts, we also noted that the use of plant and mushroom extracts as enzyme sources presents a number of advantages compared to purified enzymes: higher interference thresholds, better stability, and lower cost [24–26]. In our opinion, the study of the crude extracts' properties and the study of the possibilities of their inclusion in sol-gel films will contribute to the development of methods for the field determination of biochemically important analytes.

Historically, sol-gel films for optical applications were prepared on glass slides, which then could be put inside the cuvettes. However, a more efficient and practical way to create an optical biocomposite sensor is also described [27,28]: the sol-enzyme mixture is put directly into the dispensable polystyrene cuvettes and the film is formed on the cuvette inner side. This approach provides an easy way to fixate the film position in relation to the optical path and also to precisely measure the amount of the enzyme and sol.

The goal of this work was to develop methods for the synthesis of transparent sol-gel films doped with the most well-studied enzymes (horseradish peroxidase and mushroom tyrosinase), as well as with crude banana extract as a source of polyphenol oxidase, on the inner surface of plastic cuvettes with the use of tetraethoxysilane and organic alkoxides or silicon glycerate as precursors, studying the effect of immobilization on the activity of these enzymes, and evaluating the analytical performance of synthesized sol-gel films.

2. Results

2.1. Synthesis of Sol-Gel Films with Immobilized Enzymes and Banana Extract

The validity of the sol-gel route presented in this work to preserve the enzyme activity during the immobilization process has been studied on two of the enzymes that are most widely used in analytical methods: horseradish peroxidase (HRP) and mushroom tyrosinase (MT), and also the widely used crude plant extract—banana extract (BE) as a source of polyphenol oxidase [23]. Most sol-gel immobilization studies use HRP [9,10,15,18,19,21]; polyphenol oxidase (tyrosinase) is studied less frequently [11–14,29]. We have optimized the sol-gel matrix using HRP, and then studied the performance of HRP, MT, and BE in the selected conditions. We have developed two approaches: using sol-gel films based on mixtures of TEOS with derivatives (i.e., Ormosils) and using glycerol precursors; then we studied the effect of immobilization on the activity of enzymes and the possibility of analytical use of the synthesized films.

2.1.1. Sol-Gel Films Based on Alkoxide Precursors (TEOS Films)

Sol-gel synthesis consists of successively carrying out the following stages: hydrolysis of precursors, polymerization (transformation of a sol into a gel) and, if necessary, drying of the gels under different conditions. Typically, hydrolysis is carried out in the presence of a related alcohol; when using TEOS, in the presence of ethanol. The properties of sol-gel materials depend on the nature of the precursors, the ratio of components in the hydrolyzing mixture, the nature of the gelation catalyst, and special additives to control the porosity of the materials. To obtain sol-gel materials doped with analytical reagents, we have developed a synthesis scheme [30,31], based on the hydrolysis of tetraethoxysilane in an aqueous-ethanol medium in the presence of hydrochloric acid as a catalyst and cetylpyridinium chloride as a pore former [32]. This scheme was used as the basis for the development of a method for the synthesis of sol-gel films doped with enzymes.

Studies show that when enzymes are included in Ormosils, i.e., sol-gel materials obtained from modified alkoxide precursors, their activity decreases to a lesser extent than in the case of standard TEOS materials. Most often, methyl- and amino-derivatives of tetraethoxysilane are used as precursors for the immobilization of enzymes in sol–gel materials [16], while phenyl derivatives are used much less frequently.

According to the literature, adding the enzyme solution to the already formed sol and not to the hydrolyzing mixture seems like a promising approach. This method was proposed using sodium silicate as a precursor [19]. We decided to take this approach when we used tetraethoxysylane (TEOS) and its mixtures with phenyltriethoxysilane (PhTEOS) and 3-aminopropyltriethoxysilane (AmTEOS) as sol-gel precursors. These TEOS-based sol-gel matrices were studied using horseradish peroxidase (HRP) as a model enzyme. HRP is widely used in enzyme immobilization studies and comparing the results of the present study using HRP was likely to be easier.

In order to select the conditions for the synthesis of transparent HRP-containing sol-gel films on the inner surface of plastic cuvettes, the influence of the nature and ratio of precursors, concentrations of hydrochloric acid and cetylpyridinium chloride was studied. TEOS, AmTEOS, and PhTEOS were used as precursors. The mixtures of precursors with water were prepared with 3:1 ratio of precursors:water and were mixed under the influence of ultrasound with a sound energy density of 0.24–0.38 W/mL at the hydrolyzing stage. The transparent homogeneous sols were formed in 45–90 min. The resulting sols are stable and can be stored at +4 °C for a week. To obtain gels, the sols were mixed with HRP solution in a buffer (pH 6.0) in a ratio of 1:0.8. The loss of fluidity of this mixture (i.e., gel formation) occurs after 2–10 min, depending on the composition of the sol. To prepare the sol-gel films, 0.5–0.9 mL of the mixture was placed in cuvettes, distributed on one of the inner sides, and after 2–10 min, films with an approximate thickness of 1.5–2.5 mm were formed. The stability of the sol-gel films, the enzymatic activity of the immobilized enzymes, and their storage stability were evaluated to choose the synthesis conditions.

When 1.0–4.5% w/w AmTEOS was added to TEOS, the films lost transparency compared to films prepared with only TEOS. For TEOS-AmTEOS films, the 10–100 fold increase in the concentration of hydrochloric acid led to the increase in transparency, but it was accompanied by significant reduction in the gelation time (less than 1 min), which made it difficult to obtain films by our method.

The use of TEOS-PhTEOS mixtures with a PhTEOS content of 1–10% w/w made it possible to obtain films that are transparent in the visible light range. Films produced at PhTEOS contents greater than 2% cracked on the next day after the preparation, but at 1–2% w/w they remained stable and did not crack. When using mixtures of TEOS-AmTEOS-PhTEOS with 1% w/w of PhTEOS and 0.4% w/w of AmTEOS, a lack of film transparency was observed, which led us to stop using AmTEOS and to concentrate on studying TEOS-PhTEOS mixtures.

In the literature, enzyme-doped sol-gel materials obtained by various methods are usually characterized by the retention of the immobilized enzyme in the matrix and its activity [9,19,21]. We investigated the effect of PhTEOS content on enzyme retention using HRP as a model enzyme. Under the described above conditions, HRP-doped films were synthesized with different PhTEOS content in the mixture of precursors (0, 1, and 2%)—HRP-PhTEOS0, HRP-PhTEOS1, and HRP-PhTEOS2 films. Enzyme retention was studied by determining the enzyme activity in buffer solutions obtained by washing the sol-gel films. The results are shown in Figure 1. The introduction of PhTEOS into the precursor mixture increased the retention of peroxidase by the sol-gel matrix. Thus, it can be concluded that peroxidase retention is significantly improved when PhTEOS is introduced into the matrix, and the enzyme is washed out slightly less from films containing 1% PhTEOS than from the films containing 2% PhTEOS. The obtained HRP retention values in the films after three washes are shown in Table 1. Comparison with literature data shows that the enzyme washing out values in our experiments are somewhat greater than for the previously proposed methods [19,21]. However, our film preparation method is simple, involves the use of commercially available precursors, and preserves significant activity of the immobilized enzyme.

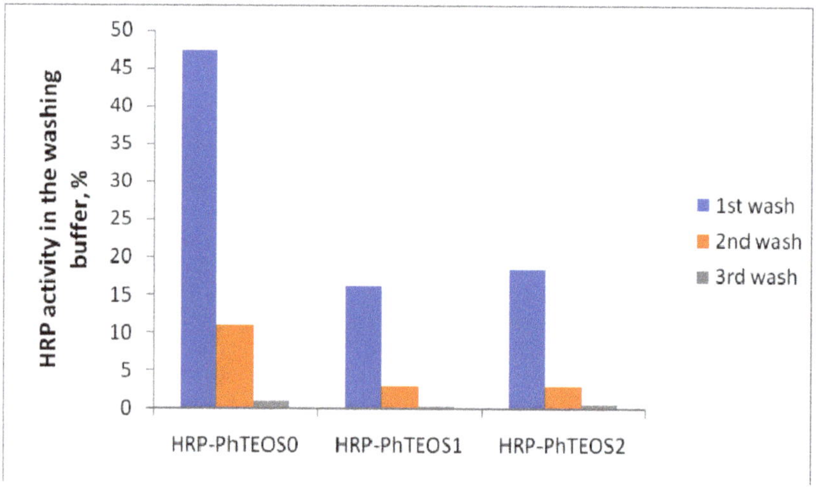

Figure 1. HRP activity (% in relation to the initial TEOS-PhTEOS sol-gel film activity) in the washing buffer solutions (pH 6.0).

Table 1. Retention of HRP in various sol-gel films after washing.

Precursor(s), Biocomposite Sol-Gel Material Preparation Procedure	Retention, %	Reference
TMOS, enzyme is added to the hydrolyzing mixture	76–90	[21]
PGS, enzyme is added to the hydrolyzing mixture	83–95	
Sodium silicate, ion-exchange elimination of sodium at the sol formation stage, enzyme is added to the sol	100	[19]
TEOS, enzyme is added to the sol	41	Present work
TEOS and 1% PhTEOS, enzyme is added to the sol	81	
TEOS and 2% PhTEOS, enzyme is added to the sol	79	

It is widely known that the activity of an entrapped enzyme is usually only a fraction of its activity in free solution [9]. The relative activity of the immobilized enzyme was calculated as a percentage of the activity of a similar amount of the enzyme in solution. The initial rates comparison allowed calculating the film loaded enzymes' relative activity (Table 2). The relative activities were in the range of 6.6–7.4%, which is similar to the results observed for other sol-gel matrices described in the literature [9,19]. The highest relative activity was observed for 1% PhTEOS sol-gel film. For further experiments, we chose a film obtained by adding 1% PhTEOS to a mixture of precursors—HRP-PhTEOS1.

Table 2. Relative activity of the immobilized enzymes (% of the activity of the same amount of native enzyme) for the different sol-gel films.

Enzyme	Precursor(s), Biocomposite Sol-Gel Material Preparation Procedure	Enzyme Relative Activity, % (n = 3, P = 0.95)	Reference
HRP	Sodium silicate, ion-exchange elimination of sodium at the sol formation stage, enzyme is added to the sol	7.2	[19]
	TEOS, enzyme sorption on the nascent sol-gel film on the glass slide	11.7 ± 0.5	[9]
	TEOS and PhTEOS, enzyme is added to the sol, 1. 0% PhTEOS 2. 1% PhTEOS 3. 2% PhTEOS	6.6 ± 0.5 7.4 ± 0.6 6.9 ± 0.5	Present work
	SPG, enzyme is added to the hydrolyzing mixture	3.4 ± 0.4	
MT	TEOS and PhTEOS, enzyme is added to the sol, 1% PhTEOS	11.2 ± 1.0	
	SPG, enzyme is added to the hydrolyzing mixture	15.2 ± 2.2	
BE	TEOS and PhTEOS, enzyme is added to the sol,1% PhTEOS	10.8 ± 0.9	
	SPG, enzyme is added to the hydrolyzing mixture	8.5 ± 0.5	

HRP—horseradish peroxidase, MT—mushroom tyrosinase, BE—crude banana extract.

Three types of films were prepared for the following studies using this sol-gel matrix: HRP-PhTEOS1 with immobilized HRP, MT-PhTEOS1 with immobilized MT, BE-PhTEOS1 with immobilized BE. The relative activities were also calculated for MT and BE (Table 2) and they were similar to the HRP activities obtained in our study and other works [9,19]. No available data on the relative activity of immobilized tyrosinase or crude extracts were found in the literature.

HRP-PhTEOS1, MT-PhTEOS1, and BE-PhTEOS1 films were used to study the kinetics of enzymatic reactions.

2.1.2. Sol-Gel Films Based on Silicon Polyethylene Glycol (SPG Films)

Another approach for creating biocomposite sol-gel films is the employment of silicon polyethylene glycol (SPG). SPG rapidly hydrolyzes and forms gels in aqueous media without the need for any catalyst, such as hydrochloric acid, to form silica hydrogels, which are transparent, and physically stable [21–23]. This approach was tested with many biological molecules, such as peroxidase, catalase, various oxidases, etc. [21]. We decided to synthesize glycerol-containing-precursors based sol-gel films and compare them to our TEOS-PhTEOS films.

Unlike alkoxide-based films, no ethanol is generated when using SPG, so SPG films are easier to prepare. In order to obtain SPG films, the precursor was mixed with a solution of the enzyme/extract in a pH 6.0 buffer solution in a ratio of 1:2. To form sol-gel films, 0.6 mL of the mixture was placed in cuvettes, distributed on one of its inner sides, and after 90 min, films with an approximate thickness of 1.5–2.5 mm were formed. Gelation occurred in the absence of a catalyst, and film formation took longer than in the case of TEOS-PhTEOS films.

Three types of films were prepared for the following studies using this sol-gel matrix: HRP-SPG with immobilized HRP, MT-SPG with immobilized MT, BE-SPG with immobilized BE. The relative activity of enzymes in these films is given in Table 2, and it is comparable to TEOS-PhTEOS based films. HRP-SPG, MT-SPG, and BE-SPG films were also used to study the kinetics of enzymatic reactions.

2.2. Study of the Kinetics of Enzymatic Reactions in the Presence of Sol-Gel Films Doped with HRP, MT, and BE

The effect of the sol-gel process on the enzyme activity was investigated by comparing the kinetic parameters (Michaelis constants) of the reactions catalyzed by native and immobilized enzyme. The initial rates of the HRP, MT, and BE catalyzed reactions were measured as the absorbance increase over time. The Michaelis constants were obtained through the fitting of the data to Michaelis–Menten kinetic analysis using Lineweaver–Burk plots. The Michaelis constants were calculated for hydrogen peroxide in the presence of constant TMB concentration (0.009%) in the case of HRP and for caffeic acid in the cases of MT and BE.

The enzymes included in the sol-gel films obtained in this work retain their activity, and the kinetics of the reactions catalyzed can be fitted to the Michaelis–Menten equation. Figure 2 shows, for example, kinetic curves for different concentrations of hydrogen peroxide in the presence of HRP-PhTEOS1 and HRP-SPG films.

For the evaluation of the immobilized enzymes' properties, we studied their interaction with substrates (hydrogen peroxide in the case of HRP and caffeic acid in the case of MT and BE) and calculated the kinetic parameters of the enzymatic reaction (Michaelis constants). Figure 3 shows the dependence of the reaction rate on the concentration of hydrogen peroxide in the presence of HRP-PhTEOS1 and HRP-SPG films. When Lineweaver–Burk coordinates are used, these dependencies become linear and allow the calculation of the Michaelis constants (Table 3). Both in our experiments and in the literature data [11,18,19] the Michaelis constant values (K_M) of the immobilized enzymes were higher than those of the native enzymes, indicating the presence of partitioning and diffusional effects in the pores of the sol-gel matrix. Table 3 shows that, when using SPG-based films, an even greater increase in K_M values is observed, meaning that such films are better fit for the determination of high concentrations of substrates. This can be explained by the greater steric hindrance because of bulkier sol-gel precursor molecules. The obtained data indicate that for all the studied sol-gel films, the inclusion of HRP, MT, and BE does not hinder their enzymatic activity and allows their use for enzymatic reactions. TEOS-PhTEOS-based films seem to be more promising for the development of methods for determining low contents of analytes-substrates.

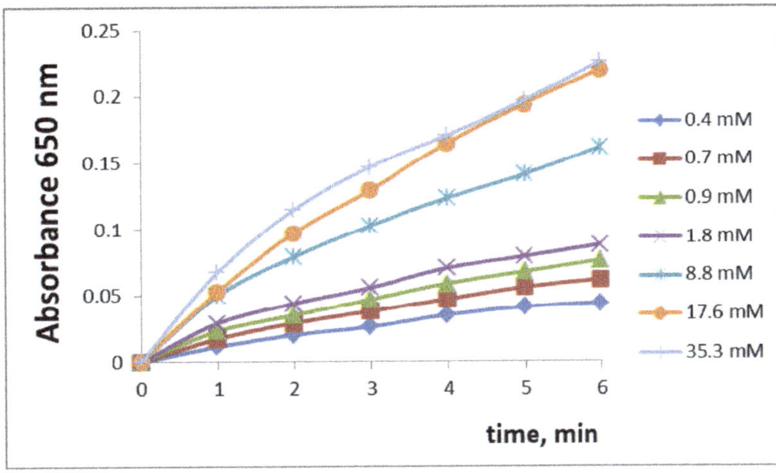

(a)

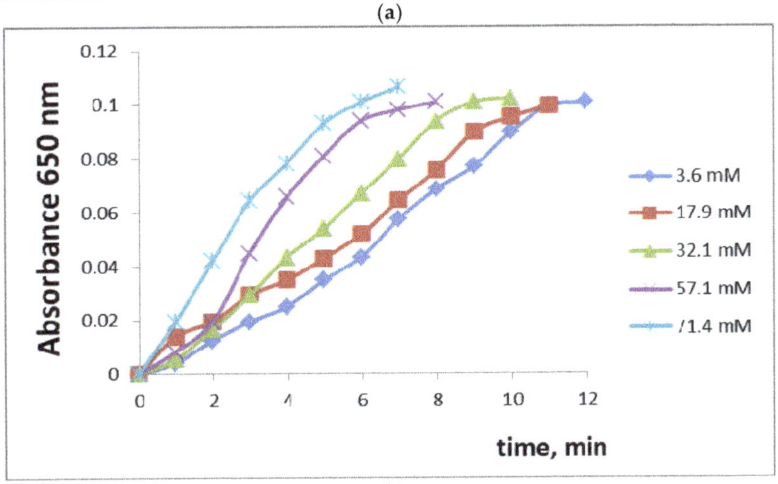

(b)

Figure 2. Kinetic curves of the TMB oxidation (absorbance at 650 nm over time) in the presence of different hydrogen peroxide concentrations (indicated in the legends) and HRP-PhTEOS1 (**a**) or HRP-SPG (**b**) sol-gel films.

We have studied crude banana extract in the present immobilization study, because earlier we have established that crude plant extracts have higher interference thresholds than purified enzymes [24]. There are data indicating that crude extracts are more robust and endure sol-gel immobilization better: in some cases, the extract can withstand the immobilization procedure that inhibits the corresponding purified enzyme activity [11]. In the present study, we observed that for crude banana extract the Michaelis constant remained almost the same after the immobilization in PhTEOS1 film (2.4 mM in solution vs. 2.8 mM in film). A similar effect was described earlier for desert truffle tyrosinase extract [11]: the Michaelis constant even slightly decreased upon immobilization (0.5 mM in solution vs.0.2 mM in film). This can be possibly explained by the presence of other plant cell fragments in the crude extracts which create a better environment for the enzymes inside the sol-gel matrices.

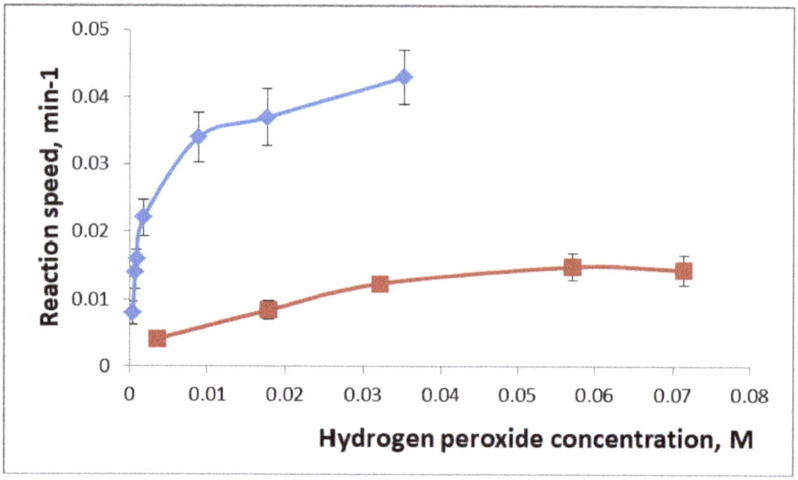

Figure 3. The dependence of the reaction speed (min^{-1}) in the presence of HRP-PhTEOS1 film (blue) and HRP-PGS film (red) on the concentration of hydrogen peroxide (n = 3).

Table 3. The influence of immobilization in sol-gel film on the Michaelis constants (K_M).

Enzyme Source	Film Precursor(s)	K_M, mM		K_M Ratio	Reference
		Solution	Film		
Horseradish peroxidase	Na silicate	0.163	0.985	6	[19]
	TEOS	0.55	2.38	4.3	[18]
	TEOS+ 1% PhTEOS	0.4	1.4	3.5	Present work
	SPG	0.4	9.9	24.8	
Mushroom tyrosinase	TEOS + colloidal silica	0.9	no activity	-	[11]
	TEOS+ 1% PhTEOS	1.1	4.1	3.7	Present work
	SPG	1.1	10.1	9.2	
Desert truffle extract	TEOS + colloidal silica	0.5	0.2	0.4	[11]
Banana extract	TEOS+ 1% PhTEOS	2.4	2.8	1.2	Present work
	SPG	2.4	8.0	3.3	

Based on the enzyme kinetics study of the immobilized enzymes we have chosen HRP-PhTEOS1, MT-PhTEOS1, and BE-PhTEOS1 films for the analytical application.

2.3. Analytical Application HRP-PhTEOS1, MT-PhTEOS1, and BE-PhTEOS1 Films

We studied the possibility of the analytical use of the proposed sol-gel films doped with enzymes and banana extract. Hydrogen peroxide was used as analyte for HRP-PhTEOS1 film in the presence of TMB, and caffeic acid was used for MT-PhTEOS1 and BE-PhTEOS1 films. Immobilized enzymes catalyze the corresponding chromogenic reactions: hydrogen peroxide reduction with TMB oxidation and caffeic acid oxidation by air oxygen. The reaction rates were used as an analytical signal; the dependence of the absorbance on time was studied for different concentrations of analytes. Analytical ranges—analyte concentration ranges with a linear dependence of the reaction rate on analyte concentrations—are given in Table 4. Comparison of detection limits (LOD) for the sol-gel film encapsulated enzymes and banana extract, with LODs for non-immobilized enzymes and extract, demonstrate only 2–3 fold loss of sensitivity (Table 4). Such effect is likely attributed to steric hindrances arising during immobilization. The simplicity of determinations using sol-gel films doped with enzymes and banana extract should be noted: it is simply needed to place 3.0 mL of a sample in a cuvette containing a sol-gel film (in the case of determining hydrogen peroxide, 0.4 mL of 0.08% TMB solution should also be added), and monitor the change in absorbance at 650 nm for the determination of hydrogen peroxide and 400 nm for the determination of caffeic acid. Such measurements can be carried out using various portable photometers, which opens the prospect of mass analyses for the determination of biochemically active analytes in the field.

Table 4. Analytical parameters of the procedures using HRP-PhTEOS1, MT-PhTEOS1, and BE-PhTEOS1 sol-gel films.

	Enzyme Source	Analytical Range, mM	LOD, mM ($n = 3$)
Hydrogen peroxide	HRP	0.05–0.5	0.02
	HRP-PhTEOS1	0.2–3.5	0.06
Caffeic acid	MT	0.25–10.0	0.08
	MT-PhTEOS1	0.5–10	0.15
	BE	0.5–10.0	0.2
	BE-PhTEOS1	2.0–10.0	0.6

In this paper, to demonstrate the analytical capabilities of sol-gel films, we present the results of the total polyphenol content determination in coffee (in caffeic acid equivalents) using an immobilized banana extract (BE-PhTEOS1 film). Total polyphenol content determination is often used in food quality control [24].

The recovery study of the caffeic acid determination using BE-PhTEOS1 film shows that the RSD values are comparable to those for banana extract in solution and equal 7–10% ($n = 3$).

The results of TPC determination in coffee compared with the results of independent methods are given in Table 5.

Table 5. Results of TPC determination in coffee using BE-PhTEOS1 film and by independent methods ($n = 3$, $P = 0.95$).

Found, mg/g			
BE—PhTEOS 1		BE	Folin's Reagent
Method of Standard Addition	Using the Calibration Curve		
114 ± 18	118 ± 36	110 ± 30	120 ± 10

The good agreement between the different procedures indicates the good accuracy of the TPC determination with BE-PhTEOS1 film. No significant difference was found between the four values using Student test ($p > 0.28$ for all the pairs).

The stability and lifetime of the immobilized BE was investigated by measuring the sol-gel film activity using 2.0 mM caffeic acid solution. 95% activity of immobilized BE was retained after 2 months storage at +4 °C (BE solution lost its activity after 4 days). 90% activity of immobilized BE was retained after 2 weeks storage at +25 °C.

These novel enzyme-doped silica matrixes provide promising platforms for development of various on-site analytical procedures. In the present work we proposed a procedure using crude plant extract immobilized in TEOS-PhTEOS sol-gel film (BE-PhTEOS1 film) for spectrophotometric determination of total polyphenol content using a standard curve of caffeic acid. The detection limit for caffeic acid equals 0.7 mM, while LOD values of other enzymatic methods of TPC determination lie in the 0.01–0.5 mM range [24]. However, the sol-gel films are ready-to-use and offer the possibility of storage at a room temperature. Using crude extract as the enzyme source in these sol-gel materials allows low-cost analysis which makes the process suitable for wide screening tests.

3. Conclusions

We have chosen the conditions for the synthesis of sol-gel films doped with HRP, MT, and BE using TEOS-PhTEOS mixture or SPG as precursors, on the inner side surface of polystyrene cuvettes. When using TEOS and PhTEOS precursors, the film preparation consists of two stages: the preparation of the sol under the influence of ultrasound for 90 min, which leads to the evaporation of a significant part of the formed alcohol, and the subsequent mixing of the sol with an enzyme solution. In the case of SPG films, the enzyme solution is mixed directly with the precursor. Basing on the study of the activity of the immobilized enzymes and the immobilized extract, we have concluded that for both types of films, the enzymatic activity is preserved, and the kinetics of the catalyzed reactions can be described by the Michaelis–Menten equation. The relative activity of the immobilized enzymes is comparable for both types of films and is about 10% of the activity of the non-immobilized enzyme. Thus, the preservation of enzyme activity in the proposed procedures is comparable to those described in the literature, which can also sometimes be significantly more complicated in execution.

When HRP and MT are included in alkoxide-based films, the Michaelis constants increase 3–4 fold, and in SPG-based films—10–20 fold. Compared to these purified enzymes, the crude banana extract demonstrates that it can better withstand the effect of immobilization: for BE the Michaelis constant almost does not change in the alkoxide-based films, and it increases only 3 fold in SPG-based films.

The analytical capabilities of sol-gel films doped with enzymes and banana extract are demonstrated: the analytical range for the hydrogen peroxide determination is 0.2–3.5 mM using HRP-PhTEOS1 film in the presence of TMB, and the analytical ranges for caffeic acid determination are 0.5–10.0 mM and 2.0–10.0 mM using MT-PhTEOS1 and BE-PhTEOS1 films, respectively. The sensitivity of the determination is decreased only 2–3 fold compared to non-immobilized enzymes, while the use of disposable cuvettes with a sol-gel film on the inner side surface greatly simplifies the determination procedure and makes it possible to carry out the determination in field conditions.

BE-PhTEOS1 films have been used to determine the total polyphenol content of coffee in caffeic acid equivalents. The lifetime of BE-PhTEOS1 is 2 months at +4 °C storage and 2 weeks at +25 °C storage, which is a significant improvement of the shelf life compared to the non-immobilized enzymes and extracts.

4. Materials and Methods

4.1. Materials

Tetraethoxysilane (TEOS), phenyltriethoxysilane (PhTEOS), 3-aminopropyltriethoxysilane (AmTEOS), and hydrogen peroxide were purchased from Sigma-Aldrich (St. Louis, MO, USA). Silicon polyethylene glycol (SPG) was synthesized under the supervision of Dr. T.G. Khonina according to the method described earlier [22].

Caffeic acid, cetylpyridinium chloride, and 3,3′,5,5′-tetramethylbenzidine (TMB) were purchased from Acros Organics (Carlsbad, CA, USA). Phosphate buffer (pH 6.0) was prepared using sodium monophosphate and potassium diphosphate.

Horseradish peroxidase (HRP) (250 U/mg) was purchased from Biozyme laboratories (UK). Mushroom tyrosinase (MT) from *A. niger* (3900 U/mg) was purchased from Sigma (St. Louis, MO, USA).

Banana extract (BE) was prepared similarly to [24]: 100.0 g of homogenized banana pulp tissue was stirred in 200.0 mL of phosphate buffer (pH 6.0) at 0 °C for 30 min, and then filtered twice through a paper filter. The protein content of the banana extract was determined by the Biuret method. Total protein content equaled 3.8 mg/mL for banana pulp crude extract. The activity of the crude banana extract used in this work has been determined by comparing the reaction speed of catechol oxidation in the presence of the crude extract and the commercial mushroom tyrosinase. Crude banana extract activity was found to be 292 ± 6 U/mL (n = 3, P = 0.95).

Polystyrene cuvettes (10 × 10 × 45 mm) with caps were purchased from Sarstedt (Numbrecht, Germany).

4.2. Synthesis of Sol-Gel Films Doped with Mushroom Tyrosinase, Horseradish Peroxidase and Banana Extract

4.2.1. Sol-Gel Films Based on TEOS and Its Mixtures with PhTEOS and AmTEOS (TEOS film)

In vials, a certain volume of AmTEOS or PhTEOS was added to a certain volume of TEOS. An aqueous solution of cetylpyridinium chloride and hydrochloric acid as a catalyst was added to the resulting mixture. The silane mixture: water ratio was 3:1. The mixture was stirred under the influence of ultrasound with the sound energy density of 0.28–0.34 W/mL for 90 min. In a plastic cuvette, 0.3 mL of the sol was mixed with 0.3 mL of the enzyme/crude extract solution in buffer. The cuvettes were capped, shaken, and then placed on their side. After 2–10 min, a transparent sol–gel film formed on the inner wall surface of the cuvette. The cuvettes with films were stored at +4 °C.

4.2.2. Synthesis of HRP-PhTEOS1, MT-PhTEOS1, BE-PhTEOS1 Films

1.5 mL of aqueous solution containing 0.5 mMcetylpyridinium chloride and 1.6 mM hydrochloric acid was added to 4.5 mL of TEOS and 0.05 mL of PhTEOS. The mixture was stirred under the influence of ultrasound with a sound energy density of 0.3 W/mL for 90 min. In a plastic cuvette, 0.3 mL of the sol was mixed with 0.3 mL of HRP solution in buffer (pH 6.0) to obtain HRP-PhTEOS1 film, with 0.3 mL MT solution in buffer (pH 6) to obtain MT-PhTEOS1 film, 0.3 mL of BE to obtain BE-PhTEOS1 film. The cuvettes were capped, shaken, and then placed on their side. After 2–10 min, a transparent sol–gel film formed on the inner wall surface of the cuvette. The cuvettes with films were stored at +4 °C.

4.2.3. Sol-Gel Films Based on SPG

In a cuvette 0.2 mL of SPG was mixed with 0.4 mL of HRP solution in buffer (pH 6.0) to obtain an HRP-SPG film, with 0.4 mL of MT solution in buffer (pH 6.0) to obtain a MT-SPG film, with 0.4 mL of BE to obtain a BE-SPG film. The cuvettes were capped, shaken, and then placed on their side. After 60–90 min, a transparent sol–gel film formed on the inner wall surface of the cuvette. The cuvettes with films were stored at +4 °C.

4.3. Study of the Immobilized HRP, MT, and BE Activity and Properties in Sol-Gel Films

4.3.1. Relative Activity of Immobilized Enzymes

To study the activity of HRP immobilized in sol-gel films, 3.0 mL of a hydrogen peroxide solution of various concentrations and 0.4 mL of a 0.08% TMB solution were added to the cuvette with film. Absorbance was measured at 650 nm for 10 min every 10 s. Enzyme activity was determined as the initial reaction rate. The relative activity was determined from the dependence of the HRP activity in solution on the HRP amount. This dependence was obtained by the following procedure: 3.0 mL of 8.0 mM hydrogen peroxide solution was mixed with 0.4 mL of 0.08% TMB solution and 0.4 mL of HRP solution with different amounts of enzyme, and absorbance was measured at 650 nm.

To study activity of MT and BE immobilized in sol-gel films, 3.0 mL of caffeic acid solution of various concentrations was added to the cuvette with film and the absorbance was measured at 400 nm for 10–15 min every 10 s. Enzyme activity was determined as the initial reaction rate. The relative activity was determined from the dependence of the MT/BE activity in the solution on the MT/BE amount. This dependence was obtained by the following procedure: 1.0 mL of MT/BE solution with different amounts of enzyme was added to 2.0 mL of 5.0 mM caffeic acid solution, and the absorbance was measured at 400 nm.

4.3.2. HRP Retention on PhTEOS0, PhTEOS1, PhTEOS2 Films

To study the retention of peroxidase in films of various compositions—TEOS (PhTEOS0), TEOS + 1%PhTEOS (PhTEOS1), TEOS + 2%PhTEOS (PhTEOS2)—2.0 mL of a buffer solution (pH 6.0) was added to the cuvettes with films and left for 30 min. After that, the buffer solution was decanted and its enzyme activity was determined according to the described above procedure.

4.3.3. Sol-Gel Films Stability Studies

The films doped with enzymes and banana extract were stored in closed cuvettes at +25 °C and at +4 °C; their stability was checked by measuring the activity according to the described above procedure.

4.3.4. Study of the Kinetics of Enzymatic Reactions in the Presence of Immobilized HRP, MT, and BE

To study the activity of HRP immobilized in sol-gel films, 3.0 mL of a hydrogen peroxide solution of various concentrations and 0.4 mL of a 0.08% TMB solution were added to the cuvette. Absorbance was measured at 650 nm for 10 min every 10 s. To calculate the Michaelis constant, the dependence of the reaction rate (min^{-1}) on the concentration of hydrogen peroxide was plotted in Lineweaver–Burk coordinates.

To study the activity of MT and BE immobilized in sol-gel films, 3.0 mL of caffeic acid solution of various concentrations was added to the cuvette and the absorbance was measured at 400 nm for 10–15 min every 10 s. To calculate the Michaelis constant, the dependence of the reaction rate (min^{-1}) on the concentration of caffeic acid was plotted in Lineweaver–Burk coordinates.

4.4. Calibration Curves Using HRP-PhTEOS1, MT-PhTEOS1, BE-PhTEOS1 Films

To obtain a calibration curve for hydrogen peroxide, 3.0 mL of a hydrogen peroxide solution of various concentrations and 0.4 mL of a 0.08% TMB solution were added to a cuvette with HRP-PhTEOS1 film. The difference in absorbance at 650 nm, measured after 1 and 2 min from the reaction start, was used as analytical signal.

To obtain a calibration curve for caffeic acid, 3.0 mL of caffeic acid solution of various concentrations was added to a cuvette with MT-PhTEOS1 or BE-PhTEOS1 films. The reaction rate, i.e., the rate of increase in absorbance at 400 nm, was used as analytical signal.

The limit of detection (LOD) was calculated as 3 standard deviation of the blank absorbance ($n = 3$) divided by the slope value. The limit of quantitation (LOQ) was calculated as 3·LOD.

4.5. Total Polyphenol Content Determination

1.0 g of coffee sample was mixed with 100.0 mL of boiling water, and filtered after 15 min. After cooling to the room temperature, 3.0 mL of the sample solution was added to the cuvette with the BE-PhTEOS1 film, and the reaction rate was used as the analytical signal. The total polyphenol content (TPC) in caffeic acid equivalents was determined in the treated sample using the standard addition method and using the calibration curve for caffeic acid in the range of 2.0–10.0 mM.

The procedure for TPC determination with Folin reagent was carried out similarly to [24].

4.6. Instrumentation

Sols were prepared under the ultrasound radiation using the ultrasound equipment UZH-02 (SonoTech, Russia). The sound energy density (W/mL) was defined as the ratio of the power absorbed in the reactor to the volume of liquid in the reactor. To determine the power, the time was measured until a certain mass of water was heated to a certain temperature.

Spectra of colored products of enzymatic oxidation of phenolic compounds were recorded with SPECTROstar Nano spectrophotometer (BMG Labtech, Ortenberg, Germany). Spectra were analyzed with MARS software (BMG Labtech, Ortenberg, Germany) and statistical analysis was carried out using MS Excel.

Author Contributions: Conceptualization, E.I.M.; Validation, E.I.M. and M.A.M.; Formal Analysis, M.A.M.; Investigation, M.A.M..; Resources, E.I.M.; Writing—Original Draft Preparation, M.A.M.; Writing—Review and Editing, E.I.M.; Supervision, E.I.M.; Funding Acquisition, E.I.M. All authors have read and agreed to the published version of the manuscript.

Funding: The study was funded by MedEcoTest Ltd. (Grant N 407/14).

Institutional Review Board Statement: Not applicable.

Informed Consent Statement: Not applicable.

Data Availability Statement: No data is available.

Acknowledgments: The authors would like to thank T.G. Khonina (I.Ya. Postovsky Institute of Organic Synthesis, Yekaterinburg, Russia) for providing silicon polyethylene glycol, and MSU students S. Terekhov, T. Ibragimov, A. Bolbat, and D. Anisimov for their participation in the preliminary experiments.

Conflicts of Interest: The authors declare no conflict of interest.

References

1. Braun, S.; Rappoport, S.; Zusman, R.; Avnir, D.; Ottolenghi, M. Biochemically active sol-gel glasses: The trapping of enzymes. *Mat. Lett.* **1990**, *10*, 1–5. [CrossRef]
2. Svobodova, J.; Miksik, I. Application of sol-gel modified with natural plants extracts as stationary phases in open-tubular electrochromatography. *Gels* **2022**, *8*, 198. [CrossRef] [PubMed]
3. Catauro, M.; Tranquillo, E.; Poggetto, G.; Naviglio, D.; Barrino, F. The Influence of Polymer on Fe(II)Citrate Release from Hybrid Materials Synthesized via Sol-Gel. *Macromol. Symp.* **2020**, *389*, 1900057. [CrossRef]
4. Ghuzali, N.A.M.; Noor, M.A.A.C.M.; Zakaria, F.A.; Hamidon, T.S.; Husin, M.H. Study on *Clitoriaternatea* extracts doped sol-gel coatings for the corrosion mitigation of mild steel. *App. Surf. Sci. Adv.* **2021**, *6*, 100177. [CrossRef]
5. Al-Saadi, S.; Singh Raman, R.K. Silane Coatings for Corrosion and Microbiologically Influenced Corrosion Resistance of Mild Steel: A Review. *Materials* **2022**, *15*, 7809. [CrossRef] [PubMed]
6. Jeronimo, P.C.A.; Araujo, A.N.; Conceicao, M.; Montenegro, B.S.M. Optical sensors and biosensors based on sol-gel films. *Talanta* **2007**, *72*, 13–27. [CrossRef]
7. Gupta, R.; Chaudhury, N.K. Entrapment of biomolecules in sol–gel matrix for applications in biosensors: Problems and future prospects. *Biosens. Bioelectron.* **2007**, *22*, 2387–2399. [CrossRef]
8. MacCraith, B.D.; McDonagh, C.M.; O'Keeffe, G.; McEvoy, A.K.; Butler, T.; Sheridan, F.R. Sol-gel coatings for optical chemical sensors and biosensors. *Sens. Actuators B Chem.* **1995**, *29*, 51–57. [CrossRef]
9. Crosley, M.S.; Yip, W.T. Silica sol–gel optical biosensors: Ultrahigh enzyme loading capacity on thin films via kinetic doping. *J. Phys. Chem. B* **2017**, *121*, 2121–2126. [CrossRef] [PubMed]
10. Crosley, M.S.; Yip, W.T. Kinetically Doped Silica Sol–Gel Optical Biosensors: Expanding Potential Through Dip-Coating. *ACS Omega* **2018**, *3*, 7971–7978. [CrossRef] [PubMed]
11. Gouzi, H.; Moreau, T.; Depagne, C.; Coradin, T. Immobilization of a Polyphenol Oxidase Extract from *Terfezialeonis Tul.* Desert Truffle in Multilayered Silica Films for Dopamine Biosensing. *Silicon* **2013**, *5*, 241–246. [CrossRef]
12. Leboukh, S.; Gouzi, H.; Coradin, T.; Yahia, H. An optical catechol biosensor based on a desert truffle tyrosinase extract immobilized into a sol–gel silica layered matrix. *J. Sol-Gel. Sci. Technol.* **2018**, *86*, 675–681. [CrossRef]
13. Wang, B.; Zhang, J.; Dong, S. Silica sol-gel composite film as an encapsulation matrix for the construction of an amperometric tyrosinase-based biosensor. *Biosens. Bioelectron.* **2000**, *15*, 397–402. [CrossRef]
14. Majidi, M.R.; Asadpour-Zeynali, K.; Gholizadn, S. Sol-gel biosensor based on Plant Tissue:The inhibitory effect of atrazine on polyphenol oxidase activity for determination of atrazine. *J. Chin. Chem. Soc.* **2008**, *55*, 522–528. [CrossRef]
15. Pandey, P.C.; Upadhyay, S.; Tiwari, I.; Tripathi, V.S. An ormosil-based peroxide biosensor—A comparative study on direct electron transport from horseradish peroxidase. *Sens. Actuators B Chem.* **2001**, *72*, 224–232. [CrossRef]
16. Tripathi, V.S.; Kandimalla, V.B.; Ju, H. Preparation of ormosil and its applications in the immobilizing biomolecules. *Sens. Actuators B Chem.* **2006**, *114*, 1071–1082. [CrossRef]
17. Kauffman, C.G.; Mandelbaum, R.T. Entrapment of atrazine-degrading enzymes in sol-gel glass. *J. Biotechnol.* **1996**, *51*, 219–225. [CrossRef]

18. Ferrer, M.L.; del Monte, F.; Levy, D. A Novel and Simple Alcohol-Free Sol-Gel Route for Encapsulation of Labile Proteins. *Chem. Mater.* **2002**, *14*, 3619–3621. [CrossRef]
19. Bhatia, R.B.; Brinker, C.J.; Gupta, A.K.; Singh, A.K. Aqueous Sol-Gel Process for Protein Encapsulation. *Chem. Mater.* **2000**, *12*, 2434–2441. [CrossRef]
20. Liu, D.M.; Chen, I.W. Encapsulation of protein molecules in transparent porous silica matrices via an aqueous colloidal sol–gel process. *Acta Mater.* **1999**, *47*, 4535–4544. [CrossRef]
21. Gill, I.; Ballesteros, A. Encapsulation of Biologicals within Silicate, Siloxane, and Hybrid Sol-Gel Polymers an Efficient and Generic Approach. *J. Am. Chem. Soc.* **1998**, *120*, 8587–8598. [CrossRef]
22. Lavrova, D.G.; Zvonarev, A.N.; Alferov, V.A.; Khonina, T.G.; Shadrina, E.V.; Alferov, S.V.; Ponamoreva, O.N. Biocompatible Silica-Polyethylene Glycol-Based Composites for Immobilization of Microbial Cells by Sol-Gel Synthesis. *Polymers* **2023**, *15*, 458. [CrossRef] [PubMed]
23. Brook, M.A.; Chen, Y.; Guo, K.; Zhang, Z.; Jin, W.; Deisingh, A.; Cruz-Aguado, J.; Brennan, J.D. Proteins entrapped in silica monoliths prepared from glyceroxysilanes. *J. Sol-Gel. Sci. Technol.* **2004**, *31*, 343–348. [CrossRef]
24. Morosanova, M.A.; Fedorov, A.S.; Morosanova, E.I. Crude Plant Extracts Mediated Polyphenol Oxidation Reactions in the Presence of 3-Methyl-2-Benzothiazolinone Hydrazone for the Determination of Total Polyphenol Content in Beverages. *Curr. Anal. Chem.* **2019**, *15*, 11–20. [CrossRef]
25. Morosanova, M.A.; Bashkatova, A.S.; Morosanova, E.I. Spectrophotometric and Smartphone-Assisted Determination of Phenolic Compounds Using Crude Eggplant Extract. *Molecules* **2019**, *24*, 4407. [CrossRef] [PubMed]
26. Morosanova, M.A.; Fedorova, T.V.; Polyakova, A.S.; Morosanova, E.I. *Agaricus bisporus* Crude Extract: Characterization and Analytical Application. *Molecules* **2020**, *25*, 5996. [CrossRef] [PubMed]
27. De Marcos, S.; Galindo, J.; Sierra, J.F.; Galban, J.; Castillo, J.R. An optical glucose biosensor based on derived glucose oxidase immobilised onto a sol–gel matrix. *Sens. Actuators B Chem.* **1999**, *57*, 227–232. [CrossRef]
28. Alqasaimeh, M.S.; Heng, L.Y.; Ahmad, M. A Urea Biosensor from Stacked Sol-Gel Films with Immobilized Nile Blue Chromoionophore and Urease Enzyme. *Sensors* **2007**, *7*, 2251–2262. [CrossRef]
29. Gul, I.; Ahmad, M.S.; Naqvi, S.S.; Hussain, A.; Wali, R.; Farooqi, A.A.; Ahmed, I. Polyphenol oxidase (PPO) based biosensors for detection of phenolic compounds: A Review. *J. Appl. Biol. Biotechnol.* **2017**, *5*, 072–085. [CrossRef]
30. Morosanova, E.I. Silica and silica–titania sol–gel materials: Synthesis and analytical application. *Talanta* **2012**, *102*, 114–122. [CrossRef]
31. Morosanova, E.I. Complex Formation in the Modified Xerogel Phase: Study and Application in Analysis. *Russ. J. Coord. Chem.* **2022**, *48*, 784–798. [CrossRef]
32. Morosanova, M.A.; Morosanova, E.I. Silica–titania sensor material prepared by cetylpyridinium chloride assisted sol–gel synthesis for solid phase spectrophotometric and visual test determination of propyl gallate in food samples. *Anal. Methods* **2016**, *8*, 8092–8098. [CrossRef]

gels

Article

Stimuli-Responsive Properties of Supramolecular Gels Based on Pyridyl-*N*-oxide Amides

Sreejith Sudhakaran Jayabhavan, Baldur Kristinsson, Dipankar Ghosh, Charlène Breton and Krishna K. Damodaran *

Department of Chemistry, Science Institute, University of Iceland, Dunhagi 3, 107 Reykjavík, Iceland
* Correspondence: krishna@hi.is; Tel.: +354-525-4846; Fax: +354-552-8911

Abstract: The nature of functional groups and their relative position and orientation play an important role in tuning the gelation properties of stimuli-responsive supramolecular gels. In this work, we synthesized and characterized mono-/bis-pyridyl-*N*-oxide compounds of *N*-(4-pyridyl)nicotinamide ($\mathbf{L_1}$–$\mathbf{L_3}$). The gelation properties of these *N*-oxide compounds were compared with the reported isomeric counterpart mono-/bis-pyridyl-*N*-oxide compounds of *N*-(4-pyridyl)isonicotinamide. Hydrogels obtained with $\mathbf{L_1}$ and $\mathbf{L_3}$ were thermally and mechanically more stable than the corresponding isomeric counterparts. The surface morphology of the xerogels of di-*N*-oxides ($\mathbf{L_3}$ and **diNO**) obtained from the water was studied using scanning electron microscopy (SEM), which revealed that the relative position of *N*-oxide moieties did not have a prominent effect on the gel morphology. The solid-state structural analysis was performed using single-crystal X-ray diffraction to understand the key mechanism in gel formation. The versatile nature of *N*-oxide moieties makes these gels highly responsive toward an external stimulus, and the stimuli-responsive behavior of the gels in water and aqueous mixtures was studied in the presence of various salts. We studied the effect of various salts on the gelation behavior of the hydrogels, and the results indicated that the salts could induce gelation in $\mathbf{L_1}$ and $\mathbf{L_3}$ below the minimum gelator concentration of the gelators. The mechanical properties were evaluated by rheological experiments, indicating that the modified compounds displayed enhanced gel strength in most cases. Interestingly, cadmium chloride formed supergelator at a very low concentration (0.7 wt% of $\mathbf{L_3}$), and robust hydrogels were obtained at higher concentrations of $\mathbf{L_3}$. These results show that the relative position of *N*-oxide moieties is crucial for the effective interaction of the gelator with salts/ions resulting in LMWGs with tunable properties.

Keywords: supramolecular gels; isomeric LMWGs; pyridyl *N*-oxide; stimuli-responsive systems; anion/cation-responsive gels; metallogels; cadmium sensor

Citation: Jayabhavan, S.S.; Kristinsson, B.; Ghosh, D.; Breton, C.; Damodaran, K.K. Stimuli-Responsive Properties of Supramolecular Gels Based on Pyridyl-*N*-oxide Amides. *Gels* **2023**, *9*, 89. https://doi.org/10.3390/gels9020089

Academic Editors: Francesco Caridi, Giuseppe Paladini and Andrea Fiorati

Received: 26 December 2022
Revised: 15 January 2023
Accepted: 17 January 2023
Published: 20 January 2023

1. Introduction

Soft materials based on stimuli-responsive systems [1–4] have gained widespread interest because of their tunable properties with respect to external stimuli such as electricity, light, heat, voltage, magnetic field, mechanical stress, pH, and salts/ions. Supramolecular gels based on low-molecular-weight gelators (LMWGs) [5–14] are an excellent class of stimuli-responsive materials with intriguing potential applications, such as sensors, dynamic gels, tissue engineering, and as media for crystal growth and catalysis [14–26]. The self-assembly of gelator molecules in a solvent medium leads to the formation of LMWGs with a three-dimensional fibrous network stabilized by various non-covalent interactions. The self-assembly process depends on various parameters [27–33], such as temperature, pressure, sound, solvent, functional groups, and gelator structure, which can significantly affect the gel network. The nature of the non-bonding interaction and the molecular structure of the gelator play a crucial role in gel network formation. However, predicting the mechanism of the self-assembly process and the gel structure is challenging, because the non-covalent interactions are dynamic in nature. The self-assembly process also depends on the spatial orientation of the functional groups. Understanding the role of building blocks/functional groups and the nature of the non-bonding interactions will help researchers to design smart materials based on LMWGs with intriguing properties. The formation of a one-dimensional (1D) hydrogen-bonded chain can be considered as one of the key primary interactions for the gel network formation in LMWGs [6,8,34]. The incorporation of supramolecular synthons [35] or functional groups with hydrogen-bond functionalities such as urea and amide that can assemble into a 1D array have extensively been used to generate LMWGs, which indicates the importance of the 1D hydrogen-bonded chain [6,8].

LMWGs based on amide groups as supramolecular synthons are an important class of supramolecular gels with tunable properties [36–39]. Amide-based LMWGs display complementary N–H···O=C interactions arising from the N−H donor and C=O acceptor of the amide moieties to form a one-dimensional chain, which self-assembles to a three-dimensional network (3D) via cooperative and unidirectional hydrogen bonding [40]. These 3D networks can immobilize solvent molecules to form organo/hydrogels [41–47]. The adjacent functional groups strongly influence the hydrogen-bonding synthons, and the gelation properties can be altered by introducing moieties that interact with the amide/urea groups. For example, attaching the pyridyl group to the amide moiety can change the intermolecular interaction involving urea/amide groups, which leads to a N···H–N hydrogen-bonding synthon due to the interaction between urea/amide groups and the pyridyl moiety [45,48]. The addition of pyridyl functionality has resulted in highly robust pyridyl amide LMWGs [45,48–50], at typical low gelator concentrations, which can be utilized as stimuli-responsive supramolecular systems [11,51–54]. Incorporating pyridyl amide moieties offers the possibility of synthesizing several isomers, depending on the relative position of the pyridyl nitrogen atom. Furthermore, they provide several advantages, such as ease of obtaining crystalline materials and modification of the pyridyl groups by simple organic reactions. For example, the gelation properties of bis-pyridyl LMWGs can be tuned by replacing the pyridyl group with the corresponding pyridyl-*N*-oxide [55], resulting in pyridyl-*N*-oxide LMWGs.

Compounds based on pyridyl-*N*-oxide moieties have gained widespread interest in synthetic chemistry, biochemistry, and pharmacology due to their intriguing potential applications in medicinal science [56–59]. This is presumably due to the characteristic of the N–O bond [60], which can be considered as a NO donating bond with an important contribution from the oxygen atom (ON back-donation). The substituents play an important role in the stability of the N–O bond; for example, additional stability can be achieved by adding electron-withdrawing substituents, but a reverse trend is observed for electron-donating groups [60]. Incorporating pyridyl-*N*-oxide moieties will lead to enhanced hydrogen bonding and increased solubility in water [55,61] because of the hydrogen-bonding capabilities of the pyridyl-*N*-oxide moiety. Thus, pyridyl-*N*-oxides have a great prospect as hydrogelators, but, surprisingly, the utilization of this functionality in LMWGs is unexplored [55,62,63]. The relative position of the functional groups plays a crucial role in gel network formation in pyridyl-amide/urea gelators, and compounds derived from the *N*-(4-pyridyl) moiety were proved to be superior gelators over the other positional isomers [45,49]. The *N*-oxide moieties could also play an important role in the self-assembly process and the gelation properties of LMWGs [64], and, to the best of our knowledge, the role of the relative position in the gelation properties has not been reported for pyridyl amide *N*-oxides. We have reported the gelation ability of mono-/bis-pyridyl-*N*-oxide compounds of N-(4-pyridyl)isonicotinamide (4PINA) [55] and have shown that bis-pyridyl-*N*-oxide (**diNO**, Scheme 1) displayed better gelation properties compared to mono-pyridyl-*N*-oxides. In this work, we are analyzing the role of the relative position of *N*-oxide moieties in gel network formation by comparing the gelation ability of isomeric pyridyl-*N*-oxide amides in water. The application of the isomeric mono-/bis-pyridyl-*N*-oxide amide LMWGs as sensors will be evaluated by analyzing the stimuli-responsive properties of these LMWGs towards various cations and anions.

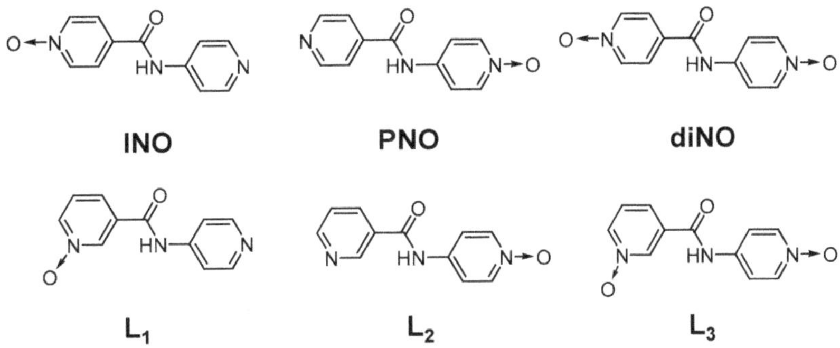

Scheme 1. Isomeric mono/di-*N*-oxides of bis-pyridyl amides.

2. Results and Discussion

We analyzed the role of positional isomers of *N*-oxide moieties in gelation properties by replacing the isonicotinic acid *N*-oxide with nicotinic acid *N*-oxide to form *N*-(4-pyridyl)nicotinamide *N*-oxide compounds (4PNANO, Scheme 1). The mono-pyridyl-*N*-oxide compounds of 4PNANO (**L_1** and **L_2**) were synthesized by reacting the corresponding *N*-oxide amine/acid (Schemes S1 and S2), but the bis-pyridyl-*N*-oxide compound was synthesized by oxidizing *N*-(4-pyridyl) nicotinamide with 3-chloroperoxybenzic acid (Scheme S3).

2.1. Gelation Experiments

The ability of **L_1–L_3** to form hydrogels was evaluated in water (1.0 wt%) or aqueous mixtures (1.0 wt/v%). In a typical experiment, 10.0 mg of the compound in 1.0 mL water/aqueous mixtures was heated in a sealed vial to obtain a clear solution, and the mixture was cooled to room temperature. The solution was left undisturbed until gelation was observed, which was confirmed by the vial inversion test. The results indicated that **L_1–L_3** did not form a gel in water at 1.0 wt%, and the experiments were performed at higher concentrations of the gelators. **L_1** and **L_3** formed gels in water at 2.0 wt%, but **L_2** did not form a gel. The minimum gelator concentration (MGC) required to form the gel network was evaluated by adding different amounts of the gelator (10.0–30.0 mg) to 1.0 mL water. The gelation experiments at various concentrations indicated that the MGC of **L_1** and **L_3** was 1.8 wt% in water. We also tested the gelation properties of **L_1-L_3** in aqueous mixtures (1:1, *v*/*v*) of high polar solvents such as MeOH, EtOH, DMF, and DMSO, and the results were similar to the experiments performed in water (Table 1). Gels were obtained in all cases for **L_1** and **L_3** at 2.0 wt/v%, indicating that the presence of cosolvents did not affect the *N*-oxide gel's self-assembly process.

Table 1. Minimum gelator concentration and T_{gel} values of **L_1** and **L_3** (2.0 wt/v%) in water and aqueous mixtures.

	L_1		L_3	
Solvents	**MGC (wt/v%)**	**T_{gel} (°C)**	**MGC (wt/v%)**	**T_{gel} (°C)**
Water	1.8	128.0	1.8	76.9
MeOH/water	2.0	77.1	2.0	69.7
EtOH/water	2.0	73.2	2.0	63.3
DMF/water	2.0	76.5	2.0	74.3
DMSO/water	2.0	87.9	2.0	94.2

2.2. Thermal Stability

The thermal stabilities of the gels were evaluated by measuring the temperature at which the gel network collapsed, and gels underwent phase transformation to a solution, which is known as gel-to-solution transition temperature (T_{gel}). A small spherical glass ball was placed on top of the preformed gels at various concentrations of the gelators in the standard vials, and the temperature was increased at a constant rate. As the temperature increased, the ball touched the bottom of the vial, and this temperature was recorded as the T_{gel}. We performed the T_{gel} experiments at the MGC of the gelators in water, and the T_{gel} values for $\mathbf{L_1}$ and $\mathbf{L_3}$ were 128.0 °C and 76.9 °C, respectively. The experiments performed at higher concentrations of $\mathbf{L_3}$ showed an increase in T_{gel} value (92.9 °C), and the results indicated that increasing the concentration of the gelators enhanced the thermal stability of the gel network. The thermal stabilities of $\mathbf{L_1}$ and $\mathbf{L_3}$ were analyzed in various aqueous solvents (Table 1), and the T_{gel} values were lower compared to the corresponding hydrogels. This was presumably due to the favorable interaction between the gelator molecules and the polar solvents, leading to enhanced solubility of the gelators in these mixed solvents, except for $\mathbf{L_3}$ in DMSO/water (1:1, *v*/*v*). We compared the MGC value and the thermal stability of $\mathbf{L_3}$ with the isomeric *N*-oxides (**diNO**) [55] to see whether the relative position of the *N*-oxide moieties affected the self-assembly process and the gel state properties. The MGC for $\mathbf{L_3}$ in water was 1.8 wt%, whereas a higher MGC was observed for isomeric **diNO** (4.0 wt%). The thermal stability of the $\mathbf{L_3}$ network in water was also higher than the isomeric **diNO**, and these results indicate that changing the isonicotinic *N*-oxide to nicotinic *N*-oxide leads to better gel network formation.

2.3. Rheology

Rheology is an important tool in studying the deformation and flow characteristics of supramolecular gels [65,66]. The mechanical strength of the gelators was analyzed by performing the amplitude- and frequency-sweep experiments on the hydrogels of $\mathbf{L_1}$ and $\mathbf{L_3}$ at 2.0 wt%. Initially, an oscillatory strain-sweep experiment was performed to evaluate the linear viscoelastic region (LVR), because the gel network undergoes reversible deformation inside LVR. The results demonstrated that $\mathbf{L_1}$ and $\mathbf{L_3}$ hydrogels had a narrow LVR, as the storage modulus G′ declined after 0.02% of the shear strain (Figure S1). An abrupt decrease in the G′ is observed at the crossover point [65,66], where the gel breaks into a viscous fluid. The crossover points for $\mathbf{L_1}$ and $\mathbf{L_3}$ hydrogels were within the range of 1.0–5.0% of the shear strain. The frequency-sweep experiments showed constant elastic (G′) and viscous (G″) moduli with a frequency range of 0.1–10.0 Hz at a constant strain of 0.02% (within LVR).

Frequency-sweep experiments were performed with $\mathbf{L_1}$ and $\mathbf{L_3}$ hydrogels (2.0 wt%) at a constant strain of 0.02% (within LVR) in a range of 0.1–10.0 Hz, which displayed constant elastic (G′) and viscous (G″) moduli under varying frequency. The hydrogel of $\mathbf{L_1}$ displayed a higher elastic modulus (~10 times, Figure 1a) compared to the $\mathbf{L_3}$ hydrogel. The enhanced mechanical stability can be correlated to the molecular structure of $\mathbf{L_1}$, which indicates that the 4-pyridyl functionality in $\mathbf{L_1}$ plays a crucial role in the thermal and mechanical strength of the gel. The role of the relative position of the *N*-oxide moieties in the mechanical strength was analyzed by performing the frequency-sweep experiments on $\mathbf{L_3}$ and isomeric **diNO** hydrogels at 4.0 wt%. The comparison of the elastic (G′) and viscous (G″) moduli of $\mathbf{L_3}$ and isomeric **diNO** gels revealed that $\mathbf{L_3}$ hydrogels displayed enhanced mechanical strength (~1.8-fold stronger) than the **diNO** hydrogels, which indicated that the mechanical strength of the gel network depended on the position of the pyridyl *N*-oxide moiety (Figure 1b).

We also performed the frequency-sweep experiments with $\mathbf{L_1}$ and $\mathbf{L_3}$ gels in DMSO/water and DMF/water (1:1, *v*/*v*) at 2.0 wt/v%, and the aqueous mixture gels of $\mathbf{L_1}$ were stronger (2–3-fold) than the hydrogel of $\mathbf{L_1}$ at 2.0 wt%. A similar trend was found for the aqueous mixture gels of $\mathbf{L_3}$ in the DMSO/water gel (~2-fold stronger than $\mathbf{L_3}$ hydrogels), but the DMF/water gel displayed similar mechanical strength as that of the hydrogel of $\mathbf{L_3}$ at 2.0 wt% (Figure S2).

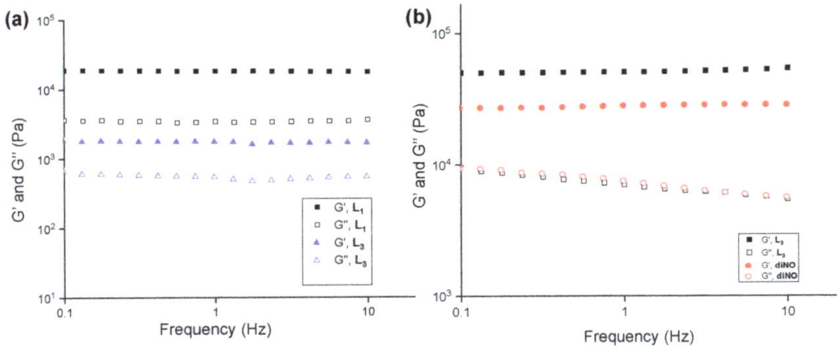

Figure 1. Frequency-sweep experiments performed on (**a**) $\mathbf{L_1}$ and $\mathbf{L_3}$ gels at 2.0 wt% and (**b**) $\mathbf{L_3}$ and **diNO** gels at 4.0 wt% in water at a constant strain of 0.02% at 20.0 °C, respectively.

2.4. *Scanning Electron Microscopy (SEM)*

SEM is an important technique to visualize the morphology of gel fibers [12,67] and can be used to distinguish the self-assembly modes in supramolecular gels by analyzing the morphology of the gel fibers [68,69]. The morphologies of the fibrous network of $\mathbf{L_1}$ and $\mathbf{L_3}$ xerogels were analyzed by performing SEM on the dried gels in water at the minimum gelator concentration (1.8 wt%). $\mathbf{L_1}$ displayed rod-like morphology with thickness ranging from 0.3–3.0 μm, but twisted rod-like morphologies were observed from $\mathbf{L_3}$ xerogels, and the thickness of the fibers was within the range of 1.0–3.0 μm (Figure S3). SEM performed at 2.0 wt% (above MGC) on $\mathbf{L_1}$ xerogels showed rod-like morphologies with thickness ranging from 0.5–4.0 μm (Figure 2a), but flake-like morphologies were observed for $\mathbf{L_3}$, with dimensions ranging from 4.0–16.0 μm (Figure 2b). The comparison of the SEM images of $\mathbf{L_3}$ and isomeric **diNO** xerogels [55] from water at 4.0 wt% revealed that the morphologies of the gels did not depend on the relative position of the pyridyl *N*-oxide moiety (Figure S4).

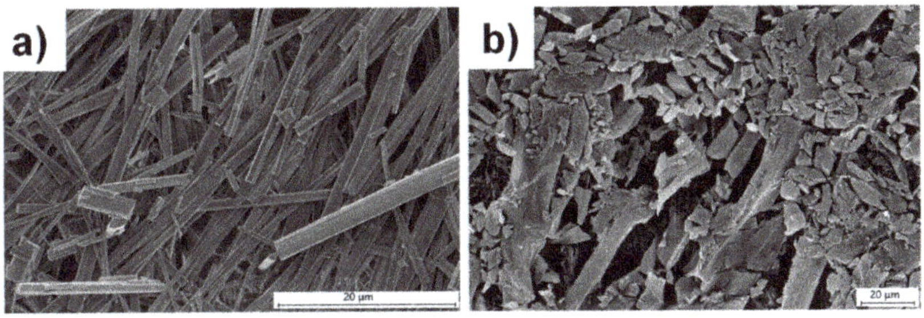

Figure 2. SEM images of xerogels in water with (**a**) $\mathbf{L_1}$ and (**b**) $\mathbf{L_3}$ at 2.0 wt%.

We also analyzed the morphologies of $\mathbf{L_1}$ and $\mathbf{L_3}$ in the aqueous mixtures of DMSO and methanol (1:1, *v*/*v*) to evaluate the effect of solvents on the morphologies of gel fibers. The xerogel of $\mathbf{L_1}$ prepared from DMSO/water and methanol/water (1:1, *v*/*v*, respectively) at 2.0 wt/v% displayed needle-shaped morphology with dimensions ranging from 0.5–6.0 μm (Figure S5). However, long rod-shaped morphology with fiber width ranging from 2.0 to 20.0 μm (Figure S6) was observed for $\mathbf{L_3}$ xerogels in DMSO/water and methanol/water (1:1, *v*/*v*, respectively) at 2.0 wt/v%. These results indicate that morphologies of the fibrous network depend on the solvent system.

2.5. Structural Analysis

2.5.1. Single Crystal X-ray Diffraction

The slow evaporation of a dilute solution of **L3** resulted in X-ray-quality single crystals of **L3**. The analysis of the crystals using a polarized light microscope indicated that needle-shaped crystals with trace amounts of plate-shaped crystals were formed. The solid-state structural data were analyzed by single-crystal diffraction analysis, which helped us to correlate the key non-bonding interactions in the solid state and the gelation properties. The structural analysis of the needle-shaped crystals revealed that **L3** crystallized in a monoclinic space group ($P2_1/c$) with a solvent water molecule (**L3**•H_2O) (Figure 3a and Table S1). The N-H moiety of the amide group displayed hydrogen-bonding interaction with the *N*-oxide moiety of the nicotinamide moieties via N–H···O interactions (Table S2), resulting in a R_2^2 (14) hydrogen-bonded dimer [70] of **L3**. The oxygen atom of the aminopyridine *N*-oxide displayed a bifurcated hydrogen bonding with the solvent water molecule via O—H···O interactions, resulting in a 1D hydrogen-bonded chain of the dimers (Figure 3b). This 1D hydrogen-bonded chain can be considered as one of the crucial factors for supramolecular gelation [6,8] in LMWGs. The comparison of the crystal structure of **L3**•H_2O with the isomeric **diNO** revealed the absence of such 1D hydrogen-bonded chains, which could be one of the factors in the better gelation ability and mechanical strength of **L3**•H_2O compared to isomeric **diNO**.

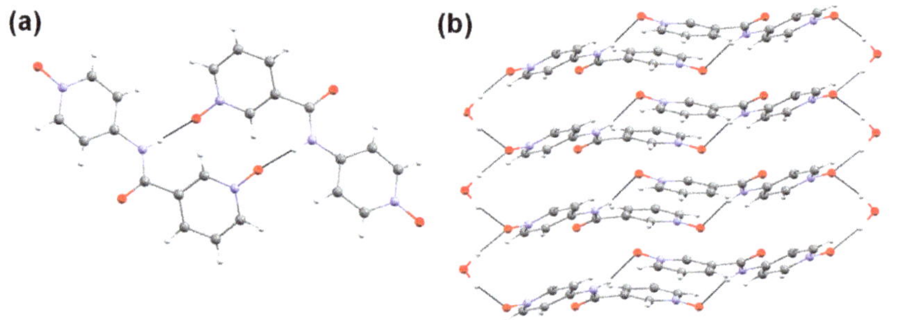

Figure 3. (**a**) Hydrogen-bonded dimer of **L3** stabilized by N–H···O interactions and (**b**) the aminopyridine *N*-oxide moiety interacting with the water molecule to form a 1D hydrogen-bonded chain of the dimers.

We also analyzed the solid-state structure of the plate-shaped crystals (minor product), which indicated that **L3** crystallized in the monoclinic space group ($P2/c$) with two water molecules (**L3**•$2H_2O$) (Figure S7 and Table S1). The molecule was planar compared to **L3**•H_2O, but a similar interaction was observed between the amide and *N*-oxide moieties, resulting in a (R_2^2) (14) hydrogen-bonded dimer [70]. However, the aminopyridine *N*-oxide moiety displayed a bifurcated hydrogen bond with two water molecules, resulting in two-dimensional porous architecture (Figure S7b), which was further stabilized by various non-bonding interactions (Table S2).

2.5.2. Powder X-ray Diffraction (PXRD)

Powder X-ray diffraction is a rapid and powerful tool to obtain an insight into the packing modes of solids. Comparing the PXRD pattern of the bulk material with the simulated pattern acquired from the single-crystal structure enabled us to evaluate the phase purity of the material [63,71]. We have shown that comparing the PXRD pattern of the xerogels with the simulated pattern of the gelator structure could provide information about the key interactions in the gel network architecture [6,12,32,55,72]. This method can be considered as an excellent strategy to correlate the self-assembly process in LMWGs despite the artefacts affecting the drying process [73]. We recorded the PXRD pattern of the **L3** crystals obtained via recrystallization from hot water (10.0 mg in 2.0 mL) and **L3** xerogel from pure water at 2.0 wt%. The PXRD pattern of the bulk crystals and the xerogel matched with the simulated pattern of the needle-shaped crystals $L_3{\bullet}H_2O$ (Figure 4), indicating that the hierarchical assembly of the xerogel network matched with the solid-state structure of $L_3{\bullet}H_2O$. The simulated pattern of $L_3{\bullet}2H_2O$ did not match with the PXRD pattern of the bulk crystals and the xerogel, which indicated that the second form did not correspond to the structure of the xerogels (Figure S8).

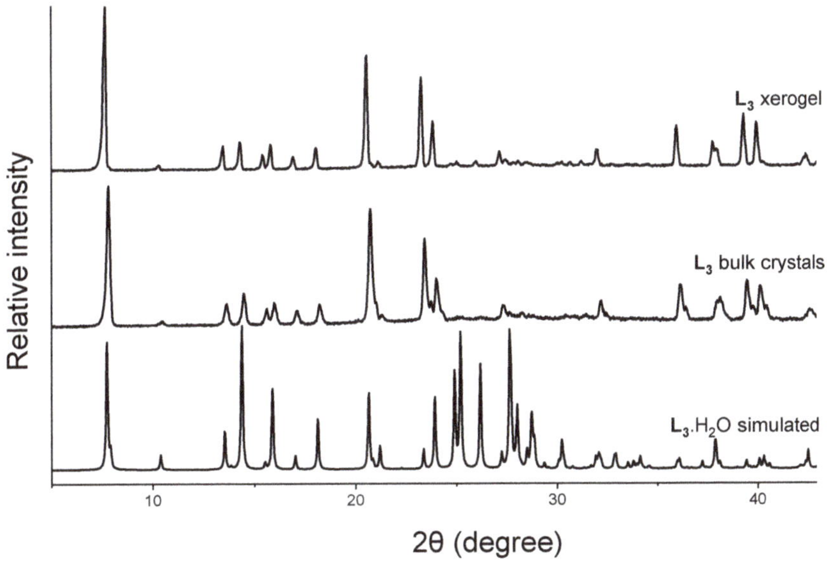

Figure 4. Comparison of the simulated pattern from single-crystal X-ray structure of $L_3{\bullet}H_2O$ with the PXRD pattern of the bulk crystals obtained from water and xerogel from water at 2.0 wt%.

2.6. Stimuli-Responsive Properties

LMWGs based on pyridyl-*N*-oxides display smart responses towards external stimuli because the molecular interactions of *N*-oxide moieties are based on partial charges, and the gelation can be turned ON/OFF in the presence of respective salts/ions. The cation/anions interact with gelator molecules constructively or destructively, depending on the electrostatic interaction and acidic/basic properties of the cations/anions [52,74]. The constructive interaction could trigger/enhance the gelation process [30,75,76], but a destructive interaction may lead to the dissolution/collapse of the gel network [27,39].

2.6.1. Anion Sensing

We have previously reported the pyridyl-*N*-oxide-based compound's sensing ability towards salts/ions [64,77] and have shown that cyanide ions can be detected using the gel-sol transition [62]. This prompted us to study the effect of salts/ions on the gelation properties of the *N*-oxide compounds of *N*-(4-pyridyl)nicotinamide. The stimuli-responsive properties of L_1 and L_3 in water were analyzed by treating the compounds at concentrations below MGC (at 1.5 wt%) with various halides of sodium and potassium ions (1.0 equiv.). Gel formation was observed for L_1 at 1.5 wt% in the presence of 1.0 equivalence of NaF, NaCl, NaBr, KF, KCl, KBr, and KI (Table S3). The experiments performed at a lower concentration of L_1 (1.2 wt%) resulted in selective gelation in the presence of 1.0 equivalence of KCl. The stimuli-responsive properties of L_3 in the presence of sodium and potassium halide salts (1.0 equiv.) displayed gel formation at 1.5 wt%. These results indicate the stimuli-responsive properties of L_1 and L_3 in water, where the salts/ions induce gel network formation (Table S3). The anion-sensing ability of L_1 and L_3 (1.5 wt%) was further analyzed in the presence of other anions, such as KNO_3, KBF_4, KPF_6, and $K_2C_2O_4$ (Table S3). Gels were obtained for all ions with **L_3**, but a colloidal solution was observed with **L_1** at 1.5 wt%. We compared the stimuli-responsive properties of **L_1** and **L_3** with isomeric *N*-oxides **INO** and **diNO**, respectively. The non-gelator **INO** (isomeric *N*-oxide of **L_1**) did not form any gels in the presence of 1.0 equivalence of sodium and potassium salts in pure water at 1.5 wt%. The isomeric *N*-oxide of **L_3 (diNO)** also failed to form gels at 1.5 wt%, which indicated that the effective interaction between the compound and the salts/ions was affected by the position of the *N*-oxide functionality. However, the **diNO** (MGC, 4.0 wt%) formed gels at a concentration below MGC (3.0 wt%) with these salts (1.0 equiv.).

The comparison of the mechanical strength of the gels (1.5 wt%) in the presence of halide salts of sodium and potassium with the hydrogels of **L_1** and **L_3** at MGC (1.8 wt%) revealed that enhanced mechanical strength was observed in both cases (Figure 5 and Table S4). A ~5.7-fold increase in the mechanical strength of **L_3** (1.5 wt%) was observed in the presence of KCl and a ~4.3-fold increase with NaCl (Figure 5b). We further studied the effect of other anions (KNO_3, KBF_4, KPF_6, and $K_2C_2O_4$) with **L_1** and **L_3** (1.5 wt%) below MGC, and **L_1** failed to form stable gels in the presence of these salts. However, the mixed gel of **L_3** and the anions displayed enhanced mechanical strength compared to the **L_3** hydrogels (Figure 5b and Table S4).

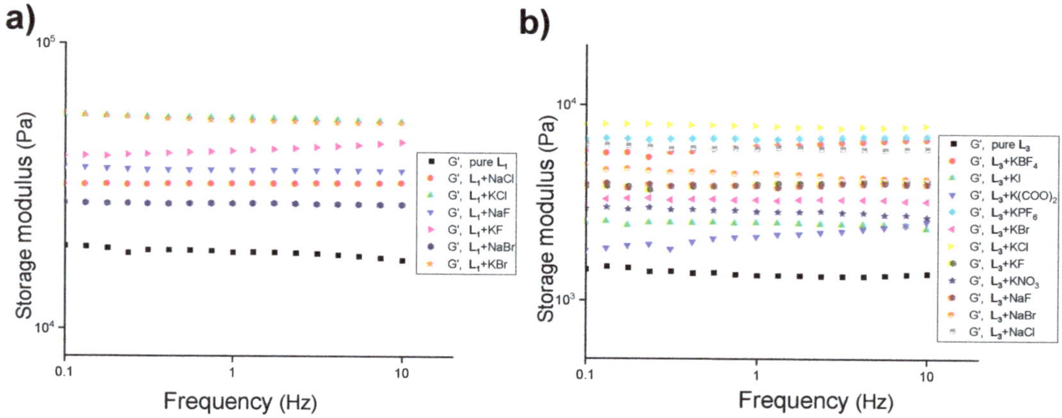

Figure 5. Frequency-sweep experiments of **L_1** and **L_3** at 1.5 wt% in water at 20.0 °C in the presence of various salts of sodium and potassium (1.0 equiv.) with a constant strain of 0.02%, (**a**) **L_1** (pure ligand at 1.8 wt%), and (**b**) **L_3** (pure ligand at 1.8 wt%).

The effect of anions on the morphology of the gel network was studied by analyzing the SEM images of the xerogels with anions. The xerogel of **L_1** and **L_3** in the presence of NaCl displayed fibrous morphologies with diameters ranging from 100 to 800 nm and 1.0 to 5.0 μm, respectively (Figure 6). The PXRD analysis of the xerogel of **L_3** hydrogels at 1.5 wt% in the presence of 1.0 equivalence of NaCl and KCl showed a similar pattern with the pure xerogel of **L_3** at 1.8 wt%, which suggested that the anions interacted with **L_3** to form a stable gel below MGC without affecting the original network (Figure S9).

2.6.2. Cation Sensing

The gelation property was observed to be independent of the size and nature of the anions, which prompted us to study the stimuli-responsive properties of **L_1** and **L_3** with chloride salts having various countercations (1.0 equiv.) at a concentration below MGC (1.5 wt%). **L_1** formed a stable gel in the presence of 1.0 equivalence CsCl, $MgCl_2$, $CaCl_2$, $SrCl_2$, and $BaCl_2$, but no gels were observed in the presence of 1.0 equivalence of NH_4Cl. On the other hand, **L_3** formed a gel in the presence of all of the above-mentioned chloride salts (Table S5). The mechanical strength of **L_1** and **L_3** with these salts was evaluated to observe the effect of cations in gel network formation. Analysis of the results indicated that the addition of chloride salts of magnesium, calcium, strontium, and barium enhanced the mechanical strength (1.5–3.0-fold) of $\mathbf{L_1}$ compared to $\mathbf{L_1}$ (1.8 wt%) hydrogel (Figure 7a and Table S6). A similar trend was observed for $\mathbf{L_3}$ gels in the presence of the chloride salts of magnesium, calcium, strontium, barium, cesium, and ammonium (Figure 7b and Table S6).

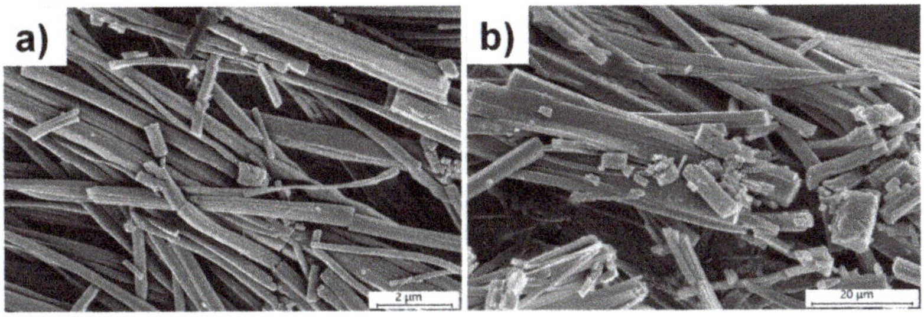

Figure 6. SEM images of xerogels obtained from pure water at 2.0 wt% in the presence of 1.0 equivalence of NaCl, (**a**) $\mathbf{L_1}$, and (**b**) $\mathbf{L_3}$.

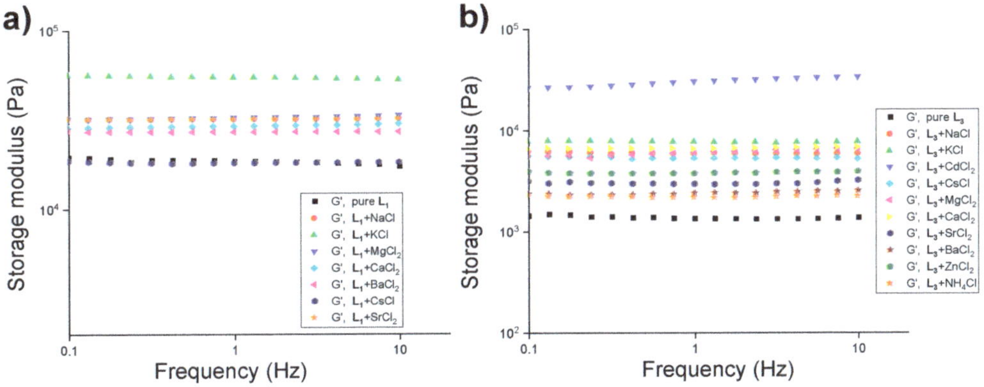

Figure 7. Frequency-sweep experiments with 1.5 wt% of the compound in the presence of various salts of chlorides (1.0 equiv.) in water at 20.0 °C with a constant strain of 0.02%, (**a**) $\mathbf{L_1}$ (pure ligand at 1.8 wt%), and (**b**) $\mathbf{L_3}$ (pure ligand at 1.8 wt%).

The cation-sensing ability of the isomeric *N*-oxide compounds (**INO** and **diNO**) were compared with $\mathbf{L_1}$ and $\mathbf{L_3}$ gels in the presence of various cations. Gelation was not observed for both **INO** and **diNO** at 1.5 wt% in the presence of $MgCl_2$ and $CaCl_2$ (1.0 equiv.) However, **diNO** formed stable hydrogel at 3.0 wt% (below MGC) with 1.0 equivalence $MgCl_2$ and $CaCl_2$. The comparison of the sensing ability of the isomeric *N*-oxides confirms that the relative position of *N*-oxide plays a crucial role in cation sensing. SEM performed on the dried gels of $\mathbf{L_1}$ and $\mathbf{L_3}$ at 2.0 wt% in the presence of $MgCl_2$ (1.0 equiv.) revealed that the morphology of the xerogels was identical to the xerogels of $\mathbf{L_1}$ and $\mathbf{L_3}$ hydrogels (2.0 wt%, Figure S10). PXRD studies with the chloride salts of magnesium and calcium (1.0 equiv) at 1.5 wt% of $\mathbf{L_3}$ showed no change compared to the powder pattern of the xerogel of $\mathbf{L_3}$ at 1.8 wt%, suggesting similar molecular packing (Figure S11). The molecular interactions of *N*-oxide moieties are based on partial charges, and the enhanced gelation property in the presence of anions/cations may be attributed to the favorable interactions between the *N*-oxide moieties and anions/cations, which is presumably due to the combination of the non-bonding and ionic interactions. The enhanced mechanical properties of $\mathbf{L_1}$ and $\mathbf{L_3}$ in the presence of various cations prompted us to study the stimuli-responsive properties of $\mathbf{L_1}$ and $\mathbf{L_3}$ with transition metal salts, which could form metallogels.

Metallogels based on transition metals are multi-responsive soft materials with intriguing potential applications in cosmetics, food processing, drug delivery, and catalysis [78,79]. Metal-based supramolecular gels are formed by the addition of metal ions to a gelator/non-gelator, and the interaction of the metal salt with the gelator could induce/enhance gelation properties to form metallogels [52,79]. However, the addition of metal salts could disrupt the key interactions in the gel network, leading to the dissolution of the gel network [52,79]. The metal-coordination-driven self-assembly plays a key role in the formation of a gel network, which is supported by various non-covalent interactions [80,81]. The presence of the coordinating functionality (pyridyl *N*-oxide moiety) and a hydrogen-bonding group (amide moiety) in $\mathbf{L_1}$ and $\mathbf{L_3}$ will make these gelators ideal candidates for metal-based supramolecular gels [52].

The effect of transition metal salts on the stimuli-responsive property of the hydrogel was studied by adding one equivalent of the metal salt ($CuCl_2$, $ZnCl_2$, or $CdCl_2$) to $\mathbf{L_1}$ or $\mathbf{L_3}$ at 1.5 wt%. Gels were not obtained for $\mathbf{L_1}$ in the presence of these transition metal chlorides. However, $\mathbf{L_3}$ gelled at 1.5 wt% in the presence of $ZnCl_2$ and $CdCl_2$ (1.0 equiv.), but no gel was observed with $CuCl_2$. We repeated the experiments by lowering the concentration of $\mathbf{L_3}$ (0.7 wt%), and selective gel formation was observed with a 1.0 equivalence of $CdCl_2$, highlighting the specific sensing of cadmium chloride at this concentration (Figure 8). The $\mathbf{L_3}$-$CdCl_2$ gelator obtained at 0.7 wt% of $\mathbf{L_3}$ and $CdCl_2$ (1.0 equiv.) was stable for several weeks, and the experiments performed at lower concentrations of $CdCl_2$ (0.5 equiv.) resulted in precipitation (Figure 8). We further analyzed the gelation behavior of $\mathbf{L_3}$ in the aqueous mixtures (1:1, v/v) of methanol, ethanol, DMF, and DMSO at 2.0 wt/v% in the presence of 1.0 equivalence of $ZnCl_2$ and $CdCl_2$. We observed gelation with $ZnCl_2$ and $CdCl_2$ in the aqueous mixtures of DMF/DMSO, but gels were formed with $CdCl_2$ in MeOH/water and EtOH/water. The gelation ability of **INO** with $ZnCl_2$ and $CdCl_2$ was previously reported by our group [62]. The isomeric **diNO** formed a precipitate below 2.5 wt% with 1.0 equivalence of $CuCl_2$ or $CdCl_2$, which highlighted the effective gelation ability of $\mathbf{L_3}$ over the isomeric gelator **diNO** in the presence of transition metals. However, hydrogels were formed at 2.5 wt% (below MGC) with 1.0 equivalence of $CuCl_2$ or $CdCl_2$, and a precipitate was formed in the presence of $ZnCl_2$.

Figure 8. Stimuli-responsive behavior of $\mathbf{L_3}$ at 0.7 wt% in the presence of $CdCl_2$.

The mechanical strength of the gels with transition metals was analyzed by performing frequency-sweep experiments, and an abrupt increase in the mechanical strength (~24.7 fold) was observed for $\mathbf{L_3}$ (1.5 wt%) with $CdCl_2$ (1.0 equiv.). We studied the mechanical strength of the gels at 2.0 wt/v% of $\mathbf{L_3}$ with $CdCl_2$ in water and a 1:1 (v/v) aqueous mixture of methanol, ethanol, DMSO, and DMF. A steep increase in the mechanical strength was observed for $\mathbf{L_3}$-$CdCl_2$ gels in water (~35.6-fold), methanol/water (~175-fold), ethanol/water (~91.7-fold), and DMF/water (~34.3-fold) with $\mathbf{L_3}$ (2.0 wt/v%) (Figure 9).

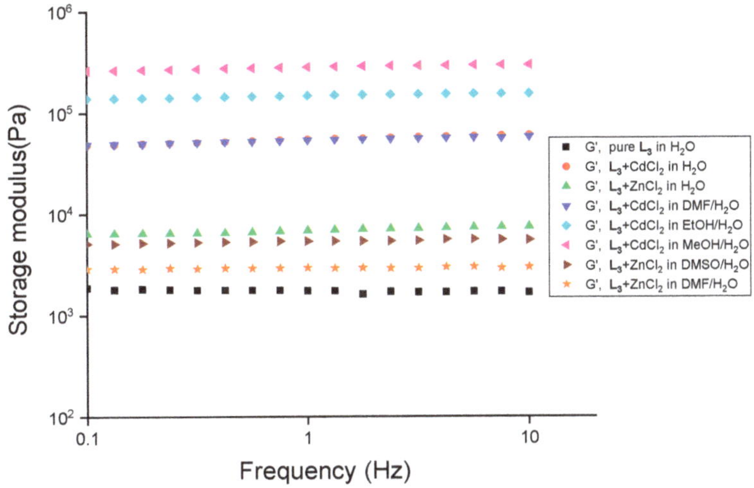

Figure 9. Frequency-sweep experiments with $\mathbf{L_3}$ at 2.0 wt% and in the presence of chloride salts of zinc and cadmium (1.0 equiv.) in various solvent mixtures at 20.0 °C with a constant strain of 0.02%.

Similarly, we observed an enhanced mechanical strength for $\mathbf{L_3}$ gels at 2.0 wt% with 1.0 equivalence of $ZnCl_2$ in water (~4.4-fold), DMF/water (~1.8-fold), and DMSO/water (~3.3-fold) (Figure 9).

The morphological analysis of the dried gels at 2.0 wt% of $\mathbf{L_3}$ with 1.0 equivalence of $CdCl_2$ from water revealed rod-shaped morphology with fiber width ranging from 0.4 to 4.0 μm (Figure S12). Long needle-shaped fibers were observed for dried gels at 0.7 wt% of $\mathbf{L_3}$ with 1.0 equivalence of $CdCl_2$, and the diameter of the fibers ranged from 1.0 to 6.0 μm (Figure 10a). The xerogels obtained from DMSO/water (1:1, *v/v*) at 2.0 wt% of $\mathbf{L_3}$ and $CdCl_2$ (1.0 equiv.) displayed similar fibrous morphology, with a fiber width range of 0.1–1.0 μm (Figure 10b).

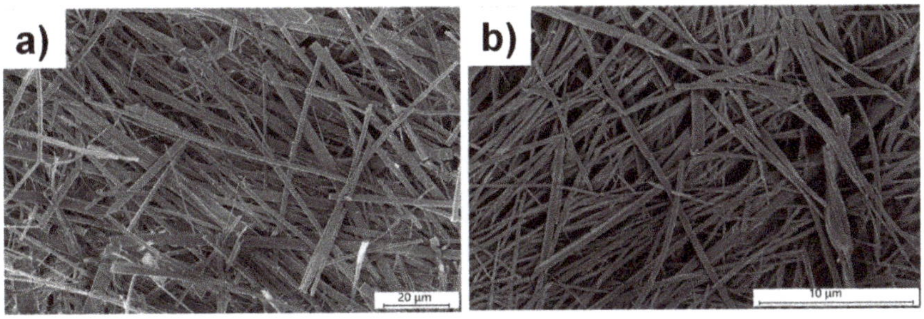

Figure 10. SEM images of xerogels obtained with $\mathbf{L_3}$ in the presence of 1.0 equivalence of $CdCl_2$ in (**a**) 0.7 wt% in pure water and (**b**) 2.0 wt/v% in DMSO/water (1:1, *v/v*).

We analyzed the powder X-ray patterns of the dried gels at 2.0 wt% and 0.7 wt% of $\mathbf{L_3}$ hydrogels in the presence of 1.0 equivalence of $CdCl_2$, respectively. Analysis of the PXRD pattern revealed a different PXRD pattern for mixed $\mathbf{L_3}$-$CdCl_2$ xerogels compared to the $\mathbf{L_3}$ xerogel at 2.0 wt% (Figure S13) in both cases. This may be attributed to the complexation of $\mathbf{L_3}$ with $CdCl_2$, resulting in the formation of a stable hydrogel at a lower ligand concentration (0.7 wt%) induced by metal-ligand-driven supramolecular self-assembly. PXRD pattern of the mixed $\mathbf{L_3}$-$CdCl_2$ xerogels at 2.0 wt/v% of $\mathbf{L_3}$ obtained from an aqueous mixture (1:1, *v/v*) of methanol, ethanol, DMF, and DMSO did not match with the xerogel of L_3, which confirmed that interaction of L_3 with $CdCl_2$ was independent of solvent composition. However, the PXRD patterns of mixed L_3-$CdCl_2$ in different solvent mixtures were not similar, because the crystal packing was affected by the nature of the solvent (Figure S13). The morphology, PXRD pattern, and gel-state properties of L_3-$CdCl_2$ gels suggest that L_3 interacts with $CdCl_2$, which may be considered as metallogels. PXRD studies with $ZnCl_2$ (1.0 equiv) at 1.5 wt% of L_3 in water and in aqueous mixtures (1:1, v/v) of DMF and DMSO matched with the xerogel of L_3 at 1.8 wt%, indicating similar solid-state structure (Figure S14). This suggests that $ZnCl_2$ exists as non-coordinated metal salts in the gel network, and the gelation properties can be explained based on the fact that metal nanoparticles and non-coordinated metal complexes are known to enhance the gelation properties of LMWGs [82–85].

3. Conclusions

The role of the relative position of functional groups in the gelation properties of LMWGs was studied by analyzing the gelation properties of isomeric mono-/bis-pyridyl-*N*-oxide compounds. We synthesized mono-/bis-pyridyl-*N*-oxide compounds of *N*-(4-pyridyl)nicotinamide (**L_1-L_3**), and the gelation properties of **L_1** and **L_3** were compared with isomeric *N*-oxide compounds (**INO** and **diNO**, respectively). Gelation tests revealed that **L_1** and **L_3** formed hydrogels, whereas **L_2** was a non-gelator, which underlines the importance of the nicotinamide *N*-oxide moiety in gel network formation. Comparing the gelation behavior of **L_1** and **L_3** with corresponding isomeric *N*-oxide compounds (**INO** and **diNO**) revealed that the relative position of the *N*-oxide moieties played a crucial role in the self-assembly process of LMWGs. SEM analysis revealed that the morphology was independent of the relative position of the pyridyl *N*-oxide moiety. Single-crystal X-ray diffraction of **L_3** revealed the existence of a 1D hydrogen-bonded chain, which was crucial for gel network formation. The solid-state structure was correlated to the xerogels to obtain an insight into the key interactions responsible for gel network formation. The stimuli-responsive properties of **L_1–L_3** were studied with various salts/ions, and anion-induced gelation was observed for **L_1** and **L_3** in the water below MGC with **L_1** and **L_3** in water. **L_3** was very versatile in nature, as it formed a gel below MGC with most of the salts or transition metal salts, which showed the cooperative interaction between the *N*-oxide and salts/ions. An abrupt increase in the mechanical property of **L_3** was observed in the presence of 1.0 equivalence of cadmium chloride, and a stable gel was formed at a very low concentration of the gelator (0.7 wt%). The effective sensing of cadmium chloride may be due to the metal-ligand coordination-driven supramolecular assembly. We showed that the salt-induced gelation depended on the nature and the position of the functional group, which will open the door for designing LMWGs based on *N*-oxide moieties with intriguing features.

4. Materials and Methods

The starting materials and solvents were obtained from Sigma-Aldrich (MEDOR ehf, Reykjavik, Iceland) and TCI-Europe (Boereveldseweg, Belgium) and utilized as provided. Deionized water was used to perform the gelation studies. The NMR spectra (^{1}H and ^{13}C, Figures S15–S20) were recorded with a Bruker Avance 400 spectrometer (Rheinstetten, Germany), and the scanning electron microscopy (SEM) images were captured with a Leo Supra 25 microscope (Carl Zeiss, Oberkochen, Germany). The mechanical strength evaluation was performed in an Anton Paar's MCR 302 (Graz, Austria) modular compact rheometer. Single-crystal X-ray diffraction (SCXRD) and powder X-ray diffraction (PXRD) experiments were carried out using a Bruker D8 venture (Karlsruhe, Germany) and PANalytical instrument (Almelo, Netherlands), respectively. We synthesized the *N*-oxide compounds by replacing the pyridyl group of *N*-(pyridin-4-yl)nicotinamide (**4PNA**) with pyridyl *N*-oxides [45]. Synthesis of 3-carboxypyridine 1-oxide [86] and 4-aminopyridine 1-oxide [87] was performed following literature and confirmed by matching the analytical data with the reported compounds. The INO and **diNO** compounds were synthesized by following the reported procedure from our previous work [55].

4.1. Synthesis of Ligands

4.1.1. Synthesis of 3-(pyridin-4-ylcarbamoyl) pyridine 1-oxide (L_1)

To a 100.0 mL two-neck round bottom flask we added 3-carboxypyridine 1-oxide (4.0 g, 30.0 mmol) with 20.0 mL of thionyl chloride, and the solution was refluxed overnight. The solvents were evaporated to dryness, and the corresponding acid chloride formed was used for the next step without further purification. Anhydrous DMF (around 25.0 mL) was added into the flask, followed by 4-aminopyridine (2.7 g, 28.7 mmol), and the mixture was cooled in an ice bath to 0 °C with constant stirring. A solution of triethylamine (4.5 mL) in 10.0 mL DMF was added dropwise to the reaction mixture at 0 °C over an hour, and the resulting yellow colloidal solution was stirred at room temperature overnight. To this mixture, water was added, resulting in a thick precipitate, which was filtered. The precipitate was then stirred in a 5.0% $NaHCO_3$ solution for 4.0 h and was filtered. The residue was washed with an excess of cold water, air-dried, and recrystallized from hot water to obtain the product. Yield 70.0%. ^1H NMR (400 MHz, DMSO-d_6) δ (ppm): δ 10.80 (s, 1H), 8.74 (s, 1H), 8.51 (d, J = 5.8 Hz, 2H), 8.43 (d, J = 6.6 Hz, 1H), 7.81 (d, J = 8.0 Hz, 1H), 7.74 (d, J = 6.0 Hz, 2H), 7.59 (t, J = 7.3 Hz, 1H). ^{13}C {^1H} NMR (101 MHz, DMSO-d_6) δ 162.65, 150.46, 145.27, 141.29, 138.02, 133.63, 126.58, 124.49, 114.11. HRMS (APCI): $C_{11}H_{10}N_3O_2$ $[M + H]^+$, 216.0768; found, 216.0765.

4.1.2. Synthesis of 4-(nicotinamido) pyridine 1-oxide (L_2)

Compound $\mathbf{L_2}$ was synthesized by following a similar procedure used for compound $\mathbf{L_1}$. Nicotonic acid (1.0 g, 8.12 mmol), thionyl chloride (20.0 mL), 4-aminopyridine 1-oxide (0.89 g, 8.12 mmol), and triethylamine (2.26 mL) in 30.0 mL DMF. Yield 68.0%. ^1H NMR (400 MHz, DMSO-d_6) δ 10.96 (s, 1H), 9.11 (s, 1H), 8.78 (dd, *J* = 4.8, 1.6 Hz, 1H), 8.32–8.28 (m, 1H), 8.19 (d, *J* = 7.5 Hz, 2H), 7.83 (d, *J* = 7.5 Hz, 2H), 7.60–7.56 (m, 1H). ^{13}C {^1H} NMR (101 MHz, DMSO-d_6) δ 164.48, 152.59, 148.79, 138.92, 136.31, 135.64, 129.84, 123.61, 116.86. HRMS (APCI): $C_{11}H_9N_3NaO_2$ $[M + Na]^+$, 238.0587; found, 238.0594.

4.1.3. Synthesis of 3-((1-oxidopyridin-4-yl) carbamoyl) pyridine 1-oxide (L_3)

To a 100.0 mL round-bottomed flask, *N*-(pyridin-4-yl) nicotinamide (2.0 g, 10.0 mmol) and MeOH (40.0 mL) were added to dissolve and stirred. 3-Chloroperoxybenzic acid (6.8 g, 40.0 mmol) was added in portions over 15.0 min to the solution and was refluxed overnight. The solution was filtered, and the precipitate was washed with 5.0% sodium bicarbonate and thrice with cold water. The precipitate was further recrystallized from hot water. Yield 55.0%. ^1H NMR (400 MHz, DMSO-d_6) δ 10.90 (s, 1H), 8.73 (s, 1H), 8.42 (m, 1H), 8.19 (d, *J* = 7.6 Hz, 2H), 7.81 (m, 1H), 7.79 (d, *J* = 7.6 Hz, 2H), 7.61–7.57 (m, 1H). ^{13}C {^1H} NMR (101 MHz, DMSO-d_6) δ 162.01, 141.31, 138.94, 137.92, 135.69, 133.50, 126.63, 124.45, 116.97. HRMS (APCI): $C_{11}H_9N_3O_2$ $[M + Na]^+$, 254.0536; found, 254.0540.

4.2. Gelation Studies

Hydrogelation ability was evaluated with all of the compounds by weighing different amounts (ranging from 10.0 to 40.0 mg) of the compound in a standard 7.0 mL vial; 1.0 mL of water was added, and the vial was sealed. The vial containing the mixture was then sonicated and slowly heated to obtain a clear solution, which was then left undisturbed for 24.0 h. Gelation was confirmed via a vial inversion test. Gelation tests were also performed in aqueous mixtures by dissolving the compound in 0.5 mL of distilled water and 0.5 mL of an appropriate solvent, and the vial was sealed. The mixture was sonicated, heated to obtain a clear solution, and checked for gelation ability. Gelation tests were performed in the presence of various salts with 15.0 mg of the compounds ($\mathbf{L_1}$ and $\mathbf{L_3}$) in 1.0 mL of water followed by the addition of 1:1 molar equivalent of an appropriate salt. The mixture was sonicated and heated after sealing the vial to obtain a transparent solution. The solution was left undisturbed, and a vial inversion test confirmed gel formation. The experiments were repeated three times to confirm the results.

4.2.1. Minimum Gelator Concentration (MGC)

MGC was performed in deionized water/aqueous mixtures by weighing various concentrations of the compounds in a standard 7.0 mL vial and adding 1.0 mL of deionized water/aqueous mixtures. The corresponding mixture was sonicated and gradually heated to dissolve the compounds, and the solution was kept at room temperature for gel formation. The minimum amount of the compound required to form a stable gel after 24.0 h was recorded as the MGC.

4.2.2. T_{gel} Experiments

The necessary amount of gelator and 1.0 mL of solvent were added to a 7.0 mL standard vial. After sonication, the mixture was heated to obtain a clear solution, and the mixture was left undisturbed to gel. A ball-drop method was used to observe the gel-to-solution transition temperature after 24.0 h (T_{gel}). A spherical glass ball was carefully positioned on top of the gel and the vial was immersed in an oil bath; a thermometer and a magnetic stirrer were equipped to check the temperature. The oil bath was gradually heated at 10.0 °C per minute. The glass ball slowly became immersed into the gel as the temperature increased, and the temperature at which the ball touched the bottom of the vial was recorded as T_{gel}.

4.3. Rheology

A 2.5 cm stainless steel parallel plate geometry configuration was used to perform the rheological measurements. In all cases, oscillatory measurements were conducted at a constant temperature of 20.0 °C. To maintain a constant temperature of 20.0 °C and to prevent solvent loss for amplitude and frequency sweeps, a Peltier temperature control hood was employed. Gels were prepared by dissolving an appropriate amount of gelator in 1.0 mL of solvent/solvent mixtures. After 24 h, amplitude-sweep experiments were carried out by adding approximately ~1.0 mL of gel to the plate. The frequency was maintained at 1.0 Hz during the amplitude sweep with log ramp strain (Y) ranges of 0.01–100%. The frequency-sweep experiments were carried out between 0.1 and 10.0 Hz within the linear viscoelasticity domain (0.02% strain). The experiments were also performed in the presence of various salts at 1:1 molar equivalent in water with a similar procedure as mentioned above.

4.4. Scanning Electron Microscopy (SEM)

The surface morphologies of the xerogels were analyzed on a Leo Supra 25 microscope. Gels of $\mathbf{L_1}$ and $\mathbf{L_3}$ were prepared in water at 2.0 wt%. We also prepared the aqueous mixture gels of $\mathbf{L_3}$ at 2.0 wt%. The gels were filtered after 24.0 h and dried under a fume hood to obtain the xerogel. A small part of xerogel was placed on a pin mount with the carbon tab on top, coated with gold for 5.0–6.0 min (~15.0 nm thickness) to avoid surface charging, and pictures were acquired at 3.0 kV with a working distance of 3–4 mm. An in-lens detector captured the SEM images. SEM of the xerogel of gelator $\mathbf{L_1}$ in the presence of sodium, magnesium, and calcium salts and $\mathbf{L_3}$ in the presence of sodium, magnesium, zinc (II), and cadmium (II) salts were also recorded.

4.5. Single-Crystal X-ray Diffraction

Crystals of compound $\mathbf{L_3}$ were obtained by the slow evaporation of 10.0 mg of the compound in 3.0 mL of water to obtain needle- and block-shaped crystals. X-ray analysis was performed on a Bruker D8 Venture (Photon100 CMOS detector) diffractometer provided with Cryostream (Oxford Cryosystems) open-flow nitrogen cryostats. The data were collected using MoKα radiation (λ = 0.71073 Å) for the plate-shaped crystals at 296(2) K and CuK$_{\alpha}$ radiation (λ = 1.542 Å) for the needle-shaped crystal at 302(2) K. Apex III software (Bruker AXS: Madison, WI, USA, 2015) was utilized for the unit cell determination, data collection, data reduction, structure solution/refinement, and empirical absorption correction (SADABS). A direct method was used for solving the structure and was refined by the full-matrix least-squares on F^2 for all data using SHELXTL version 2017/1 [88]. All non-disordered non-hydrogen atoms were refined anisotropically, and all of the hydrogen atoms were placed in the calculated positions and refined using a riding model, except for solvent water molecules. The hydrogen atoms of water molecules were located on the Fourier map and refined. The crystallographic data and hydrogen-bonding parameters are given in Tables S1 and S2 (see Supplementary Materials). The crystallographic data were deposited at the Cambridge Crystallographic Data Centre and can be obtained free of charge, and the CCDC numbers are 2226289–2226290.

4.6. Powder X-ray Diffraction (PXRD)

The bulk crystals of compound $\mathbf{L_3}$ were obtained by the slow evaporation of the solution of $\mathbf{L_3}$ (20.0 mg in 3.0 mL water). The crystals were filtered, dried in the air, and ground to a fine powder. The xerogels of $\mathbf{L_3}$ were prepared from water and the aqueous mixtures, following a similar procedure as mentioned above. We also performed PXRD of the dried gels at 2.0 wt/v% in various solvents/solvent mixtures obtained by adding different salts at 1:1 molar equivalent. All experiments were performed on a PANalytical device with a Cu anode, 2θ from 5.0 to 60.0°, and a 0.02 step size.

Supplementary Materials: The following supporting information can be downloaded at: https://www.mdpi.com/article/10.3390/gels9020089/s1, Scheme S1. Synthesis of $\mathbf{L_1}$, Scheme S2. Synthesis of $\mathbf{L_2}$, Scheme S3. Synthesis of $\mathbf{L_3}$. Figure S1. Amplitude-sweep experiments with gels of $\mathbf{L_1}$ and $\mathbf{L_2}$ (2.0 wt%) in water at 20.0 °C with a constant frequency of 1.0 Hz, Figure S2. Frequency-sweep experiments with gels of $\mathbf{L_1}$ and $\mathbf{L_3}$ (2.0 wt%) in aqueous mixtures at 20.0 °C with a constant strain of 0.02%, Figure S3. SEM images of (a) $\mathbf{L_1}$ and (b) $\mathbf{L_3}$ xerogels in water at 1.8 wt%, Figure S4. SEM images of the xerogels of (a) $\mathbf{L_3}$ and (b) **diNO** gels obtained from water at 4.0 wt%, Figure S5. SEM images of $\mathbf{L_1}$ xerogels in (a) DMSO/water (1:1, *v/v*) and (b) methanol/water (1:1, *v/v*) at 2.0 wt/v%, Figure S6. SEM images of $\mathbf{L_3}$ xerogels from (a) DMSO/water (1:1, *v/v*), and (b) methanol/water (1:1, *v/v*) at 2.0 wt/v%, Figure S7. (a) Molecular structure of $\mathbf{L_3}{\bullet}2H_2O$ and (b) two-dimensional hydrogen-bonded network with water molecules (space fill model) located in the cavity, Figure S8. Comparison of the simulated pattern of the single-crystal X-ray structure of $\mathbf{L_3}{\bullet}2H_2O$ with the PXRD pattern of the bulk crystals obtained from water, and xerogel from water at 2.0 wt%, Figure S9. PXRD pattern of xerogels obtained from the hydrogel of $\mathbf{L_3}$ at 2.0 wt% and in the presence of 1.0 equivalence of NaCl and KCl, Figure S10. SEM images of xerogels of (a) $\mathbf{L_1}$ and (b) $\mathbf{L_3}$ at 2.0 wt% obtained from water in the presence of 1.0 equivalence of $MgCl_2$, Figure S11. Comparison of the PXRD pattern of xerogels (2.0 wt%) of $\mathbf{L_3}$ hydrogels and the gels in the presence of 1.0 equivalence of $MgCl_2$ and $CaCl_2$, Figure S12. SEM images of the xerogels of $\mathbf{L_3}$ in the presence of 1.0 equivalence of $CdCl_2$ in water at 2.0 wt%, Figure S13. Comparison of the PXRD pattern of $\mathbf{L_3}$ xerogel with the PXRD pattern of the xerogels of $\mathbf{L_3}$-$CdCl_2$ mixture in various solvents, Figure S14. Comparison of the PXRD pattern of $\mathbf{L_3}$ xerogel from water with the PXRD pattern of the xerogels of the mixture ($\mathbf{L_3}$ + $ZnCl_2$) in various solvents, Figure S15. ^1H NMR spectrum of compound $\mathbf{L_1}$, Figure S16. ^{13}C NMR spectrum of compound $\mathbf{L_1}$, Figure S17. ^1H NMR spectrum of compound $\mathbf{L_2}$, Figure S18. ^{13}C NMR spectrum of compound $\mathbf{L_2}$, Figure S19. ^1H NMR spectrum of compound $\mathbf{L_3}$, Figure S20. ^{13}C NMR spectrum of compound $\mathbf{L_3}$. Table S1: Crystal data, Table S2: Hydrogen-bonding parameters, Table S3: Stimuli-responsive properties of the gelators $\mathbf{L_1}$ and $\mathbf{L_3}$: Anion sensing in water at 1.5 wt%, Table S4: Increase in G′ values of the gelators at 1.5 wt% in the presence of various sodium and potassium salts in comparison with the hydrogels (1.8 wt%), Table S5: Stimuli-responsive properties of the gelators $\mathbf{L_1}$ and $\mathbf{L_3}$: Cation sensing in water at 1.5 wt%, Table S6: Increase in G′ values of the gelators at 1.5 wt% in the presence of chloride salts of various cations in comparison with the hydrogels (1.8 wt%).

Author Contributions: Conceptualization, S.S.J., B.K., D.G. and K.K.D.; methodology, S.S.J., B.K., C.B., D.G. and K.K.D.; software, S.S.J. and D.G.; validation, S.S.J. and K.K.D.; formal analysis, S.S.J., D.G., B.K., C.B. and K.K.D.; investigation, S.S.J., D.G. and B.K.; resources, K.K.D.; data curation, K.K.D.; writing—original draft preparation, K.K.D.; writing—review and editing, S.S.J., B.K., D.G. and

K.K.D.; visualization, S.S.J. and K.K.D.; supervision, K.K.D.; project administration, K.K.D.; funding acquisition, K.K.D. All authors have read and agreed to the published version of the manuscript.

Funding: This research received no external funding.

Institutional Review Board Statement: Not applicable.

Informed Consent Statement: Not applicable.

Data Availability Statement: Not applicable.

Acknowledgments: We thank the University of Iceland Research Fund and Science Institute for funding. S.S.J. thanks the University of Iceland for the doctoral research grant. We acknowledge Sigríđur Jónsdóttir, University of Iceland, for NMR and mass spectrometry and Friđrik Magnus, University of Iceland, for powder X-ray diffraction analysis. The single-crystal X-ray diffractometer and rheometer used for this research were purchased under the Rannís Iceland infrastructure grants (150998-0031 and 191763-0031).

Conflicts of Interest: The authors declare no conflict of interest.

References

1. McConnell, A.J.; Wood, C.S.; Neelakandan, P.P.; Nitschke, J.R. Stimuli-Responsive Metal–Ligand Assemblies. *Chem. Rev.* **2015**, *115*, 7729–7793. [CrossRef] [PubMed]
2. Theato, P.; Sumerlin, B.S.; O'Reilly, R.K.; Epps, I.I.I.T.H. Stimuli responsive materials. *Chem. Soc. Rev.* **2013**, *42*, 7055–7056. [CrossRef]
3. Shigemitsu, H.; Hamachi, I. Supramolecular Assemblies Responsive to Biomolecules toward Biological Applications. *Chem. Asian J.* **2015**, *10*, 2026–2038. [CrossRef] [PubMed]
4. Segarra-Maset, M.D.; Nebot, V.J.; Miravet, J.F.; Escuder, B. Control of molecular gelation by chemical stimuli. *Chem. Soc. Rev.* **2013**, *42*, 7086–7098. [CrossRef]
5. Banerjee, S.; Das, R.K.; Maitra, U. Supramolecular gels 'in action'. *J. Mater. Chem.* **2009**, *19*, 6649–6687. [CrossRef]
6. Dastidar, P. Supramolecular gelling agents: Can they be designed? *Chem. Soc. Rev.* **2008**, *37*, 2699–2715. [CrossRef] [PubMed]
7. de Loos, M.; Feringa, B.L.; van Esch, J.H. Design and Application of Self-Assembled Low Molecular Weight Hydrogels. *Eur. J. Org. Chem.* **2005**, *2005*, 3615–3631. [CrossRef]
8. Estroff, L.A.; Hamilton, A.D. Water gelation by small organic molecules. *Chem. Rev.* **2004**, *104*, 1201–1218. [CrossRef]
9. George, M.; Weiss, R.G. Molecular Organogels. Soft Matter Comprised of Low-Molecular-Mass Organic Gelators and Organic Liquids†. *Acc. Chem. Res.* **2006**, *39*, 489–497. [CrossRef]
10. Hirst, A.R.; Escuder, B.; Miravet, J.F.; Smith, D.K. High-Tech Applications of Self-Assembling Supramolecular Nanostructured Gel-Phase Materials: From Regenerative Medicine to Electronic Devices. *Angew. Chem. Int. Ed.* **2008**, *47*, 8002–8018. [CrossRef]
11. Steed, J.W. Anion-tuned supramolecular gels: A natural evolution from urea supramolecular chemistry. *Chem. Soc. Rev.* **2010**, *39*, 3686–3699. [CrossRef] [PubMed]
12. Yu, G.; Yan, X.; Han, C.; Huang, F. Characterization of supramolecular gels. *Chem. Soc. Rev.* **2013**, *42*, 6697–6722. [CrossRef] [PubMed]
13. Chivers, P.R.A.; Smith, D.K. Shaping and structuring supramolecular gels. *Nat. Rev. Mater.* **2019**, *4*, 463–478. [CrossRef]
14. Kumar, D.K.; Steed, J.W. Supramolecular gel phase crystallization: Orthogonal self-assembly under non-equilibrium conditions. *Chem. Soc. Rev.* **2014**, *43*, 2080–2088. [CrossRef]
15. Truong, W.T.; Su, Y.; Meijer, J.T.; Thordarson, P.; Braet, F. Self-assembled gels for biomedical applications. *Chem. Asian J.* **2011**, *6*, 30–42. [CrossRef]
16. Foster, J.A.; Damodaran, K.K.; Maurin, A.; Day, G.M.; Thompson, H.P.; Cameron, G.J.; Bernal, J.C.; Steed, J.W. Pharmaceutical polymorph control in a drug-mimetic supramolecular gel. *Chem. Sci.* **2017**, *8*, 78. [CrossRef]
17. Zhu, J.; Wang, R.; Geng, R.; Zhang, X.; Wang, F.; Jiao, T.; Yang, J.; Bai, Z.; Peng, Q. A facile preparation method for new two-component supramolecular hydrogels and their performances in adsorption, catalysis, and stimuli-response. *RSC Adv.* **2019**, *9*, 22551–22558. [CrossRef]
18. Qu, R.; Shen, L.; Qu, A.; Wang, R.; An, Y.; Shi, L. Artificial Peroxidase/Oxidase Multiple Enzyme System Based on Supramolecular Hydrogel and Its Application as a Biocatalyst for Cascade Reactions. *ACS Appl. Mater. Interfaces* **2015**, *7*, 16694–16705. [CrossRef]
19. Zhang, J.; Bai, Y.; Yu, Q.; Ma, Z.; Liu, Q.; Zhou, F.; Liu, W.; Cai, M. Physicochemical and Tribological Performance of Bi-Component Supramolecular Gel Lubricants. *Adv. Mater. Interfaces* **2019**, *6*, 1801391. [CrossRef]
20. Yan, L.; Li, G.; Ye, Z.; Tian, F.; Zhang, S. Dual-responsive two-component supramolecular gels for self-healing materials and oil spill recovery. *Chem. Commun.* **2014**, *50*, 14839–14842. [CrossRef]
21. Patterson, A.K.; Smith, D.K. Two-component supramolecular hydrogel for controlled drug release. *Chem. Commun.* **2020**, *56*, 11046–11049. [CrossRef] [PubMed]
22. Buerkle, L.E.; Rowan, S.J. Supramolecular gels formed from multi-component low molecular weight species. *Chem. Soc. Rev.* **2012**, *41*, 6089–6102. [CrossRef] [PubMed]

23. Du, X.; Zhou, J.; Shi, J.; Xu, B. Supramolecular Hydrogelators and Hydrogels: From Soft Matter to Molecular Biomaterials. *Chem. Rev.* **2015**, *115*, 13165–13307. [CrossRef] [PubMed]
24. Cao, X.; Gao, A.; Hou, J.-t.; Yi, T. Fluorescent supramolecular self-assembly gels and their application as sensors: A review. *Coord. Chem. Rev.* **2021**, *434*, 213792. [CrossRef]
25. Picci, G.; Mulvee, M.T.; Caltagirone, C.; Lippolis, V.; Frontera, A.; Gomila, R.M.; Steed, J.W. Anion-Responsive Fluorescent Supramolecular Gels. *Molecules* **2022**, *27*, 1257. [CrossRef]
26. Oliveira, C.B.P.; Gomes, V.; Ferreira, P.M.T.; Martins, J.A.; Jervis, P.J. Peptide-Based Supramolecular Hydrogels as Drug Delivery Agents: Recent Advances. *Gels* **2022**, *8*, 706. [CrossRef]
27. Cui, J.; Liu, A.; Guan, Y.; Zheng, J.; Shen, Z.; Wan, X. Tuning the Helicity of Self-Assembled Structure of a Sugar-Based Organogelator by the Proper Choice of Cooling Rate. *Langmuir* **2010**, *26*, 3615–3622. [CrossRef]
28. Huang, X.; Terech, P.; Raghavan, S.R.; Weiss, R.G. Kinetics of 5α-Cholestan-3β-yl N-(2-Naphthyl)carbamate/n-Alkane Organogel Formation and Its Influence on the Fibrillar Networks. *J. Am. Chem. Soc.* **2005**, *127*, 4336–4344. [CrossRef]
29. Huang, H.; Zhu, X.; Su, L.; Wang, H.; Yang, Y. Effect of temperature on self-assembly/disassembly transition of dialkylurea supramolecular gels at high pressure. *RSC Adv.* **2013**, *3*, 11854–11859. [CrossRef]
30. Naota, T.; Koori, H. Molecules That Assemble by Sound: An Application to the Instant Gelation of Stable Organic Fluids. *J. Am. Chem. Soc.* **2005**, *127*, 9324–9325. [CrossRef]
31. Chen, S.; Fan, Y.; Song, J.; Xue, B. The remarkable role of hydrogen bond, halogen, and solvent effect on self-healing supramolecular gel. *Mater. Today Chem.* **2022**, *23*, 100719. [CrossRef]
32. Jayabhavan, S.S.; Steed, J.W.; Damodaran, K.K. Crystal Habit Modification of Metronidazole by Supramolecular Gels with Complementary Functionality. *Cryst. Growth Des.* **2021**, *21*, 5383–5393. [CrossRef]
33. Bera, S.; Basu, S.; Jana, B.; Dastidar, P. Real-time Observation of Macroscopic Helical Morphologies under Optical Microscope: A Curious Case of π–π Stacking Driven Molecular Self-assembly of an Organic Gelator Devoid of Hydrogen Bonding. *Angew. Chem. Int. Ed.* **2022**, e202216447. [CrossRef]
34. van Esch, J.H.; Feringa, B.L. New Functional Materials Based on Self-Assembling Organogels: From Serendipity towards Design. *Angew. Chem. Int. Ed.* **2000**, *39*, 2263–2266. [CrossRef]
35. Desiraju, G.R. Supramolecular Synthons in Crystal Engineering—A New Organic Synthesis. *Angew. Chem. Int. Ed.* **1995**, *34*, 2311–2327. [CrossRef]
36. Weiss, R.G.; Terech, P. (Eds.) *Molecular Gels: Materials with Self-Assembled Fibrillar Networks*; Springer: Berlin/Heidelberg, Germany, 2006; p. 978.
37. Fages, F.; Voegtle, F.; Zinic, M. Systematic design of amide- and urea-type gelators with tailored properties. *Top. Curr. Chem.* **2005**, *256*, 77–131.
38. Isare, B.; Pensec, S.; Raynal, M.; Bouteiller, L. Bisurea-based supramolecular polymers: From structure to properties1. *C. R. Chim.* **2016**, *19*, 148–156. [CrossRef]
39. Moulin, E.; Armao, J.J.; Giuseppone, N. Triarylamine-Based Supramolecular Polymers: Structures, Dynamics, and Functions. *Acc. Chem. Res.* **2019**, *52*, 975–983. [CrossRef]
40. Gale, P.A.; Busschaert, N.; Haynes, C.J.E.; Karagiannidis, L.E.; Kirby, I.L. Anion receptor chemistry: Highlights from 2011 and 2012. *Chem. Soc. Rev.* **2014**, *43*, 205–241. [CrossRef]
41. Dzolic, Z.; Cametti, M.; Dalla Cort, A.; Mandolini, L.; Zinic, M. Fluoride-responsive organogelator based on oxalamide-derived anthraquinone. *Chem. Commun.* **2007**, 3535–3537. [CrossRef]
42. Kotova, O.; Daly, R.; dos Santos, C.M.G.; Boese, M.; Kruger, P.E.; Boland, J.J.; Gunnlaugsson, T. Europium-Directed Self-Assembly of a Luminescent Supramolecular Gel from a Tripodal Terpyridine-Based Ligand. *Angew. Chem. Int. Ed.* **2012**, *51*, 7208–7212. [CrossRef] [PubMed]
43. Feng, L.; Cavicchi, K.A. Investigation of the relationships between the thermodynamic phase behavior and gelation behavior of a series of tripodal trisamide compounds. *Soft Matter* **2012**, *8*, 6483–6492. [CrossRef]
44. Mukhopadhyay, S.; Ira; Krishnamoorthy, G.; Maitra, U. Dynamics of Bound Dyes in a Nonpolymeric Aqueous Gel Derived from a Tripodal Bile Salt. *J. Phys. Chem. B* **2003**, *107*, 2189–2192. [CrossRef]
45. Kumar, D.K.; Jose, D.A.; Dastidar, P.; Das, A. Nonpolymeric Hydrogelator Derived from N-(4-Pyridyl)isonicotinamide. *Langmuir* **2004**, *20*, 10413–10418. [CrossRef]
46. Li, Z.; Cao, J.; Li, H.; Liu, H.; Han, F.; Liu, Z.; Tong, C.; Li, S. Self-assembled drug delivery system based on low-molecular-weight bis-amide organogelator: Synthesis, properties and in vivo evaluation. *Drug Delivery* **2016**, *23*, 3168–3178. [CrossRef]
47. Bradberry, S.J.; Dee, G.; Kotova, O.; McCoy, C.P.; Gunnlaugsson, T. Luminescent lanthanide (Eu(iii)) cross-linked supramolecular metallo co-polymeric hydrogels: The effect of ligand symmetry. *Chem. Commun.* **2019**, *55*, 1754–1757. [CrossRef]
48. Chandran, S.K.; Nath, N.K.; Cherukuvada, S.; Nangia, A. N–H ... N(pyridyl) and N–H ... O(urea) hydrogen bonding and molecular conformation of N-aryl-N′-pyridylureas. *J. Mol. Struct.* **2010**, *968*, 99–107. [CrossRef]
49. Kumar, D.K.; Jose, D.A.; Das, A.; Dastidar, P. First snapshot of a nonpolymeric hydrogelator interacting with its gelling solvents. *Chem. Commun.* **2005**, 4059–4061. [CrossRef]
50. Kumar, D.K.; Jose, D.A.; Dastidar, P.; Das, A. Nonpolymeric Hydrogelators Derived from Trimesic Amides. *Chem. Mater.* **2004**, *16*, 2332–2335. [CrossRef]

51. Lloyd, G.O.; Piepenbrock, M.-O.M.; Foster, J.A.; Clarke, N.; Steed, J.W. Anion tuning of chiral bis(urea) low molecular weight gels. *Soft Matter* **2012**, *8*, 204–216. [CrossRef]
52. Piepenbrock, M.-O.M.; Lloyd, G.O.; Clarke, N.; Steed, J.W. Metal- and Anion-Binding Supramolecular Gels. *Chem. Rev.* **2010**, *110*, 1960–2004. [CrossRef] [PubMed]
53. Lloyd, G.O.; Steed, J.W. Anion-tuning of supramolecular gel properties. *Nat. Chem.* **2009**, *1*, 437–442. [CrossRef] [PubMed]
54. Piepenbrock, M.-O.M.; Lloyd, G.O.; Clarke, N.; Steed, J.W. Gelation is crucially dependent on functional group orientation and may be tuned by anion binding. *Chem. Commun.* **2008**, 2644–2646. [CrossRef] [PubMed]
55. Ghosh, D.; Mulvee, M.T.; Damodaran, K.K. Tuning Gel State Properties of Supramolecular Gels by Functional Group Modification. *Molecules* **2019**, *24*, 3472. [CrossRef]
56. Reddy, L.S.; Babu, N.J.; Nangia, A. Carboxamide–pyridine N-oxide heterosynthon for crystal engineering and pharmaceutical cocrystals. *Chem. Commun.* **2006**, 1369–1371. [CrossRef]
57. Mfuh, A.M.; Larionov, O.V. Heterocyclic N-oxides-an emerging class of therapeutic agents. *Curr. Med. Chem.* **2015**, *22*, 2819–2857. [CrossRef]
58. Saikia, B.; Khatioda, R.; Bora, P.; Sarma, B. Pyridine N-oxides as coformers in the development of drug cocrystals. *CrystEngComm* **2016**, *18*, 8454–8464. [CrossRef]
59. Puttreddy, R.; Beyeh, N.K.; Ras, R.H.A.; Trant, J.F.; Rissanen, K. Endo-/exo- and halogen-bonded complexes of conformationally rigid C-ethyl-2-bromoresorcinarene and aromatic N-oxides. *CrystEngComm* **2017**, *19*, 4312–4320. [CrossRef]
60. Łukomska, M.; Rybarczyk-Pirek, A.J.; Jabłoński, M.; Palusiak, M. The nature of NO-bonding in N-oxide group. *PCCP* **2015**, *17*, 16375–16387. [CrossRef]
61. Sauvée, C.; Ström, A.; Haukka, M.; Sundén, H. A Multi-Component Reaction towards the Development of Highly Modular Hydrogelators. *Chem. Eur. J.* **2018**, *24*, 8071–8075. [CrossRef]
62. Ghosh, D.; Deepa; Damodaran, K.K. Metal complexation induced supramolecular gels for the detection of cyanide in water. *Supramol. Chem.* **2020**, *32*, 276–286. [CrossRef]
63. Ghosh, D.; Ferfolja, K.; Drabavičius, Ž.; Steed, J.W.; Damodaran, K.K. Crystal habit modification of Cu(ii) isonicotinate–N-oxide complexes using gel phase crystallisation. *New J. Chem.* **2018**, *42*, 19963–19970. [CrossRef]
64. Ghosh, D.; Bjornsson, R.; Damodaran, K.K. Role of N–Oxide Moieties in Tuning Supramolecular Gel-State Properties. *Gels* **2020**, *6*, 41. [CrossRef]
65. Goodwin, J.W.; Hughes, R.W. *Rheology for Chemists: An Introduction*; Royal Society of Chemistry: London, UK, 2008.
66. Guenet, J.-M. *Organogels: Thermodynamics, Structure, Solvent Role, and Properties*; Springer: Berlin/Heidelberg, Germany, 2016.
67. Denzer, B.R.; Kulchar, R.J.; Huang, R.B.; Patterson, J. Advanced Methods for the Characterization of Supramolecular Hydrogels. *Gels* **2021**, *7*, 158. [CrossRef]
68. Tomasson, D.A.; Ghosh, D.; Kržišnik, Z.; Fasolin, L.H.; Vicente, A.n.A.; Martin, A.D.; Thordarson, P.; Damodaran, K.K. Enhanced mechanical and thermal strength in mixed-enantiomers-based supramolecular gel. *Langmuir* **2018**, *34*, 12957–12967. [CrossRef] [PubMed]
69. Ghosh, D.; Farahani, A.D.; Martin, A.D.; Thordarson, P.; Damodaran, K.K. Unraveling the Self-Assembly Modes in Multicomponent Supramolecular Gels Using Single-Crystal X-ray Diffraction. *Chem. Mater.* **2020**, *32*, 3517–3527. [CrossRef]
70. Bernstein, J.; Davis, R.E.; Shimoni, L.; Chang, N.-L. Patterns in Hydrogen Bonding: Functionality and Graph Set Analysis in Crystals. *Angew. Chem. Int. Ed.* **1995**, *34*, 1555–1573. [CrossRef]
71. Ghosh, D.; Lebedytė, I.; Yufit, D.S.; Damodaran, K.K.; Steed, J.W. Selective gelation of N-(4-pyridyl) nicotinamide by copper (ii) salts. *CrystEngComm* **2015**, *17*, 8130–8138. [CrossRef]
72. Tómasson, D.A.; Ghosh, D.; Kurup, M.P.; Mulvee, M.T.; Damodaran, K.K. Evaluating the role of a urea-like motif in enhancing the thermal and mechanical strength of supramolecular gels. *CrystEngComm* **2021**, *23*, 617–628. [CrossRef]
73. Mears, L.L.E.; Draper, E.R.; Castilla, A.M.; Su, H.; Zhuola; Dietrich, B.; Nolan, M.C.; Smith, G.N.; Doutch, J.; Rogers, S.; et al. Drying Affects the Fiber Network in Low Molecular Weight Hydrogels. *Biomacromolecules* **2017**, *18*, 3531–3540. [CrossRef]
74. Li, L.; Sun, R.; Zheng, R.; Huang, Y. Anions-responsive supramolecular gels: A review. *Mater. Des.* **2021**, *205*, 109759. [CrossRef]
75. Fages, F. Metal coordination to assist molecular gelation. *Angew. Chem. Int. Ed.* **2006**, *45*, 1680–1682. [CrossRef] [PubMed]
76. Offiler, C.A.; Jones, C.D.; Steed, J.W. Metal 'turn-off', anion 'turn-on' gelation cascade in pyridinylmethyl ureas. *Chem. Commun.* **2017**, *53*, 2024–2027. [CrossRef] [PubMed]
77. Sudhakaran Jayabhavan, S.; Ghosh, D.; Damodaran, K.K. Making and Breaking of Gels: Stimuli-Responsive Properties of Bis (Pyridyl-N-oxide Urea) Gelators. *Molecules* **2021**, *26*, 6420. [CrossRef] [PubMed]
78. Lima, J.C.; Rodríguez, L. Supramolecular gold metallogelators: The key role of metallophilic interactions. *Inorganics* **2014**, *3*, 1–18. [CrossRef]
79. Tam, A.Y.-Y.; Yam, V.W.-W. Recent advances in metallogels. *Chem. Soc. Rev.* **2013**, *42*, 1540–1567. [CrossRef]
80. Adarsh, N.; Sahoo, P.; Dastidar, P. Is a crystal engineering approach useful in designing metallogels? A case study. *Cryst. Growth Des.* **2010**, *10*, 4976–4986. [CrossRef]
81. Häring, M.; Díaz, D.D. Supramolecular metallogels with bulk self-healing properties prepared by in situ metal complexation. *Chem. Commun.* **2016**, *52*, 13068–13081. [CrossRef]
82. Bhattacharjee, S.; Bhattacharya, S. Pyridylenevinylene based Cu2+-specific, injectable metallo(hydro)gel: Thixotropy and nanoscale metal-organic particles. *Chem. Commun.* **2014**, *50*, 11690–11693. [CrossRef]

83. Bhattacharjee, S.; Samanta, S.K.; Moitra, P.; Pramoda, K.; Kumar, R.; Bhattacharya, S.; Rao, C.N.R. Nanocomposite Made of an Oligo(p-phenylenevinylene)-Based Trihybrid Thixotropic Metallo(organo)gel Comprising Nanoscale Metal-Organic Particles, Carbon Nanohorns, and Silver Nanoparticles. *Chem. Eur. J.* **2015**, *21*, 5467–5476. [CrossRef]
84. Kumar Vemula, P.; Aslam, U.; Ajay Mallia, V.; John, G. In Situ Synthesis of Gold Nanoparticles Using Molecular Gels and Liquid Crystals from Vitamin-C Amphiphiles. *Chem. Mater.* **2007**, *19*, 138–140. [CrossRef]
85. Paul, M.; Sarkar, K.; Dastidar, P. Metallogels Derived from Silver Coordination Polymers of C3-Symmetric Tris(pyridylamide) Tripodal Ligands: Synthesis of Ag Nanoparticles and Catalysis. *Chem. Eur. J.* **2015**, *21*, 255–268. [CrossRef]
86. Schobert, R.; Biersack, B. cis-Dichloroplatinum (II) complexes with aminomethylnicotinate and -isonicotinate ligands. *Inorg. Chim. Acta* **2005**, *358*, 3369–3376. [CrossRef]
87. Gardner, J.N.; Katritzky, A.R. N-oxides and related compounds. Part V. The tautomerism of 2- and 4-amino- and -hydroxy-pyridine 1-oxide. *J. Chem. Soc.* **1957**, *875*, 4375–4385. [CrossRef]
88. Sheldrick, G. Crystal structure refinement with SHELXL. *Acta Crystallogr. Sect. C* **2015**, *71*, 3–8. [CrossRef]

gels

MDPI

Article

Optimising Soy and Pea Protein Gelation to Obtain Hydrogels Intended as Precursors of Food-Grade Dried Porous Materials

Lorenzo De Berardinis, Stella Plazzotta and Lara Manzocco *

Department of Agricultural, Food, Environmental and Animal Sciences, University of Udine, 33100 Udine, Italy
* Correspondence: lara.manzocco@uniud.it

Abstract: Dried porous materials based on plant proteins are attracting large attention thanks to their potential use as sustainable food ingredients. Nevertheless, plant proteins present lower gelling properties than animal ones. Plant protein gelling could be improved by optimising gelation conditions by acting on protein concentration, pH, and ionic strength. This work aimed to systematically study the effect of these factors on the gelation behaviour of soy and pea protein isolates. Protein suspensions having different concentrations (10, 15, and 20% w/w), pH (3.0, 4.5, 7.0), and ionic strength (IS, 0.0, 0.6, 1.5 M) were heat-treated (95 °C for 15 min) and characterised for rheological properties and physical stability. Strong hydrogels having an elastic modulus (G′) higher than 10^3 Pa and able to retain more than 90% water were only obtained from suspensions containing at least 15% soy protein, far from the isoelectric point and at an IS above 0.6 M. By contrast, pea protein gelation was achieved only at a high concentration (20%), and always resulted in weak gels, which showed increasing G′ with the increase in pH and IS. Results were rationalised into a map identifying the gelation conditions to modulate the rheological properties of soy and pea protein hydrogels, for their subsequent conversion into xerogels, cryogels, and aerogels.

Keywords: plant proteins; heat gelation; gelling behaviour; structure; pH

Citation: De Berardinis, L.; Plazzotta, S.; Manzocco, L. Optimising Soy and Pea Protein Gelation to Obtain Hydrogels Intended as Precursors of Food-Grade Dried Porous Materials. *Gels* **2023**, *9*, 62. https://doi.org/10.3390/gels9010062

Academic Editors: Francesco Caridi, Giuseppe Paladini and Andrea Fiorati

Received: 15 December 2022
Revised: 4 January 2023
Accepted: 10 January 2023
Published: 12 January 2023

1. Introduction

Xerogels, cryogels, and aerogels indicate dry porous materials produced by removing the solvent from a gel. Most studies have been carried out on the development of inorganic dried porous materials (e.g., silica and carbon-based) [1–3] to be used in a wide variety of applications, such as catalysis, environmental remediation, energy storage, and insulation [4–7]. Nevertheless, in recent years, growing interest has been focused on the development of biopolymeric-based dried porous templates, due to their biocompatibility, and non-toxic profile. Thanks to these characteristics, their application has been successfully extended to life science fields, including the biomedical and pharmaceutical sectors [8–10]. The potentialities of dried porous materials in the food sector are nowadays attracting large attention, due to their unique physico-chemical properties and techno-functionalities. Both cryogels and aerogels have been suggested as innovative delivery systems to protect bioactives and flavours during processing, storage, and digestion [11–16]. In addition, their capacity to absorb large amounts of food solvents has been identified as a key feature to modulate food structural properties [17,18]. For instance, they have been suggested as templates for oil structuring, leading to fat replacers with improved nutritional properties [16,19–21]. By contrast, as concerns xerogels, to the best of our knowledge, no applications in the food sector have been reported, despite the high potentialities of these materials have been demonstrated in other life science sectors.

To produce food-grade dried porous material, an aqueous gel is first produced by inducing the networking of the selected biopolymer in water, leading to a hydrogel [22]. To obtain a xerogel, subsequently, water is removed from the network by evaporative drying. The latter can also be performed by evaporating ethanol after substituting hydrogel water with ethanol [23,24]. The evaporative drying usually induces capillary forces during solvent removal, so that xerogels present low porosity [25]. Cryogels are instead obtained through freeze-drying, and thus by water sublimation [18]. This reduces the capillary forces, leading to materials with large pores and channels left upon the sublimation of water crystals grown during freezing [26]. Finally, aerogels are obtained by replacing the water contained in the starting gel with ethanol, followed by ethanol removal with a flow of CO_2 in the supercritical state [27]. This technique preserves the structure of the initial network, and the dried material is thus characterised by low density and high internal surface area, due to the presence of micro- and macropores [28].

Food-grade xerogels, cryogels, and aerogels can be prepared either from polysaccharides or proteins. As concerns proteins, most literature studies focus on animal ones (e.g., whey, egg white, casein, gelatin) [29,30], while studies on the development of dried porous templates from plant proteins are limited to a few works exploiting silk fibroin, patatins, and soy proteins [31–37]. The interest in plant-based products is constantly growing due to their low environmental impact, low cost, and the possibility of being obtained from food industry wastes, in a circular economy perspective [38–40]. For these reasons, plant proteins are ideal candidates for developing sustainable dried porous materials intended as innovative ingredients for the food sector. However, the production of plant-based xerogels, cryogels, and aerogels is rather challenging. This is mainly due to the poor gelling properties of vegetable proteins compared to their animal counterparts. Protein gelation is commonly induced by heat treatment, during which the protein chains unfold, exposing their reactive groups, which subsequently drive protein reassembling in a 3D network. Although both covalent (i.e., S-S bridges) and weak interactions (i.e., hydrophobic interactions, hydrogen bonds, and electrostatic interactions) play an important role in the formation and stabilisation of protein gels [41], the availability of free -SH groups available for covalent stabilisation is known to lead to stronger gels. The possibility to obtain strong hydrogels is pivotal in determining their suitability in the conversion into dried porous materials, since the stronger the gel, the higher its capacity to structurally withstand the subsequent drying steps. In this regard, plant proteins present a lower number of -SH groups as compared to animal ones [42]. Moreover, the extraction process performed to isolate the protein fraction from the vegetable matrix, where it is intimately embedded in fiber–protein complexes, is known to induce structural modifications in the protein chains, further reducing gelling properties [43]. Nevertheless, several factors, including protein concentration, pH, and ionic strength, can be properly modulated to improve the plant protein gelling capacity. In this regard, the increase in protein concentration usually leads to a denser protein network, accounting for the formation of firmer gels that better maintain the original volume upon water removal [41]. When gelation occurs at a pH approaching the isoelectric point (pI), globular and strongly aggregated protein structures are formed, mostly driven by hydrophobic interactions [44,45]. At a pH far above or below the pI, instead, proteins form a fine-stranded network, as a result of the presence of surface charges which prevent intimate protein aggregation [46]. For example, aerogels derived from gels prepared near protein pI have been shown to present higher structural stability during drying, associated with lower density and higher pore sizes as compared to aerogels prepared far from the pI [28,47]. Gelation properties are also affected by ionic strength (IS). The increase in IS reduces electrostatic repulsive forces among protein chains, favouring the formation of a stronger network. For instance, the elastic modulus of pea protein gels was increased by 12 times by adding 0.3 M NaCl [41]. However, beyond a salt concentration threshold, specific for each protein (usually >2.0 M), a weakening of the hydrogel structure is commonly observed, due to salt-induced stabilisation of the protein structure, which suppresses protein unfolding during gelation [48,49].

This work aimed to systematically study the effect of gelation conditions on the physical properties of plant protein-based hydrogels, with the final aim of identifying the conditions leading to hydrogels suitable for the development of dried porous materials. For this purpose, soy and pea proteins were selected as the protein sources widely used as alternatives to animal proteins. Aqueous suspensions containing increasing amounts of soy and pea protein isolates (SPI and PPI) at different pH (3.0, 4.5 pI, 7.0) and IS (0.0, 0.6, 1.5 M) were heat-treated to induce gelation. The obtained hydrogels were characterised for rheological properties and physical stability, and the results were rationalised into a gelation map.

2. Results and Discussion

2.1. *Effect of Protein Type and Concentration*

SPI and PPI solutions were prepared at increasing concentrations from 10 to 20% (w/w) at pH 7.0, and thermally treated. Table 1 reports the appearance of the obtained SPI and PPI samples.

Table 1. Appearance, elastic (G′), loss modulus (G″), loss tangent (tan δ), and water-holding capacity (WHC) of soy protein isolate (SPI) and pea protein isolate (PPI) systems obtained after heat treatment of protein solutions at 10, 15, and 20% w/w; at pH 7.0; and 0.0 ionic strength.

Protein	Concentration (%, w/w)	Appearance	$G' \times 10^2$ (Pa)	$G'' \times 10^2$ (Pa)	Tan δ	WHC
SPI	10	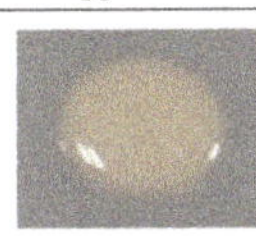	N.D.	N.D.	N.D.	N.D.
	15		3.94 ± 0.24 [c]	0.77 ± 0.04 [c]	0.19 ± 0.01 [b]	89.51 ± 3.56 [a]
	20		47.57 ± 1.61 [a]	6.60 ± 0.24 [a]	0.14 ± 0.01 [c]	99.60 ± 0.44 [a]
PPI	10		N.D.	N.D.	N.D.	N.D.
	15		N.D.	N.D.	N.D.	N.D.
	20	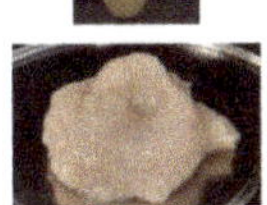	13.64 ± 0.24 [b]	3.13 ± 0.05 [b]	0.23 ± 0.01 [a]	75.19 ± 0.03 [b]

N.D.: not determined, since the system did not gel. [a, b, c]: means indicated by different letters in the same column are significantly different ($p < 0.05$).

As expected, for both SPI and PPI, the increase in protein concentration resulted in a visible increase in system structuring [50,51]. At a given protein concentration, SPI always led to a more structured system as compared to PPI, so a minimum protein concentration of 15 and 20% (w/w) was required to form a semi-solid system by using SPI and PPI, respectively (Table 1). This difference was also confirmed by the rheological analysis. Supplementary Figure S1 reports the frequency sweep test results for SPI and PPI hydrogels obtained from 20% (w/w) protein solutions.

For both proteins, G′ higher than G″ and parallel to G″ was obtained, indicating the formation of gel systems [52]. The moduli of the PPI gel showed a higher frequency dependence (higher slope) than those of the SPI gels. The latter showed a negligible frequency dependence, indicating that a stronger gel structure was obtained; SPI gels also presented rheological moduli higher than those of the PPI gel, and a lower loss tangent (tan δ) (Table 1). These results confirm the higher gelling ability of SPI as compared to PPI. In agreement with the literature [53,54], this difference between SPI and PPI gelation properties can be attributed to the different compositions of the globulin fraction of the considered proteins. Soybean globulins are mainly represented by glycinin (11S) and β-conglycinin (7S), which present higher solubility than pea ones (legumin 11S and vicilin 7S). As a result, a higher protein fraction would remain homogeneously suspended during the gelation of soy proteins [43,53]. Moreover, soybean globulins have been previously demonstrated to present a threshold gelling concentration lower than pea ones [55].

The higher strength of the gel obtained with SPI rather than PPI was also related to an improvement in gel stability, as shown by the higher WHC values (Table 1). The increased density network obtained by increasing protein concentration was actually able to retain more water, due to the better distribution of the solvent in the 3D structure, as well as to the higher number of protein residues available for the interaction with water [56].

2.2. Effect of pH

The precursor protein solutions were adjusted to pH 3.0, 4.5, and 7.0 and thermally treated. Independently of the pH, self-standing gelled systems were only obtained at 15 and 20% (w/w) SPI concentrations and at a 20% (w/w) PPI concentration. As representative examples, Table 2 reports the appearance and the rheological parameters of the hydrogels obtained from the SPI and PPI solutions at 20% (w/w) protein concentration and adjusted at the different pH values.

Table 2. Appearance, storage modulus (G′), loss modulus (G″), loss tangent (tan δ), and water-holding capacity (WHC) of soy protein isolate (SPI) and pea protein isolate (PPI) hydrogels at 20% protein concentration at pH 3.0 and 4.5, and 0.0 ionic strength.

Protein	pH	Appearance	G′ × 10^2 (Pa)	G″ × 10^2 (Pa)	Tan δ	WHC
SPI	3.0		30.14 ± 2.78 [a]	4.10 ± 0.46 [a]	0.14 ± 0.01 [c]	99.80 ± 0.09 [a]
	4.5		24.32 ± 0.66 [b]	3.36 ± 0.85 [b]	0.14 ± 0.01 [c]	99.68 ± 0.07 [a]
PPI	3.0		8.52 ± 1.26 [c]	2.38 ± 0.31 [c]	0.28 ± 0.01 [a]	67.07 ± 0.73 [b]
	4.5	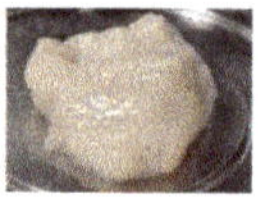	7.89 ± 0.59 [d]	1.99 ± 0.73 [d]	0.25 ± 0.01 [b]	75.00 ± 2.99 [b]

[a, b, c, d]: means indicated by different letters in the same column are significantly different ($p < 0.05$).

Similar to data achieved at pH 7.0 (Table 1), also at pH 3.0 and 4.5, SPI led to higher system structuration as compared to PPI. At pH 4.5, which is close to protein pI, a particulate gel, otherwise known as a microgel, was obtained with both proteins [57,58]. Proteins actually show a higher tendency towards aggregation in the isoelectric region, where the net charge is low, and thus protein–protein interactions are promoted with the formation of spherical particles, which, at a high protein concentration, can randomly associate into larger self-supporting hydrogels [58]. By contrast, at pH values away from the pI, where strong electrostatic repulsions are present, the gels present a fine-stranded structure.

For both proteins, the decrease in pH from 7.0 (Table 1) to 3.0 (Table 2) caused a significant decrease in system structuration, as evidenced by the rheological parameters. In fact, not only both moduli showed lower values for gels prepared at pH 3.0 as compared to those obtained at neutral pH, but they also presented a slightly higher frequency dependence. In this regard, Supplementary Figure S2 shows the effect of the pH change on the frequency sweep results of PPI gels prepared at 20% (w/w) protein concentration at pH 3.0 and 7.0. A significant decrease in gel strength was instead observed upon adjusting the protein solution at pH 4.5 (Tables 1 and 2). This can be attributed to the different microstructure of the hydrogels obtained at different pHs. In particular, microgelled systems obtained near the pI are stabilised by weak surface interactions among spherical protein aggregates, which can easily flow one on the other [59]. By contrast, at pHs far from the pI (pH 3.0 and 7.0), stranded gel structures are obtained, stabilised by numerous disulphide bridges and weak-interaction entanglement regions, thus accounting for the higher resistance to mechanical perturbation [58]. Moreover, in the isoelectric region, protein solubility is minimised, resulting in a significant decrease in well-solubilised protein fractions able to efficaciously interlink in a 3D gel network [53].

For both SPI and PPI, pH had a negligible effect on gel stability, as indicated by the comparable WHC values (Tables 1 and 2). This is probably due to the counterbalancing effect of the high protein concentration on the effect of pH. In other words, the effect of the different gel architectures induced by pH would be made negligible in the presence of a high protein concentration, which would increase the network density, thus allowing a high solvent retention [53].

2.3. *Effect of Ionic Strength*

The precursor protein solutions were added with different NaCl amounts to modulate the ionic strength (IS) of the system. As representative examples of the effect of this parameter at low protein concentrations, Table 3 shows the appearance of systems obtained upon the thermal treatment of 10% (w/w) SPI and 15% (w/w) PPI solutions, at pH 7.0, and having 0.6 and 1.5 M IS.

Although the final system showed an evident phase separation, as compared to the system with no salt added (Table 1), which showed a liquid-like homogeneous structure, the increase in IS resulted in a local gelling effect with the formation of a microgel-like structure. This effect can be traced back to the shielding effect of salt ions of the protein surface charge, favouring protein aggregation [60]. The positive effect of the IS increase on SPI and PPI gelling properties was also observed at a higher protein concentration. In this regard, Table 3 reports the appearance and the rheological parameters of the hydrogels obtained from 20% (w/w) SPI and PPI solutions at pH 7.0, at 0.6 and 1.5 M IS. As compared to the gels obtained without salt addition (Table 1), the increase in IS resulted in particulate gels, well-evident in the case of the PPI-based systems (Table 3). This was due to the changes induced by the increase in IS in the gel microstructure, which shifted from a fine-stranded structure (low IS) to a particulate structure (high IS) [22]. NaCl concentration increase also caused a considerable increase in both SPI and PPI gel strength, as indicated by the increase in G' values (Tables 1 and 3), as shown in Supplementary Figure S3, which reports the frequency sweep results for PPI gels at 20% (w/w) protein concentration at 0.0 and 1.5 M IS. The presence of Na^+ ions actually promotes protein–protein interactions during gelation, due to the reduction of the repulsive electrostatic interactions between protein chains [51].

Moreover, the increase in IS is known to promote the so-defined "salting-in" effect, i.e., the increase in the solubility of globulins, which are the main protein fraction of both SPI and PPI [61]. A higher IS thus results in higher availability of well-hydrated proteins available for networking during gelation [51,62].

Table 3. Appearance, storage modulus (G′), loss modulus (G″), loss tangent (tan δ), and water-holding capacity (WHC) of soy protein isolate (SPI) and pea protein isolate (PPI) hydrogels at 10, 15, or 20% (w/w) protein concentrations at 0.6 and 1.5 M ionic strength.

Protein	Concentration (%, w/w)	Ionic Strength (M)	Appearance	$G' \times 10^2$ (Pa)	$G'' \times 10^2$ (Pa)	Tan δ	WHC
SPI	10	0.6		N.D.	N.D.	N.D.	N.D.
		1.5		N.D.	N.D.	N.D.	N.D.
	20	0.6		58.21 ± 7.10 [b]	9.97 ± 1.19 [b]	0.17 ± 0.01 [b]	97.95 ± 0.70 [a]
		1.5		115.51 ± 46.08 [a]	22.40 ± 8.52 [a]	0.19 ± 0.01 [a]	89.24 ± 3.17 [b]
PPI	15	0.6	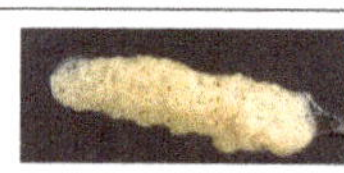	N.D.	N.D.	N.D.	N.D.
		1.5		N.D.	N.D.	N.D.	N.D.
	20	0.6		48.48 ± 0.29 [b]	11.52 ± 0.07 [b]	0.24 ± 0.01 [c]	84.76 ± 3.56 [b]
		1.5	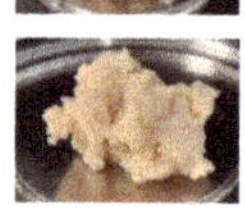	51.57 ± 6.38 [b]	12.56 ± 1.64 [b]	0.24 ± 0.01 [c]	91.96 ± 0.02 [c]

N.D.: not determined, since the system did not gel. [a, b, c]: means indicated by different letters in the same column are significantly different ($p < 0.05$).

IS also affected gel stability. In the case of the SPI gels, WHC decreased with IS, despite the higher gel strength (Table 3). Similar results were found for gels from both soy [63,64] and egg white proteins [65–68] and can be attributed to the microstructural changes induced by the presence of ions. In this regard, Munialo et al. [69] have demonstrated that a gel with an evenly distributed fine-stranded network, obtained at low IS, generally presents higher WHC as compared to particulate gels, obtained at high IS, where water is less tightly trapped. Likewise, Maltais et al. [70] and Urbonaite, et al. [71,72] reported an inverse correlation between aggregate size and WHC, with larger aggregates resulting in lower WHC. On the contrary, in the case of PPI hydrogels, the increase in IS promoted an increase in the WHC. It can be inferred that, in this case, the increased gel structural properties obtained upon NaCl addition (Tables 1 and 3) prevailed over the microstructural changes induced by the IS increase.

2.4. Gelation Map

Collected data were further elaborated and rationalised in order to obtain a gelation map (Figure 1), which is useful to have an immediate view of the gelation performances of SPI and PPI under the considered conditions.

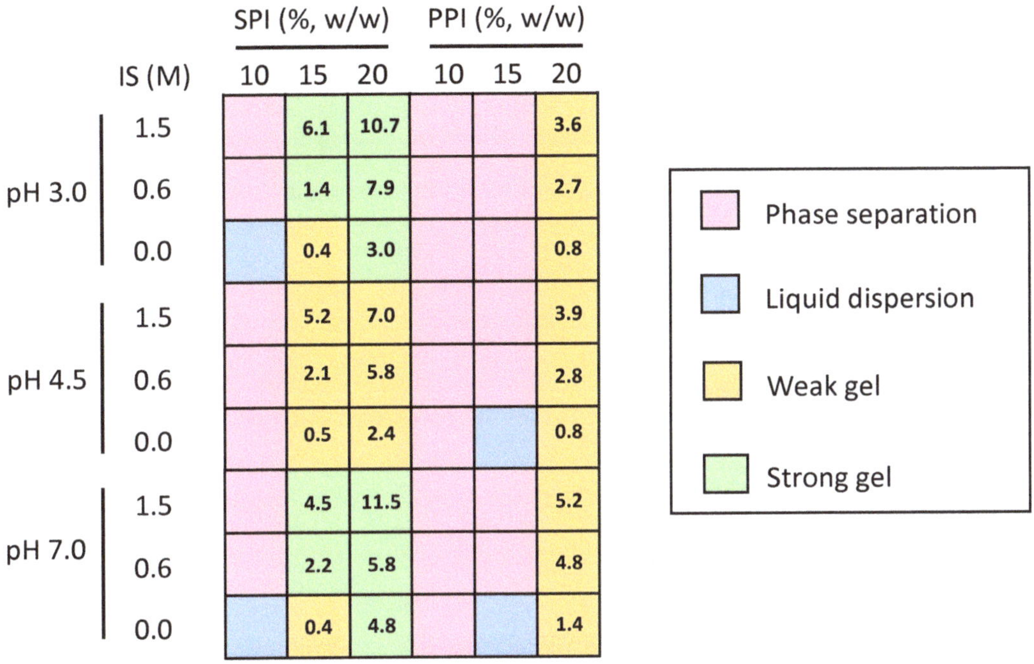

Figure 1. Gelation map of soy (SPI) and pea protein isolate (PPI) at increasing protein concentration (%, w/w), pH, and ionic strength (IS). The mean values of elastic modulus ($G' \times 10^3$ Pa) of the gelled systems are also reported within cells.

The obtained map clearly highlights the complex effect of protein type, pH, IS, and their combination on the sample structure. For example, the higher gelation propensity of SPI as compared to PPI is immediately visible, as well as the higher structuration obtained far away from the protein isoelectric region pH or increasing the IS. This map represents a useful tool to identify optimal conditions leading to SPI and PPI gels presenting the desired physical properties. In particular, the conditions allowing for the preparation of hydrogels presenting a network strong enough to withstand the conversion into xerogels, cryogels, and aerogels can be identified. Moreover, additional considerations can be drawn, with the aim of optimising the production process of these dried porous materials. For example, at pH 3.0 or 7.0, in view of minimising the consumption of SPI, and thus raw material costs, while also maintaining a strong gel structure, the possibility to reduce the SPI concentration from 20 to 15% (w/w) while increasing the ionic strength can be identified. Similarly, in the case of PPI, it is immediately evident how only weak gels can be obtained at 20% concentration.

3. Conclusions

The results collected in this study show that the gelling behaviour of vegetable proteins is highly dependent on both the protein nature and formulation parameters (protein concentration, pH, ionic strength). In particular, hydrogel strength can be enhanced by choosing soy proteins over pea ones, as well as avoiding the isoelectric region and increasing the ionic strength. The obtained gelation map can be considered a useful tool to identify the optimal conditions to produce soy and pea protein hydrogels with physical properties suitable for the subsequent conversion into xerogels, cryogels, and aerogels.

The results obtained in this research, although relevant to soy and pea protein isolates solely, clearly indicate the potential of plant proteins as interesting precursors for the production of food-grade and plant protein-based dried porous materials. Further studies are therefore required to investigate the correlation between the physical and techno-functional properties of the precursor hydrogel and the resulting dried materials. In this regard, different drying processes such as evaporative drying, freeze-drying, and supercritical drying can be applied to convert the obtained hydrogels into xerogels, cryogels, and aerogels, respectively. At the same time, a comprehensive characterisation of the dried templates obtained thereof could be performed. The latter should include the physical characterisation of the materials (e.g., SEM microstructure, BET surface area, porosity) but also their interaction properties with food fluids (oil, water) to obtain a first insight into their applicability as innovative food ingredients.

4. Materials and Methods

4.1. Soy and Pea Protein Solution Preparation

Aqueous solutions presenting different ionic strength (IS), 0.6 and 1.5 M, were prepared by adding NaCl (Sigma Aldrich, Milan, Italy) in deionised water (System advantage A10®, Millipore S.A.S, Molsheim, France). Deionised water without the addition of NaCl was considered to have an IS equal to 0.0 M. Aqueous solutions were added with 10, 15, or 20% (w/w) of soy (SPI) or pea (PPI) protein isolates (Myprotein, Manchester, England). The suspensions were subjected to high shear mixing at $1120\times g$ for 1 min (Polytron PT-MR3000, Kinematica AG, Littau, Switzerland), and pH was adjusted to 3.0, 4.5, and 7.0 by adding 1 M NaOH or HCl.

4.2. Heat Treatment

To induce gelation, soy and pea protein suspensions were transferred in 50 mL-sealed falcon tubes and subjected to thermal treatment in a water bath (95 °C for 15 min), followed by cooling in an ice bath (0 °C for 15 min). The heat-treated samples were then stored at 4 °C for 48 h, until analysis.

4.3. Image Acquisition

Images were captured with a digital camera (EOS 550D, Canon, Milano, Italy) in an image acquisition cabinet (Immagini & Computer, Bareggio, Italy). The digital camera was positioned in an adjustable stand positioned at 45 cm from the samples and enlightened by 4×100 W frosted photographic floodlights, in a position allowing minimum shadow and glare.

4.4. Rheological Properties

Hydrogel rheological properties were tested using an RS6000 Rheometer (Thermo Scientific RheoStress, Haake, Germany), equipped with a Peltier system for temperature control. The analysis was performed with a parallel plate geometry, with a gap of 2.0 mm at 20 °C. Hydrogels were cut into cylinders with 2 mm of height and 20 mm of diameter. The linear viscoelastic region (LVR) was determined using an oscillatory sweep test (0.01 to 1000 Pa at 1 Hz frequency). The frequency sweep tests were carried out increasing the frequency from 0.1 to 20 Hz, at stress values selected in the LVR.

4.5. Physical Stability

The physical stability of hydrogels was evaluated based on their water-holding capacity (*WHC*). Hydrogels were accurately weighed (W_1) and transferred into 1.5 mL-Eppendorf microcentrifuge tubes, and then centrifugated at 15,000× g for 15 min at 4 °C (D3024, DLAB, Scientific Europe S.A.S, Schiltigheim, France). The supernatant was then removed, and the samples were weighed again (W_2). The *WHC* was determined according to Equation (1).

$$WHC = \frac{W_1 - (W_1 - W_2)}{W_1} \cdot 100 \quad (1)$$

4.6. Data Analysis

Data are expressed as the mean ± standard deviation of at least three measurements resulting from two replicates. The statistical analysis was performed using the program R version 4.1.2 (The R Foundation for Statistical Computing, Vienna, Austria). The homogeneity of the variance was evaluated with Bartlett tests, a one-way ANOVA was applied, and the difference between the averages was assessed by the post-hoc Tukey test ($p < 0.05$).

Supplementary Materials: The following supporting information can be downloaded at: https://www.mdpi.com/article/10.3390/gels9010062/s1, Figure S1: Elastic (G′) and viscous (G″) modulus of soy (SPI) and pea (PPI) hydrogels obtained from 20% (w/w) protein solutions. Figure S2. Elastic (G′) and viscous (G″) modulus of pea protein isolate (PPI) hydrogels obtained from 20% (w/w) protein solutions at pH 3.0 and 7.0. Figure S3. Elastic (G′) and viscous (G″) modulus of pea protein isolate (PPI) hydrogels obtained from 20% (w/w) protein solutions at 0.0 and 1.5 M ionic strength.

Author Contributions: L.D.B.: Investigation, Formal Analysis, Visualisation, Writing—Original Draft; S.P.: Conceptualisation, Methodology, Writing—Review and Editing, Supervision; L.M.: Conceptualisation, Resources, Writing—Review and Editing, Supervision, Funding Acquisition. All authors have read and agreed to the published version of the manuscript.

Funding: This research received no external funding.

Institutional Review Board Statement: Not applicable.

Informed Consent Statement: Not applicable.

Data Availability Statement: Data available on request to the corresponding author.

Acknowledgments: The authors thank Tijana Bjelogrlic for helping in the sample preparation and analysis.

Conflicts of Interest: The authors declare no conflict of interest.

References

1. Khoshnevis, H.; Mint, S.M.; Yedinak, E.; Tran, T.Q.; Zadhoush, A.; Youssefi, M.; Pasquali, M.; Duong, H.M. Super High-Rate Fabrication of High-Purity Carbon Nanotube Aerogels from Floating Catalyst Method for Oil Spill Cleaning. *Chem. Phys. Lett.* **2018**, *693*, 146–151. [CrossRef]
2. Di Luigi, M.; Guo, Z.; An, L.; Armstrong, J.N.; Zhou, C.; Ren, S. Manufacturing Silica Aerogel and Cryogel through Ambient Pressure and Freeze Drying. *RSC Adv.* **2022**, *12*, 21213–21222. [CrossRef] [PubMed]
3. Ptaszkowska-Koniarz, M.; Goscianska, J.; Bazan-Wozniak, A.; Pietrzak, R. Amine-Modified Carbon Xerogels as Effective Carbon-Based Adsorbents of Anionic Dye from Aqueous Solutions. *Materials* **2022**, *15*, 5736. [CrossRef] [PubMed]
4. Li, Z.; Yang, L.; Cao, H.; Chang, Y.; Tang, K.; Cao, Z.; Chang, J.; Cao, Y.; Wang, W.; Gao, M.; et al. Carbon Materials Derived from Chitosan/Cellulose Cryogel-Supported Zeolite Imidazole Frameworks for Potential Supercapacitor Application. *Carbohydr. Polym.* **2017**, *175*, 223–230. [CrossRef] [PubMed]
5. Sakuma, W.; Yamasaki, S.; Fujisawa, S.; Kodama, T.; Shiomi, J.; Kanamori, K.; Saito, T. Mechanically Strong, Scalable, Mesoporous Xerogels of Nanocellulose Featuring Light Permeability, Thermal Insulation, and Flame Self-Extinction. *ACS Nano* **2021**, *15*, 1436–1444. [CrossRef]
6. Kudaibergenov, S.E. Physicochemical, Complexation and Catalytic Properties of Polyampholyte Cryogels. *Gels* **2019**, *5*, 8. [CrossRef]
7. Pham, T.H.; Jung, S.H.; Kim, Y.J.; Kim, T.Y. Adsorptive Removal and Recovery of Organic Pollutants from Wastewater Using Waste Paper-Derived Carbon-Based Aerogel. *Chemosphere* **2021**, *268*, 129319. [CrossRef]

8. Križman, K.; Novak, S.; Kristl, J.; Majdič, G.; Drnovšek, N. Long-Acting Silk Fibroin Xerogel Delivery Systems for Controlled Release of Estradiol. *J. Drug. Deliv. Sci. Technol.* **2021**, *65*, 102701. [CrossRef]
9. Sharma, M.; Tavares, A.P.M.; Nunes, J.C.F.; Singh, N.; Mondal, D.; Neves, M.C.; Prasad, K.; Freire, M.G. Hybrid Alginate–Protein Cryogel Beads: Efficient and Sustainable Bio-Based Materials to Purify Immunoglobulin G Antibodies. *Green Chem.* **2020**, *22*, 2225–2233. [CrossRef]
10. Goimil, L.; Santos-Rosales, V.; Delgado, A.; Évora, C.; Reyes, R.; Lozano-Pérez, A.A.; Aznar-Cervantes, S.D.; Cenis, J.L.; Gómez-Amoza, J.L.; Concheiro, A.; et al. ScCO_2-Foamed Silk Fibroin Aerogel/Poly(ε-Caprolactone) Scaffolds Containing Dexamethasone for Bone Regeneration. *J. CO2 Util.* **2019**, *31*, 51–64. [CrossRef]
11. Raschip, I.E.; Fifere, N.; Varganici, C.D.; Dinu, M.V. Development of Antioxidant and Antimicrobial Xanthan-Based Cryogels with Tuned Porous Morphology and Controlled Swelling Features. *Int. J. Biol. Macromol.* **2020**, *156*, 608–620. [CrossRef] [PubMed]
12. Kleemann, C.; Schuster, R.; Rosenecker, E.; Selmer, I.; Smirnova, I.; Kulozik, U. In-Vitro-Digestion and Swelling Kinetics of Whey Protein, Egg White Protein and Sodium Caseinate Aerogels. *Food Hydrocoll.* **2020**, *101*, 105534. [CrossRef]
13. Selmer, I.; Karnetzke, J.; Kleemann, C.; Lehtonen, M.; Mikkonen, K.S.; Kulozik, U.; Smirnova, I. Encapsulation of Fish Oil in Protein Aerogel Micro-Particles. *J. Food Eng.* **2019**, *260*, 1–11. [CrossRef]
14. Schroeter, B.; Yonkova, V.P.; Goslinska, M.; Orth, M.; Pietsch, S.; Gurikov, P.; Smirnova, I.; Heinrich, S. Spray Coating of Cellulose Aerogel Particles in a Miniaturized Spouted Bed. *Cellulose* **2021**, *28*, 7795–7812. [CrossRef]
15. Koshy, S.T.; Zhang, D.K.Y.; Grolman, J.M.; Stafford, A.G.; Mooney, D.J. Injectable Nanocomposite Cryogels for Versatile Protein Drug Delivery. *Acta Biomater.* **2018**, *65*, 36–43. [CrossRef]
16. Plazzotta, S.; Alongi, M.; De Berardinis, L.; Melchior, S.; Calligaris, S.; Manzocco, L. Steering Protein and Lipid Digestibility by Oleogelation with Protein Aerogels. *Food Funct.* **2022**, *13*, 10601–10609. [CrossRef]
17. Mallepally, R.R.; Bernard, I.; Marin, M.A.; Ward, K.R.; McHugh, M.A. Superabsorbent Alginate Aerogels. *J. Supercrit. Fluids* **2013**, *79*, 202–208. [CrossRef]
18. Manzocco, L.; Plazzotta, S.; Powell, J.; de Vries, A.; Rousseau, D.; Calligaris, S. Structural Characterisation and Sorption Capability of Whey Protein Aerogels Obtained by Freeze-Drying or Supercritical Drying. *Food Hydrocoll.* **2022**, *122*, 107117. [CrossRef]
19. Plazzotta, S.; Calligaris, S.; Manzocco, L. Structure of Oleogels from κ-Carrageenan Templates as Affected by Supercritical-CO_2-Drying, Freeze-Drying and Lettuce-Filler Addition. *Food Hydrocoll.* **2019**, *96*, 1–10. [CrossRef]
20. Plazzotta, S.; Jung, I.; Schroeter, B.; Subrahmanyam, R.P.; Smirnova, I.; Calligaris, S.; Gurikov, P.; Manzocco, L. Conversion of Whey Protein Aerogel Particles into Oleogels: Effect of Oil Type on Structural Features. *Polymers* **2021**, *13*, 4063. [CrossRef]
21. Plazzotta, S.; Calligaris, S.; Manzocco, L. Structural Characterization of Oleogels from Whey Protein Aerogel Particles. *Int. Food Res. J.* **2020**, *132*, 109099. [CrossRef]
22. Brodkorb, A.; Croguennec, T.; Bouhallab, S.; Kehoe, J.J. Heat-Induced Denaturation, Aggregation and Gelation of Whey Proteins. In *Advanced Dairy Chemistry*, 4th ed.; Springer: Manhattan, NY, USA, 2016; Volume 1B, pp. 155–178. [CrossRef]
23. Manzocco, L.; Valoppi, F.; Calligaris, S.; Andreatta, F.; Spilimbergo, S.; Nicoli, M.C. Exploitation of κ-Carrageenan Aerogels as Template for Edible Oleogel Preparation. *Food Hydrocoll.* **2017**, *71*, 68–75. [CrossRef]
24. Yamasaki, S.; Sakuma, W.; Yasui, H.; Daicho, K.; Saito, T.; Fujisawa, S.; Isogai, A.; Kanamori, K. Nanocellulose Xerogels with High Porosities and Large Specific Surface Areas. *Front. Chem.* **2019**, *7*, 316. [CrossRef] [PubMed]
25. Buchtová, N.; Budtova, T. Cellulose Aero-, Cryo- and Xerogels: Towards Understanding of Morphology Control. *Cellulose* **2016**, *23*, 2585–2595. [CrossRef]
26. Baudron, V.; Gurikov, P.; Smirnova, I.; Whitehouse, S. Porous Starch Materials via Supercritical- and Freeze-Drying. *Gels* **2019**, *5*, 12. [CrossRef] [PubMed]
27. García-González, C.A.; Alnaief, M.; Smirnova, I. Polysaccharide-Based Aerogels—Promising Biodegradable Carriers for Drug Delivery Systems. *Carbohydr. Polym.* **2011**, *86*, 1425–1438. [CrossRef]
28. Betz, M.; García-González, C.A.; Subrahmanyam, R.P.; Smirnova, I.; Kulozik, U. Preparation of Novel Whey Protein-Based Aerogels as Drug Carriers for Life Science Applications. *J. Supercrit. Fluids* **2012**, *72*, 111–119. [CrossRef]
29. Selmer, I.; Kleemann, C.; Kulozik, U.; Heinrich, S.; Smirnova, I. Development of Egg White Protein Aerogels as New Matrix Material for Microencapsulation in Food. *J. Supercrit. Fluids* **2015**, *106*, 42–49. [CrossRef]
30. Kleemann, C.; Selmer, I.; Smirnova, I.; Kulozik, U. Tailor Made Protein Based Aerogel Particles from Egg White Protein, Whey Protein Isolate and Sodium Caseinate: Influence of the Preceding Hydrogel Characteristics. *Food Hydrocoll.* **2018**, *83*, 365–374. [CrossRef]
31. Arboleda, J.C.; Hughes, M.; Lucia, L.A.; Laine, J.; Ekman, K.; Rojas, O.J. Soy Protein-Nanocellulose Composite Aerogels. *Cellulose* **2013**, *20*, 2417–2426. [CrossRef]
32. Andlinger, D.J.; Bornkeßel, A.C.; Jung, I.; Schröter, B.; Smirnova, I.; Kulozik, U. Microstructures of Potato Protein Hydrogels and Aerogels Produced by Thermal Crosslinking and Supercritical Drying. *Food Hydrocoll.* **2021**, *112*, 106305. [CrossRef]
33. Yetiskin, B.; Okay, O. High-Strength and Self-Recoverable Silk Fibroin Cryogels with Anisotropic Swelling and Mechanical Properties. *Int. J. Biol. Macromol.* **2019**, *122*, 1279–1289. [CrossRef] [PubMed]
34. Marin, M.A.; Mallepally, R.R.; McHugh, M.A. Silk Fibroin Aerogels for Drug Delivery Applications. *J. Supercrit. Fluids* **2014**, *91*, 84–89. [CrossRef]
35. Li, Y.; Wang, S.; Zhang, G.; Liu, X.; Liu, H.; He, Y.; Zhu, D. Morphological and Structural Changes in Thermally-Induced Soybean Protein Isolate Xerogels Modulated by Soybean Polysaccharide Concentration. *Food Hydrocoll.* **2022**, *133*, 107967. [CrossRef]

36. Andlinger, D.J.; Schlemmer, L.; Jung, I.; Schroeter, B.; Smirnova, I.; Kulozik, U. Hydro- and Aerogels from Ethanolic Potato and Whey Protein Solutions: Influence of Temperature and Ethanol Concentration on Viscoelastic Properties, Protein Interactions, and Microstructure. *Food Hydrocoll.* **2022**, *125*, 107424. [CrossRef]
37. Cheng, K.; Tao, X.; Qi, Z.; Yin, Z.; Kundu, S.C.; Lu, S. Highly Absorbent Silk Fibroin Protein Xerogel. *ACS Biomater. Sci. Eng.* **2021**, *7*, 3594–3607. [CrossRef]
38. de Boer, J.; Aiking, H. On the Merits of Plant-Based Proteins for Global Food Security: Marrying Macro and Micro Perspectives. *Ecol. Econ.* **2011**, *70*, 1259–1265. [CrossRef]
39. Prag, A.A.; Henriksen, C.B. Transition from Animal-Based to Plant-Based Food Production to Reduce Greenhouse Gas Emissions from Agriculture—The Case of Denmark. *Sustainability* **2020**, *12*, 8228. [CrossRef]
40. Machovina, B.; Feeley, K.J.; Ripple, W.J. Biodiversity Conservation: The Key Is Reducing Meat Consumption. *Sci. Total Environ.* **2015**, *536*, 419–431. [CrossRef]
41. Sun, X.D.; Arntfield, S.D. Gelation Properties of Salt-Extracted Pea Protein Induced by Heat Treatment. *Int. Food Res. J.* **2010**, *43*, 509–515. [CrossRef]
42. Yin, X.; Cheng, H.; Wusigale; Dong, H.; Huang, W.; Liang, L. Resveratrol Stabilization and Loss by Sodium Caseinate, Whey and Soy Protein Isolates: Loading, Antioxidant Activity, Oxidability. *Antioxidants* **2022**, *11*, 647. [CrossRef] [PubMed]
43. Nicolai, T.; Chassenieux, C. Heat-Induced Gelation of Plant Globulins. *Curr. Opin. Food Sci.* **2019**, *27*, 18–22. [CrossRef]
44. Chen, N.; Lin, L.; Sun, W.; Zhao, M. Stable and pH-Sensitive Protein Nanogels Made by Self-Assembly of Heat Denatured Soy Protein. *J. Agric. Food Chem.* **2014**, *62*, 9553–9561. [CrossRef] [PubMed]
45. Ako, K.; Nicolai, T.; Durand, D.; Brotons, G. Micro-Phase Separation Explains the Abrupt Structural Change of Denatured Globular Protein Gels on Varying the Ionic Strength or the pH. *Soft Matter* **2009**, *5*, 4033–4041. [CrossRef]
46. Renkema, J.M.S.; Gruppen, H.; van Vliet, T. Influence of pH and Ionic Strength on Heat-Induced Formation and Rheological Properties of Soy Protein Gels in Relation to Denaturation and Their Protein Compositions. *J. Agric. Food Chem.* **2002**, *50*, 6064–6071. [CrossRef]
47. Fitzpatrick, S.E.; Staiger, M.P.; Deb-Choudhury, S.; Ranford, S. Protein-Based Aerogels: Processing and Morphology. *RSC Green Chem.* **2018**, *201*, 67–102. [CrossRef]
48. Hermansson, A.M. Aggregation and Denaturation Involved In Gel Formation. *ACS Symp. Ser.* **1979**, *5*, 81–103. [CrossRef]
49. Hermansson, A.M. Soy Protein Gelation. *J. Am. Oil Chem. Soc.* **1986**, *63*, 658–666. [CrossRef]
50. KangII, J.; Matsumura, Y.; Mori, T. Characterization of Texture and Mechanical Properties of Heat-Induced Soy Protein Gels. *J. Am. Oil Chem. Soc.* **1991**, *68*, 339–345. [CrossRef]
51. Chen, N.; Zhao, M.; Chassenieux, C.; Nicolai, T. The Effect of Adding NaCl on Thermal Aggregation and Gelation of Soy Protein Isolate. *Food Hydrocoll.* **2017**, *70*, 88–95. [CrossRef]
52. Osen, R.; Toelstede, S.; Wild, F.; Eisner, P.; Schweiggert-Weisz, U. High Moisture Extrusion Cooking of Pea Protein Isolates: Raw Material Characteristics, Extruder Responses, and Texture Properties. *J. Food Eng.* **2014**, *127*, 67–74. [CrossRef]
53. Shand, P.J.; Ya, H.; Pietrasik, Z.; Wanasundara, P.K.J.P.D. Physicochemical and Textural Properties of Heat-Induced Pea Protein Isolate Gels. *Food Chem.* **2007**, *102*, 1119–1130. [CrossRef]
54. Hua, Y.; Cui, S.W.; Wang, Q.; Mine, Y.; Poysa, V. Heat Induced Gelling Properties of Soy Protein Isolates Prepared from Different Defatted Soybean Flours. *Int. Food Res. J.* **2005**, *38*, 377–385. [CrossRef]
55. O'Kane, F.E.; Happe, R.P.; Vereijken, J.M.; Gruppen, H.; van Boekel, M.A.J.S. Heat-Induced Gelation of Pea Legumin: Comparison with Soybean Glycinin. *J. Agric. Food Chem.* **2004**, *52*, 5071–5078. [CrossRef] [PubMed]
56. Kuhn, K.R.; Cavallieri, Â.L.F.; da Cunha, R.L. Cold-Set Whey Protein Gels Induced by Calcium or Sodium Salt Addition. *Int. J. Food Sci. Technol.* **2010**, *45*, 348–357. [CrossRef]
57. Nicolai, T. Formation and Functionality of Self-Assembled Whey Protein Microgels. *Colloids Surf. B.* **2016**, *137*, 32–38. [CrossRef]
58. Nicolai, T.; Durand, D. Controlled Food Protein Aggregation for New Functionality. *Curr. Opin. Colloid Interface Sci.* **2013**, *18*, 249–256. [CrossRef]
59. van der Linden, E.; Venema, P. Self-Assembly and Aggregation of Proteins. *Curr. Opin. Colloid Interface Sci.* **2007**, *12*, 158–165. [CrossRef]
60. Chen, N.; Zhao, M.; Chassenieux, C.; Nicolai, T. Structure of Self-Assembled Native Soy Globulin in Aqueous Solution as a Function of the Concentration and the pH. *Food Hydrocoll.* **2016**, *56*, 417–424. [CrossRef]
61. Kumar, S.; Nussinov, R. Relationship between Ion Pair Geometries and Electrostatic Strengths in Proteins. *Biophys. J.* **2002**, *83*, 1595–1612. [CrossRef]
62. Jiang, J.; Xiong, Y.L.; Chen, J. pH Shifting Alters Solubility Characteristics and Thermal Stability of Soy Protein Isolate and Its Globulin Fractions in Different pH, Salt Concentration, and Temperature Conditions. *J. Agric. Food Chem.* **2010**, *58*, 8035–8042. [CrossRef] [PubMed]
63. Puppo, M.C.; Lupano, C.E.; Añón, M.C. Gelation of Soybean Protein Isolates in Acidic Conditions. Effect of pH and Protein Concentration. *J. Agric. Food Chem.* **1995**, *43*, 2356–2361. [CrossRef]
64. Puppo, M.C.; Añón, M.C. Structural Properties of Heat-Induced Soy Protein Gels As Affected by Ionic Strength and pH. *J. Agric. Food Chem.* **1998**, *46*, 3583–3589. [CrossRef]
65. Li, J.; Li, X.; Wang, C.; Zhang, M.; Xu, Y.; Zhou, B.; Su, Y.; Yang, Y. Characteristics of Gelling and Water Holding Properties of Hen Egg White/Yolk Gel with NaCl Addition. *Food Hydrocoll.* **2018**, *77*, 887–893. [CrossRef]

66. Li, J.; Wang, C.; Zhang, M.; Zhai, Y.; Zhou, B.; Su, Y.; Yang, Y. Effects of Selected Phosphate Salts on Gelling Properties and Water State of Whole Egg Gel. *Food Hydrocoll.* **2018**, *77*, 1–7. [CrossRef]
67. Croguennec, T.; Nau, F.; Brulé, G. Influence of pH and Salts on Egg White Gelation. *J. Food Sci.* **2002**, *67*, 608–614. [CrossRef]
68. Khemakhem, M.; Attia, H.; Ayadi, M.A. The Effect of pH, Sucrose, Salt and Hydrocolloid Gums on the Gelling Properties and Water Holding Capacity of Egg White Gel. *Food Hydrocoll.* **2019**, *87*, 11–19. [CrossRef]
69. Munialo, C.D.; van der Linden, E.; de Jongh, H.H.J. The Ability to Store Energy in Pea Protein Gels Is Set by Network Dimensions Smaller than 50nm. *Food Res. Int.* **2014**, *64*, 482–491. [CrossRef]
70. Maltais, A.; Remondetto, G.E.; Gonzalez, R.; Subirade, M. Formation of Soy Protein Isolate Cold-Set Gels: Protein and Salt Effects. *J. Food Sci.* **2005**, *70*, C67–C73. [CrossRef]
71. Urbonaite, V.; de Jongh, H.H.J.; van der Linden, E.; Pouvreau, L. Origin of water loss from soy protein gels. *J. Agric. Food Chem.* **2014**, *62*, 7550–7558. [CrossRef]
72. Urbonaite, V.; de Jongh, H.H.J.; van der Linden, E.; Pouvreau, L. Water holding of soy protein gels is set by coarseness, modulated by calcium binding, rather than gel stiffness. *Food Hydrocoll.* **2015**, *46*, 103–111. [CrossRef]

gels

MDPI

Article

Pd-Loaded Cellulose NanoSponge as a Heterogeneous Catalyst for Suzuki–Miyaura Coupling Reactions

Laura Riva [1], Gloria Nicastro [1], Mingchong Liu [1], Chiara Battocchio [2], Carlo Punta [1,3] and Alessandro Sacchetti [1,*]

1 Department of Chemistry, Materials, and Chemical Engineering "G. Natta" and INSTM Local Unit, Politecnico di Milano, 20131 Milan, Italy
2 Department of Science, Roma Tre University, Via della Vasca Navale 79, 00146 Rome, Italy
3 Istituto di Scienze e Tecnologie Chimiche, "Giulio Natta" (SCITEC), National Research Council-CNR, 20131 Milan, Italy
* Correspondence: alessandro.sacchetti@polimi.it; Tel.: +39-0223993017

Abstract: The (eco)design and synthesis of durable heterogeneous catalysts starting from renewable sources derived from biomass waste represents an important step for reducing environmental impacts of organic transformations. Herein, we report the efficient loading of Pd(II) ions on an eco-safe cellulose-based organic support (CNS), obtained by thermal cross-linking between TEMPO-oxidized cellulose nanofibers and branched polyethyleneimine in the presence of citric acid. A 22.7% *w/w* Pd-loading on CNS was determined by the ICP-OES technique, while the metal distribution on the xerogel was evidenced by SEM–EDS analysis. XPS analysis confirmed the direct chelation of Pd(II) ions by means of the high number of amino groups present in the network, so that further functionalization of the support with specific ligands was not necessary. The new composite turned to be an efficient heterogeneous pre-catalyst for promoting Suzuki–Miyaura coupling reactions between aryl halides and phenyl boronic acid in water, obtaining yields higher than 90% in 30 min, by operating in a microwave reactor at 100 °C and with just 2% *w/w* of CNS-Pd catalyst with respect to aryl halides (4.5‰ for Pd). At the end of first reaction cycle, Pd(II) ions on the support resulted in being reduced to Pd(0) while maintaining the same catalytic efficiency. In fact, no leaching was observed at the end of reactions, and five cycles of recycling and reusing of CNS-Pd catalyst provided excellent results in terms of yields and selectivity in the desired products.

Keywords: nanocellulose; Suzuki–Miyaura coupling; heterogeneous catalysis; sustainable catalyst; nanocellulose-based xerogels; green chemistry

Citation: Riva, L.; Nicastro, G.; Liu, M.; Battocchio, C.; Punta, C.; Sacchetti, A. Pd-Loaded Cellulose NanoSponge as a Heterogeneous Catalyst for Suzuki–Miyaura Coupling Reactions. *Gels* **2022**, *8*, 789. https://doi.org/10.3390/gels8120789

Academic Editor: Pavel Gurikov

Received: 11 November 2022
Accepted: 29 November 2022
Published: 2 December 2022

1. Introduction

Pd-catalyzed Suzuki–Miyaura cross-coupling is one of the most investigated C-C bond formation reactions, widely applied for the synthesis of complex molecules, including pharmaceuticals, semiconductors, supramolecular structures, and pesticides [1–4].

Many efforts have been devoted over the years to the design and synthesis of Pd(II)-precatalysts by selecting proper ligands such as *N*-heterocyclic carbenes [5–7], phosphine [8,9], palladacycles [10], the PEPPSI (pyridine-enhanced precatalyst preparation) system [11], and allyl-based ligands [12], in order to improve the efficiency of this catalysis under homogeneous conditions.

Ligands guarantee excellent donor abilities, high steric hindrance, and, in most cases, the stabilization of a Pd(0) reduced form, which is considered to be the active species once generated in situ [13].

While most of these approaches allow operation at low catalyst loadings, under mild conditions, and even in green solvents, including water [14,15], they all suffer from the limits related to homogeneous catalysis, which can be summarized in the direct costs for catalysts' synthesis, and the indirect ones for their efficient recovery and reuse, which is also a key issue for the environmental impact of the process.

For these reasons, in recent years many solutions have been proposed for the immobilization of Pd(II)-precatalysts onto heterogeneous networks [16–19], also opening the route to continuous-flow synthetic processes [20,21]. However, in most cases the pre-functionalization of solid supports with proper ligands is necessary to guarantee high efficiency in fixing Pd(II), minimizing leaching phenomena [22,23]. Moreover, the increasing demand for bio-based materials derived from renewable sources, such as biomass waste, in the framework of the circular economy, pushes towards the design of new solutions as heterogenous supports for organometallic catalysis, capable of minimizing synthetic steps and providing sustainable solutions with low environmental impact.

In this context, in recent years we designed and developed a microporous cellulose-based nanosponge (CNS), having TEMPO ((2,2,6,6-Tetramethylpiperidin-1-yl)oxyl)-oxidized cellulose nanofibers and 25 kDa branched polyethylenimine (bPEI) as main components [24]. The high porosity of the system derived from the freeze-drying process followed for converting the original hydrogel-like suspension of the two polymers into the resulting xerogel, with ice crystals acting as pores' templates, while its chemical stability was guaranteed by the thermally induced (~100 °C) formation of amide bonds between the carboxyl groups of the oxidized nanocellulose and the amine groups of the polyamine. Further optimization of pristine formulations by addition of citric acid (CA) allowed the nanostructure to have higher mechanical resistance [25] and to better fix bPEI, improving the eco-safety [26–29] and the sustainability [30] of the material. More recently, the nanoporosity of the material was revealed by small-angle neutron scattering (SANS) investigation [31] and by FTIR-ATR analysis of the H-bond network, which evidenced water nanoconfinement in the nanostructure [32,33].

CNS have found ample application in different fields, including wastewater remediation [29,34], sensing [35,36], as drug-delivery systems [25], and as heterogenous catalysts for promoting amino-catalyzed organic reactions [37]. Very recently, we were inspired by the high heavy-metal adsorption efficiency of CNS exploited in wastewater treatment [24,29]. This property had to be ascribed to the strong chelating action of primary, secondary, and tertiary amino groups, provided by the presence of bPEI in the xerogel network. Inspired by this behavior, we envisioned the opportunity to consider these materials as suitable organic heterogeneous supports for transition metal ions, opening the synthesis of a new class of heterogeneous organometallic catalysts for organic reactions. In a first attempt, we confirmed this hypothesis by designing CNS-Cu- and CNS-Zn-loaded catalysts, which were successfully used to promote the synthesis of aromatic acetals [38].

With these premises, herein we propose CNS as easy-to-prepare biobased and sustainable supports for Pd(II) ions. As CNS do not require a pre-functionalization by specific ligands to efficiently trap Pd(II), the environmental and economic impact of these systems is minimized. The new CNS-Pd(II) composite turned out to be an efficient and stable heterogeneous precatalyst towards the Suzuki–Miyaura cross-coupling reaction.

2. Results and Discussion

2.1. CNS-Pd Synthesis and Characterization

CNS-Pd was synthesized as reported in Scheme 1, following a two-step procedure: (i) the production of CNS and (ii) the loading of Pd(II) ions on the resulting xerogel. For the first step, TOCNF and bPEI were mixed in a 1:2 weight ratio in deionized water in the presence of 18% of CA with respect to primary amino groups of bPEI. The resulting hydrogel was transferred in molds and underwent a freeze-drying process, providing a highly porous xerogel. The latter was heated in an oven at ~100 °C in order to impart chemical stability and mechanical resistance to the final material and allowing the formation of amidic bonds by dehydration between the carboxylic groups of TOCNF and CA, and the primary amino groups of bPEI [25]. In the second step, the nanosponge was ground in a mortar before use to increase the superficial area and consequently the Pd-sorption efficiency. Pd loading was performed by soaking the CNS material in a saturated $PdCl_2$ solution, running several loading cycles.

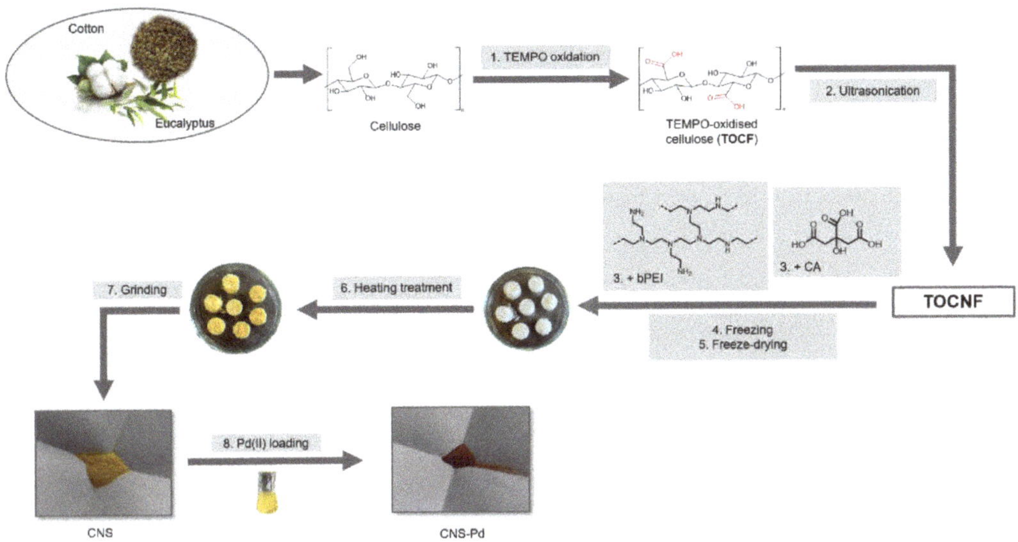

Scheme 1. Preparation of CNS-Pd.

A complete chemical and morphological characterization of CNS has already been reported in the literature. More specifically, evidence of amide bonding was shown by FTIR [32,34] and ^{15}N CP-MAS solid-state NMR analyses [25]; microporosity of the system was detected by scanning electron microscopy (SEM) images and better-investigated by microcomputed tomography quantitative analysis, resulting in about 70% of the bulk material [25,29]; finally, nanoporosity of the xerogel was revealed by a small-angle neutron scattering (SANS) [31] study and an investigation of water nanoconfinement in the network by ATR-FTIR [32,34]. We thus proceeded with an in-deep characterization of the new Pd-loaded system. Scanning Electron Microscopy/Energy Dispersive Spectroscopy (SEM–EDS) analysis revealed a homogeneous distribution of the metal on ground CNS (Figure 1).

Figure 1. SEM images of CNS-Pd (**A**) and distribution of Pd on the catalyst's surface, obtained with EDS (**B**).

An EDS absorption spectrum is also reported in Figure 2, where the Pd and Cl^- signals can be clearly observed.

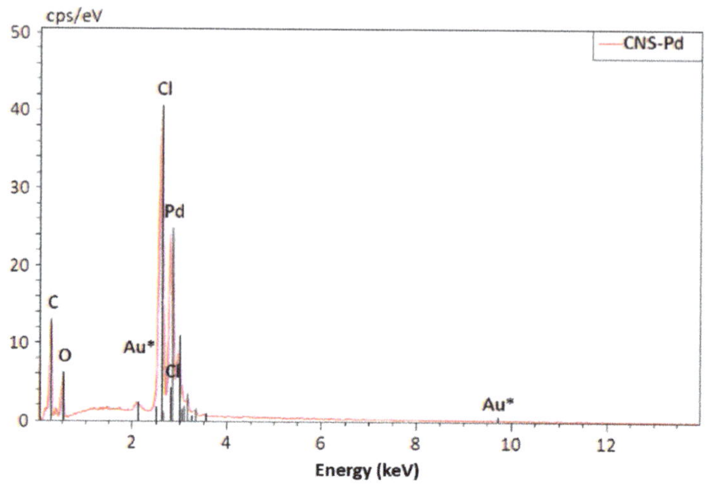

Figure 2. EDS absorption spectrum of CNS-Pd.

A quantification of the amount of Pd(II) actually loaded on CNS-Pd was measured by ICP-OES analysis, obtaining a value of 22.7% *w/w*, corresponding to 2.13 $mmol_{Pd}/g_{CNS}$.

To obtain information on the structural features of the CNS-Pd catalyst, X-ray photoelectron spectroscopy (XPS) studies were performed on the pristine catalyst as well as on the catalyst recovered after the first catalytic cycle (see Section 2.6 for reaction conditions). The pristine CNS matrix was also investigated to provide useful references for the assignment of Pd3d spectral features. Measurements were carried out at C 1s, N 1s, O 1s, and Pd 3d core levels. A complete collection of core-level binding energy (BE), full width at half-maxima (FWHM) values, and proposed assignments is reported in Table S1 in the Supporting Information; here, the Pd3d spin-orbit components will be discussed with particular attention since they are of major interest for the assessment of the Pd–CNS interaction in the catalyst and for the investigation of its role in the catalytic reaction. C1s and N1s spectral features will also be briefly discussed since the reproducibility of such signals confirms the stability of the catalyst molecular structure upon use.

C1s and N1s spectra collected on the pristine CNS and on the CNS-Pd catalyst before and after the catalytic process show analogous features, in excellent accordance with the chemical composition of CNS; in more detail, C1s spectra are composite and at least four spectral components can be individuated by applying a peak fitting procedure, assigned, respectively, to aliphatic C atoms (BE = 285.00 eV), C atoms bonded to N or O in C-N, C-O functional groups (286.3 eV), O-C-O or C=O carbons (287.5 eV), and carboxylic COOH functional groups (288.9 eV) [39–43]. The relative amount of each species is well-reproducible in the three samples, as reported in the column "atomic ratios" in Table S1. C1s spectra of CNS-Pd pristine and recovered from the catalysis reaction are reported in Figure 3A,B. As for the N 1s spectra, a main signal is always found at about 400 eV, as expected for N atoms in the polyamine [43]. At higher BE values, a signal of very low intensity is also observed, probably due to oxidized nitrogen atoms belonging to impurities of the CNS.

Pd3d spectra show the presence of several contributions in different BE positions for the pristine and recovered catalysts, as reported in Figure 3C (pristine CNS-Pd) and D (recovered CNS-Pd).

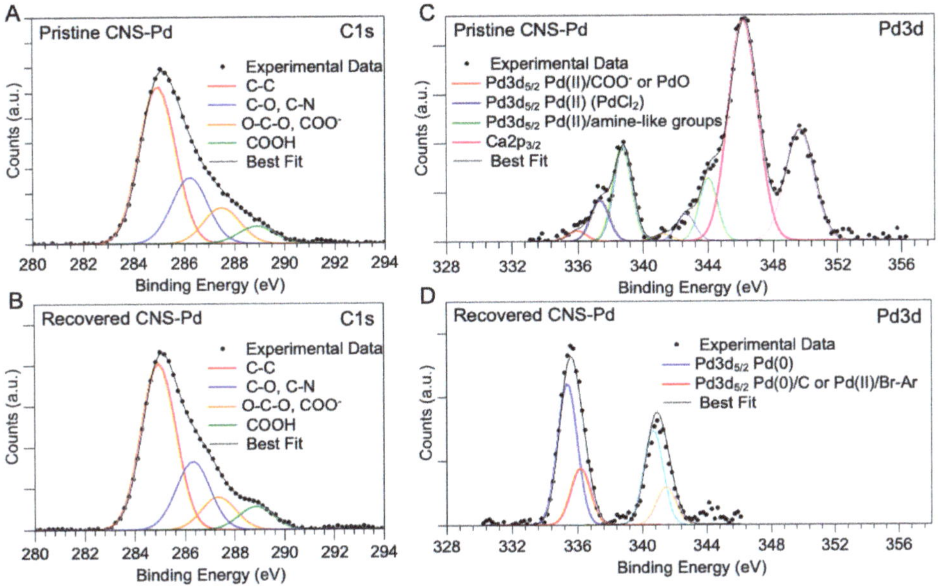

Figure 3. (**A**) XPS C 1s core-level spectra acquired on pristine CNS and (**B**) CNS recovered after catalysis; (**C**) XPS Pd 3d core-level spectra acquired on CNS before and (**D**) after catalysis.

The Pd3d spectrum of pristine CNS-Pd shows the presence of Pd(II) ions only; however, by applying the peak-fitting procedure, it is possible to point out at least three spin-orbit pairs whose BE positions are indicative of different chemical environments. The signal at lower BE ($Pd3d_{5/2}$ BE = 337.38 eV) can be associated with PdO [43] or, as suggested by some authors, with Pd(II) ions coordinating carboxylic groups [44]. A less-suitable assignment could be atomic Pd(0) interacting with amorphous carbon, as suggested by Bertolini et al. [45]. Moreover, the low-BE component is also less intense (about 7% of all Pd contribution) in the pristine CNS-Pd Pd3d spectrum. The second feature ($Pd3d_{5/2}$ component at 338.77 eV BE) is indicative of ionic Pd(II) in $PdCl_2$ [43], as also confirmed by the position of the Cl2p signal ($Cl2p_{3/2}$ BE = 198.40 eV, see Table S1) [46]. Finally, the most intense signal, at higher BE value ($Pd3d_{5/2}$ BE = 340.19 eV, 64% of all the palladium in the catalyst), is due to Pd(II) ions interacting with the amine-like nitrogens of the polyamine moieties in coordination compounds, similarly to $Pd(NH_3)_4Cl_2$ [43].

In Figure 3C, a Ca2p signal arising from Ca(II) ions contained as impurities in the CNS matrix is observed and partially superimposed to the 3/2 Pd3d spin-orbit components; an analogous signal is also observed in pure-pristine CNS (see Supporting Information, Table S1). The Ca2p signal is not found in the recovered catalyst, probably due to repeated washing with ultrapure water.

After recovering, the Pd signal measured for CNS-Pd catalyst suggests a different composition. The most-intense spectral component (71% of Pd species) is shifted at low BE values ($Pd3d_{5/2}$ BE = 335.46 eV) and it is now indicative of reduced Pd(0) atoms. At higher BE values, a less intense signal can be observed ($Pd3d_{5/2}$ BE = 338.20); from the literature, this minor (29%) contribution could be due to Pd(0) atoms interacting with an amorphous carbon matrix [45] or Pd(II) ions coordinating halogenated organic molecules (Br-containing reactant residues, for example, or reaction intermediates and byproducts) [47,48]. Measurements at the Br3d core level reveal a small number of bromine atoms covalently bonded to C ($Br3d_{5/2}$ BE = 68.80 eV) in the recovered sample, supporting the latter assignment (see Table S1).

2.2. *Suzuki–Miyaura Reaction Optimization*

The reaction between 4-Br-anisole (**1a**) and phenylboronic acid (**2**) in water (Scheme 2) was selected as the model for the optimization of the conditions. CNS-Pd precatalyst was used in 2–10% *w*/*w* with KOH (2 eq) as base. In addition, to facilitate the solubilization of the reagents, the phase-transfer agent TBAB (0.6 eq) was used. ^{1}H-NMR in $CDCl_3$ with acetonitrile as internal standard was used to calculate the reaction conversion during the reaction optimization process. In all cases, a complete selectivity toward the desired product could be observed, without the formation of any undesired by-products.

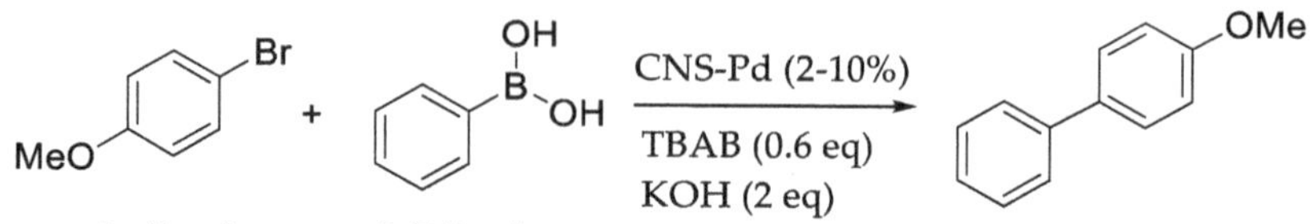

1a (1 eq) **2** (1.5 eq) **3a**

Scheme 2. Suzuki–Miyaura reference reaction.

The influence of reaction time and temperature was first investigated with the use of a microwave reactor, according to the principle of "time–temperature equivalence", shortening the reaction time under the condition of increasing temperature. For comparison, conventional heating was also considered.

As reported in Table 1, after working at room temperature (T = 25 °C) with long reaction time (48 h), the reaction yield was not satisfactory (entry 1). As expected, at higher temperatures yields increased and shorter reaction times could be applied, going from 80 °C, 30 min (entry 2), with a yield of 57% to 100 °C, 20 min (entry 3), with a yield of 72%. Finally, applying a temperature of 100 °C for a reaction time of 30 min (entry 4), a satisfying 92% yield was achieved. It has to be noticed that in the same conditions but with the use of conventional heating, a very low 19% yield was obtained (entry 5), thus highlighting the unique role and the high effectiveness of microwave irradiation in this catalytic system.

Table 1. Time and temperature reaction optimization [a].

Entry	T (°C)	T (h)	Yield (%)
1	25 (RT)	48	19
2	80	0.5	57
3	100	0.33	72
4	100	0.5	92
5 [b]	100	0.5	19

[a] Reaction conditions: 0.268 mmol of **1a** (1 eq), 0.161 mmol of TBAB (0.6 eq), 0.563 mmol of KOH (2 eq), and 1 mg (2% *w*/*w*) catalyst in 2.5 mL of water under MW irradiation. [b] Conventional heating instead of MW irradiation was used.

In order to achieve higher yields, the amount of catalyst was increased (Table 2). No significant results were obtained. Indeed, going from 2% (entry 2) to 5% (entry 3) and 10% *w*/*w* (entry 4), the yields did not substantially improve. As expected, in the absence of catalyst (entry 1) and in the presence of a metal-free catalyst (CNS, entry 5), no conversion was observed. According to these results, the 2% *w*/*w* catalyst loading was taken as the optimal condition. It is important to highlight that according to the measured loading of palladium on CNS, the 2% *w*/*w* of catalyst corresponds to a very low 0.45% *w*/*w* amount of Pd, making this system very efficient in terms of metal loading in the reaction.

The role of the base was also considered, and different inorganic bases, namely, NaOAc, Na_2CO_3, K_2CO_3, and KOH, were tested in these conditions. Results are reported in Table 3.

Table 2. Catalyst loading optimization [a].

Entry	CNS-Pd (%)	Pd (%)	Yield (%)
1	0	0.00	0
2	2	0.45	92
3	5	1.14	95
4	10	2.27	94
5 [b]	10	-	0

[a] Reaction conditions: 0.268 mmol of **1a** (1 eq), 0.161 mmol of TBAB (0.6 eq), 0.563 mmol of KOH (2 eq), and the listed amount of catalyst in 2.5 mL of water under MW irradiation (T = 100 °C, 30 min). The *w/w* percentage of catalyst is referred to as **1a**. [b] metal free CNS was used.

Table 3. Role of the base in the reaction [a].

Entry	Base	Yield (%)
1	-	15
2	NaOAc	54
3	Na_2CO_3	25
4	K_2CO_3	25
5	KOH	92

[a] Reaction conditions: 0.268 mmol of **1a** (1 eq), 0.161 mmol of TBAB (0.6 eq), 0.563 mmol (2 eq) of base, and 2% *w/w* of catalyst in 2.5 mL of water under MW irradiation (T = 100 °C, 30 min).

The results obtained show that, in base-free conditions (entry 1), the pH is probably not alkaline enough to promote the reduction of Pd (II) to Pd (0), thus obtaining poor yields [49]. With the use of NaOAc, Na_2CO_3, or K_2CO_3 (entry 2, 3, and 4), an adequate alkaline environment could not be provided, with a consequent unsatisfactory activation of the catalytic cycle, resulting again in low yields [49]. The only base that gave good results was the strong base KOH, able to give a yield of 92% (entry 5).

Finally, the use of the phase-transfer agent was explored. As the coupling reaction is run in water and under heterogeneous conditions, producing water-insoluble products from organic reagents, tetrabutylammonium bromide (TBAB) was added to the reaction mixture. Optimization of the amount of TBAB in the reaction was performed by gradually reducing it from the starting 0.6 to 0.15 equivalents, obtaining the results shown in Table 4.

Table 4. Role of the phase-transfer catalyst in the reaction [a].

Entry	TBAB Equivalents	Conversion (%)
1	0.60	92
2	0.30	96
3	0.15	97
4	0	90
5 [b]	0	60
6 [b]	0.15	95

[a] Reaction conditions: 0.268 mmol of 1a (1 eq), TBAB, 0.563 mmol (2 eq) of KOH, and 2% *w/w* of catalyst in 2.5 mL of water under MW irradiation (T = 100 °C, 30 min). [b] A 10× scale-up of the reference reaction.

The results underline that the reaction yield was somewhat higher when using small amounts of TBAB (97% with 0.15 eq, entry 3). It can be also noticed that, without using TBAB, acceptable yields can be obtained (90%, entry 4). The role of TBAB turned out to be crucial when scaling up the reaction. In fact, when performing a ×10 reaction scale-up (entry 5 and 6), the presence of TBAB was essential in increasing the efficiency of the reaction in the presence of large quantities of polar solvent. In this case, without the use of TBAB, a low 60% yield was indeed obtained.

After all this screening, the optimized conditions were defined as follows: 100 °C, 30 min, 2% *w/w* catalyst, corresponding to 0.45% *w/w* of palladium, KOH as base, and 0.15 equivalents of TBAB. For the model Suzuki–Miyaura reaction, in these conditions a TON = 1.2×10^2 and a TOF = 6.5×10^{-2} s^{-1} could be calculated. These values are reasonably good and comparable to many industrial catalytic processes, but are obtained by operating in the presence of an eco-friendly cellulose-derived palladium support. In fact, while we are aware that higher TONs can be obtained under homogeneous conditions, these values are comparable or even better than those of most of the examples reported in the literature for fixing Pd on heterogeneous supports [16–18,20–23]. Moreover, in this case, the choice of the support falls on a bio-based system derived from waste biomass.

To verify the reliability of this catalytic system, a complete mass recovery of the reaction after column chromatography purification was performed. Reaction conditions, work-up, and purification are described in detail in the Section 4 and with this test we were able to verify that, in the optimized conditions after the purification process, a 95% yield could be achieved, calculated by weighting the white crystalline powder obtained after the whole process.

2.3. Suzuki–Miyaura Substrate Scope

After the definition of the optimized conditions, different substrates were tested in the reaction. All the experiments are reported in Table 5.

Table 5. Suzuki–Miyaura cross-couplings of ArX and $PhB(OH)_2$ catalyzed by CNS-Pd [a].

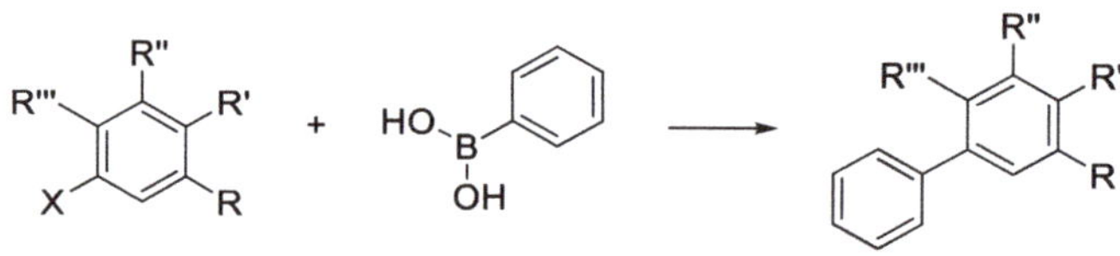

1a-i **2** **3a-i**

Entry	ArX	R	R′	R″	R‴	X	Conversion (%)
1	**1a**	-H	$-OCH_3$	-H	-H	-Br	99
2	**1b**	-H	-H	$-OCH_3$	-H	-Br	98
3	**1c**	-H	-H	-H	$-OCH_3$	-Br	54
4 [c]	**1c**	-H	-H	-H	$-OCH_3$	-Br	38
5 [b]	**1d**	-H	$-CH_3$	-H	-H	-Br	72
6 [b]	**1e**	-H	-CHO	-H	-H	-Br	13
7 [b,c]	**1e**	-H	-CHO	-H	-H	-Br	80
8	**1f**	-H	-H	-H	-H	-Br	93
9	**1g**	-H	$-NH_2$	-H	-H	-Br	76
10	**1h**	-CN	-H	-H	-H	-Br	99
11	**1i**	-H	-H	-H	-H	-Cl	55

[a] Reaction conditions: 0.268 mmol of **1a** (1 eq), TBAB (0.15 eq), 0.563 mmol (2 eq) of KOH, and 2% *w*/*w* of catalyst in 2.5 mL of water under MW irradiation (T = 100 °C, 30 min). [b] A 0.6 eq value of TBAB. [c] A 10% *w*/*w* value of catalyst.

An early analysis on the position of the methoxyl substituent was carried out. As can be seen from the results (entry 1, 2, and 3), with the optimized reaction conditions it was possible to obtain excellent yields with the substituent in para- and meta-position. In the case of the methoxy substituent in the ortho- position (entry 3), we observed a decrease in yield, probably due to the steric effect of the substituent in the position adjacent to the halogen involved in the reaction mechanism. The test with more catalyst (10% *w*/*w*, entry 4) further confirmed the low conversion, supporting the hypothesis of the steric effect of the substituent.

After the position analysis, a study with structurally different substituents was then performed. In the absence of substituents on the benzene ring of the organ halide (entry 8), a satisfactory yield of more than 90% was obtained. In the presence of electron-donating activating substituents on the benzene ring in para-position, such as $-CH_3$ (entry 5) and $-NH_2$ (entry 9), good yields were achieved, in particular a yield of 72% with p-Br-toluene and 76% with p-Br-aniline. In the presence of an electron-withdrawing group, such as

-CHO (entry 6), a dramatic drop in reaction efficiency could be observed, reaching a yield around only 13%. When the reaction was conducted increasing the catalyst amount to 10% *w*/*w*, we noticed an increase in the yield, reaching a good value (80%, entry 7), finding that the reduced reactivity of the starting aril-halide could be balanced by increasing the amount of catalyst, thus achieving good conversions and maintaining selectivity toward the desired product. In contrast, surprisingly, the presence of the electron-withdrawing group -CN in the ortho-position (entry 10) gave an extremely high yield (99%), comparable with the yield obtained with p-Br-anisole (entry 1). By changing the halogen involved in the coupling reaction and considering Cl-benzene (entry 11) instead of Br-benzene (entry 8), we observed a decrease in yield from 93% to 55.5%. Once again, these values are in line with those reported for several heterogeneous supports [16–18,20–23].

2.4. Multiple Cross-Coupling Suzuki–Miyaura Reactions

After studying the effect of substituents, we then performed tests to confirm that CNS-Pd could also be used for multiple couplings. For this study, we selected 1,4-dibromobenzene (Scheme 3) and 1,3,5-tribromobenzene (Scheme 4) as organ halides, increasing the amount of phenylboronic acid (2.5 equivalents for **1j** and 3.5 equivalents for **1k**) and slightly changing the reaction conditions (reaction time increased to 1.5 h for **1j** and to 3 h for **1k**).

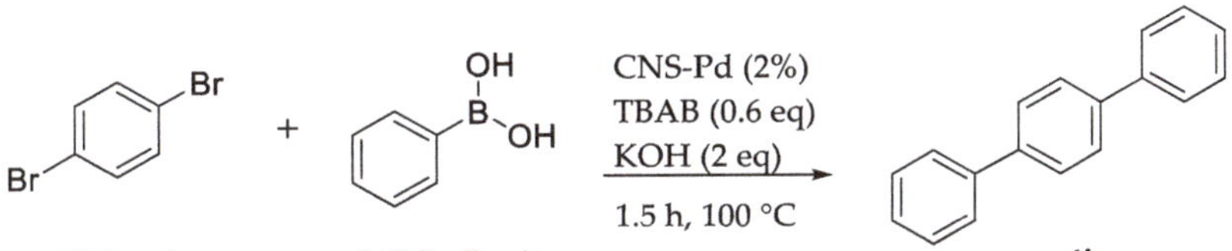

Scheme 3. Reaction of double cross-coupling with 1,4-dibromobenzene.

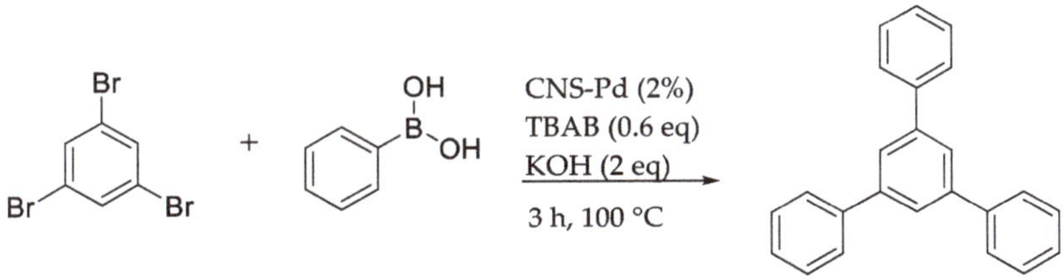

Scheme 4. Reaction of triple cross-coupling with 1,3,5-tribromobenzene.

In these reaction conditions, we were able to obtain high selectivity (>95%) towards the product **4j** in the reaction with 1,4-dibromobenzene (Scheme 3), reaching a conversion of 55%. Increasing the phenylboronic acid amount (from 2.5 to 5 equivalents compared to **1j**) and maintaining the same reaction conditions (1.5 h, 30 °C), we were able to reach a very high conversion (90%) without losing selectivity toward product **4j**.

In the reaction with 1,3,5-tribromobenzene (Scheme 4), under the conditions of 3 h, at 100 °C, with 3.5 equivalents of **2** compared with **1k** reagent, we obtained a very high selectivity (>95%) towards product **5k**, with a conversion of 82%, confirming the catalytic efficiency of CNS-Pd also in multiple cross-coupling reactions.

2.5. Leaching Tests

Tests were performed to evaluate the leaching of Pd from the CNS-Pd catalyst under the reaction conditions (see Section 4). The amount of Pd in solution before and after the reaction was evaluated by means of ICP-OES analysis. After one reaction cycle, only 7% of the initial amount of metal (0.016 mg when 1 mg of CNS-Pd is used) was released in the solution. No significant metal loss was observed after a second cycle, thus confirming that, after stabilization, Pd remains anchored on the heterogeneous support, paving the way for the possible recycling of CNS-Pd.

2.6. Recyclability Tests

To evaluate the possibility of reusing CNS-Pd several times without losing catalytic activity, reusability tests were performed between **1a** and **2** to give **3a** under the optimized conditions. Five consecutive reaction runs were realized and, after each cycle, the reaction yield was evaluated by ^{1}H-NMR with acetonitrile as internal standard to evaluate the efficiency of the reused catalytic system. Results are reported in Table 6.

Table 6. Results from Recyclability tests [a].

Entry	Cycle Number	Yield (%)
1	I	97
2	II	99
3	III	98
4	IV	99
5	V	96

[a] Reaction conditions: 0.268 mmol of **1a** (1 eq), TBAB (0.15 eq), 0.563 mmol (2 eq) of KOH, and 2% *w/w* of catalyst in 2.5 mL of water under MW irradiation (T = 100 °C, 30 min).

As can be observed, CNS-Pd can be considered a heterogeneous catalyst with good reusability characteristics, as, after five cycles, more than 96% conversion was still achieved without losing selectivity towards product **3a**, demonstrating the long catalyst life.

3. Conclusions

In this work, cellulose-based nanosponges (CNS) were prepared. TEMPO-oxidized cellulose nanofibers and branched polyethyleneimine were cross-linked together in the presence of citric acid to obtain a micro- and nanoporous structure and this latter was then loaded with Pd(II) to obtain a potential heterogeneous catalyst for Suzuki–Miyaura coupling reactions. The morphology and structure of the material were characterized by various SEM–EDS and ICP-OES analyses. The CNS-Pd system was also thoroughly investigated by XPS analysis to evaluate the behavior of the metal in the catalytic activity. An optimization study on the heterogeneous reaction was conducted, reaching optimal condition for the obtainment of high-rate yields. The catalytic recycling ability of the material and the catalytic effect for different substituents were also examined, confirming the possibility of reusing this sustainable catalyst several times.

4. Materials and Methods

All of the reagents were purchased from Merck (Darmstadt, Germany). Cotton linter was obtained from Bartoli paper factory (Capannori, Lucca, Italy). Deionized water was produced within the laboratories with a Millipore Elix® Deionizer with Progard® S2 ion exchange resins (Merck KGaA, Darmstadt, Germany). All 1H-NMR spectra were recorded on a 400 MHz Brüker (Billerica, MA, USA) NMR spectrometer. Microwave reactions were conducted in a Biotage® Initiator+ (Uppsala, Sweden). Other equipment used in the procedures includes a Branson SFX250 Sonicator (Emerson Electric Co., Ferguson, MI, USA), a SP Scientific BenchTop Pro Lyophilizer (SP INDUSTRIES, 935 Mearns Road, Warminster, UK), a Büchi Rotavapor® R-124 8 (Flawil, Switzerland), and a Thermotest Mazzali laboratory oven (Monza, Italy). Scanning electron microscopy (SEM) was performed using a variable pressure instrument (SEM Cambridge Stereoscan 360) at 100/120 pA with a detector BSD. The operating voltage was 15 kV with an electron beam current intensity of 100 pA. The focal distance was 9 mm. The EDS analysis was performed using a Bruker Quantax 200 6/30 instrument (Billerica, MA, USA). The metal concentrations were measured by

ICPOES atomic emission spectroscopy using a Perkin Elmer Optima 3000 SD spectrometer (Wellesley, MA, USA).

4.1. TEMPO-Oxidized Cellulose (TOC) Production and Titration and Synthesis of Cellulose NanoSponges (CNS)

TEMPO-Oxidized Cellulose (TOC) production and titration were performed according to a procedure previously reported in the literature [34,50,51]. After TOC synthesis and characterization, TEMPO-Oxidized Cellulose Nanofibers (TOCNF) for the production of CNS were produced by suspending 3.5 g of TOC in 0.14 L of deionized water, adding 0.210 g of NaOH pellets. The suspension was ultrasonicated with a Branson SFX250 Sonicator, achieving a homogeneous suspension of TOCNF. This latter was then acidified with 12 N HCl (2 mL), filtered under vacuum on a Büchner funnel, and washed with deionized water (250 mL) reaching neutral pH. Then, two aqueous solutions of anhydrous citric acid (CA) (1.792 g in 10 mL) and 25 kDa branched polyethylenimine (bPEI) (7 g of bPEI in 10 mL) were mixed with the TOCNF solution, while continuously stirring until reaching a white and homogeneous hydrogel and obtaining a final TOCNF concentration of 3% w/w. The mixture was then placed in well plates used as molds, quickly frozen at −20 °C, frozen-dried for 48 h at −52 °C, 140 µbar using an SP Scientific BenchTop Pro Lyophilizer, and finally thermally treated in the laboratory oven performing a heating ramp from a temperature of 55 °C to a maximum of 98 °C. This temperature was kept for 16 h. At the end of the process, CNS was ground with a mortar and then washed with deionized water (500 mL) to remove the excess bPEI. This procedure for the synthesis of CNS was described in our previous works [37,38].

4.2. Preparation of the Catalyst

After washing, 275 mg of CNS was ground into a fine powder. In the meantime, a saturated acidic solution of $PdCl_2$ (101 mg in 50 mL of 0.1 M HCl) was prepared. Ground CNS was soaked in this solution until complete adsorption of Pd (10 min), repeating the adsorption cycle three times in the same conditions. After every sorption cycle, Pd-loaded CNS was filtered on a Buchner funnel, washed with 100 mL of deionized water and 50 mL of ethanol, and, finally, left to dry in open air, obtaining the catalyst CNS-Pd.

4.3. Catalyst Characterization

Scanning Electron Microscopy (SEM) was performed using a variable-pressure instrument (SEM Cambridge Stereoscan 360) at 100/120 Pa with a detector VPSE. The operating voltage was 20 kV with an electron beam current intensity of 150 pA. The focal distance was 8 mm. The specimens were analyzed in High Vacuum mode after metallization.

EDS analysis was performed using a Bruker Quantax 200 6/30 instrument.

Inductively Coupled Plasma Optical Emission Spectroscopy (ICP-OES) analysis was performed on solid materials to determine the percentage of palladium in CNS-Pd. ICP-OES was conducted using a Perkin Elmer Optima 3000 SD spectrometer and the samples were treated with nitric acid (HNO_3) to completely dissolve the organic portion of the material before the determination of the metal concentration.

X-ray photoelectron spectroscopy (XPS) measurements were carried out using a custom designed spectrometer, described in previous studies [52] and equipped with a non-monochromatized Mg Ka X-ray source (1253.6 eV pass energy 25 eV, step 0.1 eV). For this experiment, photoelectrons emitted by C1s, O1s, N1s, Pd3d, Cl2p, Br3d, and B1s core levels were detected on powder samples of the pristine and recovered CNS-Pd catalyzer. All spectra were energy-referenced to the C 1s signal of aliphatic C atoms at 285.00 eV binding energy (BE) [53]. Atomic ratios were calculated from peak intensities using Scofield's cross-section values [54]. Curve-fitting analysis was performed using Gaussian profiles as fitting functions after subtraction of a polynomial background. For qualitative data, the BE values were mostly referred to from the NIST database [43].

4.4. Suzuki–Miyaura Reaction

In a 5 mL microwave vial, CNS-Pd (1–5 mg, 2–10% *w*/*w*), KOH (0.536 mmol, 2 eq), phenylboronic acid (0.402 mmol, 1.5 eq), and 2.5 mL of water as solvent were added. Then, tetrabutylammonium bromide (TBAB) (0.0402–0.1608 mmol, 0.15–0.6 eq) and reagent **1a–k** (0.268 mmol, 1 eq) were put in the vial (all the percentages and equivalents are referred to as reagents **1a–k**). The reaction was performed under microwave irradiation in a temperature range of 40–130 °C and for reaction times between 10 min and 3 h.

4.5. Reaction Work-Up and Product Purification

At the end of the reaction, 2 mL of ethyl acetate was added into the reaction mixture and stirred for 10 min. The solution was then filtered in a glass straw equipped with cotton to remove the catalyst and the reaction vial was washed three times with 2 mL of water and three times with 2 mL of ethyl acetate. The organic and aqueous phases were transferred in an extractor funnel and 5 mL of 0.1 N HCl solution was added. The aqueous phase was then extracted three times with 15 mL of ethyl acetate and all the organic phases were collected together, and then washed once with 10 mL of NaOH 0.1 N and twice with 10 mL of a saturated solution of NaCl. Then, the organic phase was collected and anhydrified with Na_2SO_4. Lastly, the final organic solution was then filtered off and the solvent was removed under vacuum to obtain a crude for the NMR analysis. The purification of the product was performed with flash column chromatography using a solvent mixture of hexane and ethyl acetate in a 95:5 ratio.

Yields were calculated by means of ^{1}H-NMR analysis using acetonitrile (ACN) as internal standard. The analysis was performed on the crude reaction without purification process. For the standard reaction (4-Br-anisole as reagent **A**), the yield calculated by internal standard technique with the ^{1}H-NMR was further confirmed with the yield obtained with mass recovery after purification with column chromatography, as described in the previous paragraph.

4.6. Leaching Test

In a 20 mL microwave vial, CNS-Pd (12 mg), KOH (360 mg), and 15 mL of water as solvent were added. Then, TBAB (312 mg) was put in the vial. The reaction was performed under microwave irradiation at 100 °C for 30 min. Once finished, the reaction was filtered off on a Buchner funnel to remove the solid and the solution was analyzed through ICP-OES analysis to quantify the Pd released in the reaction environment.

4.7. Reusability Test

Re-usability tests were conducted on the reference reaction with *p*-Br-anisole (Scheme 2). In a 5 mL microwave vial, CNS-Pd (10 mg, 10% *w*/*w*), KOH (1.072 mmol, 2 eq), phenylboronic acid (98 mg), and 5 mL of water as solvent were added. Then, TBAB (0.804 mmol, 1.5 eq) and *p*-Br-anisole (0.536 mmol, 1 eq) were put in the vial. The reaction was performed under microwave irradiation at 100 °C for 30 min. At the end of the reaction time, the vial was centrifuged, and the reaction solvent was removed. Three cycles of washing of CNS-Pd inside the vial were then carried out as described: a 5 mL measure of AcOEt was added into the reaction vial, and the solution was left stirring for 10 min and subsequently centrifuged. The supernatant was removed, taking care not to remove the catalyst, and the washing was repeated twice more. At the end of the washing passages, CNS-Pd was allowed to dry inside the vial, and this latter was then used for a new reaction cycle. This procedure was repeated five times.

4.8. NMR Product Characterization

Internal Standard (ACN): ^{1}H NMR (400 MHz, $CDCl_3$) δ 2.04 (s, 3H); **Product 3a (4-Methoxybiphenyl):** ^{1}H NMR (400 MHz, $CDCl_3$) δ 7.55–7.50 (m, 4H), 7.40 (t, *J* = 7.5, 2H), 7.29 (t, *J* = 7.3, 1H), 6.97 (d, *J* = 8.8, 2H), 3.84 (s, 3H) [55]; **Product 3b (3-Methoxybiphenyl):** ^{1}H NMR (400 MHz, $CDCl_3$) δ 7.61 (d, *J* = 7.5, 2H), 7.45 (t, *J* = 7.6, 2H), 7.39–7.35 (m, 2H),

7.20 (d, J = 7.7, 1H), 7.15 (t, J = 2.0, 1H), 6.92 (dd, J = 8.2, 2.3, 1H), 3.87 (s, 3H) [56]; **Product 3c (2-Methoxybiphenyl):** ^{1}H NMR (400 MHz, $CDCl_3$) δ 7.54 (d, J = 8.3, 2H), 7.41 (t, J = 7.6, 2H), 7.34–7.31 (m, 3H), 7.03 (t, J = 7.5, 1H), 6.99 (d, J = 8.5, 1H), 3.81 (s, 3H) [56]; **Product 3d (4-Methylbiphenyl):** ^{1}H NMR (400 MHz, $CDCl_3$): δ = 7.64 (d, J = 8.0 Hz, 2H), 7.54 (d, J = 8.0 Hz, 2H),7.47 (t, J = 7.2, 2H), 7.37 (t, J = 7.2 Hz, 1H), 7.28 (d, J = 8.0 Hz, 2H), 2.44 (s, 3H) [55]; **Product 3e (Biphenyl-4-carbaldehyde):** ^{1}H NMR (400 MHz, $CDCl_3$): δ = 9.96 (s, 1H), 7.86 (d, J = 8.0 Hz, 2H), 7.66 (d, J = 8.0 Hz, 2H), 7.55 (d, J = 8.4 Hz, 2H), 7.40 (t, J = 7.2 Hz, 2H), 7.34 (t, J = 7.2 Hz, 1H) [55]; **Product 3f (Biphenyl):** ^{1}H NMR (400 MHz, $CDCl_3$): δ = 7.54 (d, J = 8.0 Hz, 4H), 7.39 (t, J = 8.0 Hz, 4H), 7.31–7.27 (m, 2H) [55]; **Product 3g (4-Aminobiphenyl):** ^{1}H NMR (400 MHz, $CDCl_3$, TMS) δ 7.52 (d, J = 7.9 Hz, 2H, CH), 7.41–7.36 (m, 4H, CH), 7.25 (t, J = 7.3 Hz, 1H, CH), 6.73 (d, J = 8.5 Hz, 2H, CH), 3.68 (s, 2H) [57]; **Product 3h (2-Cynobiphenyl):** ^{1}H NMR (400 MHz, $CDCl_3$): δ (ppm) 7.76 (dd, J = 8.0Hz, 1.2Hz, 1H), 7.64 (td, J = 7.6 Hz, 1.2Hz, 1H), 7.58–7.54 (m, 2H), 7.54–7.40 (m, 5H) [58]; **Product 4j (1,1′:4′,1″-terphenyl):** ^{1}H NMR (400 MHz, Chloroform-d) δ 7.70 (s, 4H), 7.69–7.64 (m, 4H), 7.50–7.45 (m, 4H), 7.41–7.35 (m, 2H) ppm [43]; **Product 5k (1,3,5-Triphenylbenzene):** ^{1}H NMR (400 MHz, $CDCl_3$): δ = 7.71 (s, 3H), 7.63 (d, J = 7.2 Hz, 6H), 7.42 (t, J = 8.0 Hz, 6H), 7.31 (t, J = 6.8 Hz, 3H) [59].

Supplementary Materials: The following supporting information can be downloaded at: https://www.mdpi.com/article/10.3390/gels8120789/s1. Table S1: XPS data analysis results: core-level binding energy (BE), full width at half-maxima (FWHM) values, internal atomic ratios, and proposed assignments. Reference [60] is cited in the supplementary materials.

Author Contributions: Conceptualization, C.P. and A.S.; methodology, L.R., G.N., C.B. and M.L.; validation, L.R., G.N., C.B. and M.L.; formal analysis, investigation, L.R., G.N., C.B. and M.L.; resources, A.S. and C.P.; data curation, A.S.; writing—original draft preparation, L.R.; writing—review and editing, C.P., A.S. and C.B.; supervision, A.S. and C.P. All authors have read and agreed to the published version of the manuscript.

Funding: This research received no external funding.

Institutional Review Board Statement: Not applicable.

Informed Consent Statement: Not applicable.

Conflicts of Interest: The authors declare no conflict of interest.

References

1. Molnár, Á. Efficient, Selective, and Recyclable Palladium Catalysts in Carbon-Carbon Coupling Reactions. *Chem. Rev.* **2011**, *111*, 2251–2320. [CrossRef] [PubMed]
2. Roy, D.; Uozumi, Y. Recent Advances in Palladium-Catalyzed Cross-Coupling Reactions at Ppm to Ppb Molar Catalyst Loadings. *Adv. Synth. Catal.* **2018**, *360*, 602–625. [CrossRef]
3. D'Alterio, M.C.; Casals-Cruañas, È.; Tzouras, N.V.; Talarico, G.; Nolan, S.P.; Poater, A. Mechanistic Aspects of the Palladium-Catalyzed Suzuki-Miyaura Cross-Coupling Reaction. *Chem. A Eur. J.* **2021**, *27*, 13481–13493. [CrossRef] [PubMed]
4. Bhattacherjee, D.; Rahman, M.; Ghosh, S.; Bagdi, A.K.; Zyryanov, G.V.; Chupakhin, O.N.; Das, P.; Hajra, A. Advances in Transition-Metal Catalyzed Carbonylative Suzuki-Miyaura Coupling Reaction: An Update. *Adv. Synth. Catal.* **2021**, *363*, 1597–1624. [CrossRef]
5. Shi, S.; Nolan, S.P.; Szostak, M. Well-Defined Palladium(II)-NHC Precatalysts for Cross-Coupling Reactions of Amides and Esters by Selective N-C/O-C Cleavage. *Acc. Chem. Res.* **2018**, *51*, 2589–2599. [CrossRef]
6. Sau, S.C.; Hota, P.K.; Mandal, S.K.; Soleilhavoup, M.; Bertrand, G. Stable Abnormal N-Heterocyclic Carbenes and Their Applications. *Chem. Soc. Rev.* **2020**, *49*, 1233–1252. [CrossRef]
7. Wang, K.; Fan, R.; Wei, X.; Fang, W. Palladacyclic N-Heterocyclic Carbene Precatalysts for Transition Metal Catalysis. *Green Synth. Catal.* **2022**, *3*, 327–338. [CrossRef]
8. Wei, X.; Xue, B.; Handelmann, J.; Hu, Z.; Darmandeh, H.; Gessner, V.H.; Gooßen, L.J. Ylide-Functionalized Diisopropyl Phosphine (PrYPhos): A Ligand for Selective Suzuki-Miyaura Couplings of Aryl Chlorides. *Adv. Synth. Catal.* **2022**, *364*, 3336–3341. [CrossRef]
9. Martin, R.; Buchwald, S.L. Palladium-Catalyzed Suzuki-Miyaura Cross-Coupling Reactions Employing Dialkylbiaryl Phosphine Ligands. *Acc. Chem. Res.* **2008**, *41*, 1461–1473. [CrossRef]
10. Bruneau, A.; Roche, M.; Alami, M.; Messaoudi, S. 2-Aminobiphenyl Palladacycles: The "Most Powerful" Precatalysts in C-C and C-Heteroatom Cross-Couplings. *ACS Catal.* **2015**, *5*, 1386–1396. [CrossRef]

11. Valente, C.; Çalimsiz, S.; Hoi, K.H.; Mallik, D.; Sayah, M.; Organ, M.G. The Development of Bulky Palladium NHC Complexes for the Most-Challenging Cross-Coupling Reactions. *Angew. Chem. Int. Ed.* **2012**, *51*, 3314–3332. [CrossRef]
12. Espinosa, M.R.; Doppiu, A.; Hazari, N. Differences in the Performance of Allyl Based Palladium Precatalysts for Suzuki-Miyaura Reactions. *Adv. Synth. Catal.* **2020**, *362*, 5062–5078. [CrossRef] [PubMed]
13. Hazari, N.; Melvin, P.R.; Beromi, M.M. Well-Defined Nickel and Palladium Precatalysts for Cross-Coupling. *Nat. Rev. Chem.* **2017**, *1*, 1–17. [CrossRef]
14. Orecchia, P.; Petkova, D.S.; Goetz, R.; Rominger, F.; Hashmi, A.S.K.; Schaub, T. Pd-Catalysed Suzuki-Miyaura Cross-Coupling of Aryl Chlorides at Low Catalyst Loadings in Water for the Synthesis of Industrially Important Fungicides. *Green Chem.* **2021**, *23*, 8169–8180. [CrossRef]
15. Abi Fayssal, S.; Naret, T.; Buendia, J.; Labattut, A.; Huc, V.; Martini, C.; Schulz, E. Synthesis, Catalytic Activity and Comparative Leaching Studies of Calix[8]Arene-Supported Pd-NHC Complexes for Suzuki-Miyaura Cross-Couplings. *Adv. Synth. Catal.* **2022**, *364*, 947–957. [CrossRef]
16. Jin, H.; Zhang, C.; Liu, P.; Ge, X.; Zhou, S. Covalent Organic Framework-Supported Pd Nanoparticles: An Efficient and Reusable Heterogeneous Catalyst for Suzuki–Miyaura Coupling Reactions. *Appl. Organomet. Chem.* **2022**, *36*, 1–10. [CrossRef]
17. Gao, M.; Wang, J.; Shang, W.; Chai, Y.; Dai, W.; Wu, G.; Guan, N.; Li, L. Zeolite-Encaged Palladium Catalysts for Heterogeneous Suzuki-Miyaura Cross-Coupling Reactions. *Catal. Today* **2022**, *in press*. [CrossRef]
18. Labulo, A.H.; Martincigh, B.S.; Omondi, B.; Nyamori, V.O. Advances in Carbon Nanotubes as Efficacious Supports for Palladium-Catalysed Carbon–Carbon Cross-Coupling Reactions. *J. Mater. Sci.* **2017**, *52*, 9225–9248. [CrossRef]
19. Hoshiya, N.; Isomura, N.; Shimoda, M.; Yoshikawa, H.; Yamashita, Y.; Iizuka, K.; Tsukamoto, S.; Shuto, S.; Arisawa, M. Development of a Recyclable and Low-Leaching Palladium Catalyst Supported on Sulfur-Modified Gallium Arsenide (001) for Use in Suzuki-Miyaura Coupling. *ChemCatChem* **2009**, *1*, 279–285. [CrossRef]
20. Monguchi, Y.; Ichikawa, T.; Yamada, T.; Sawama, Y.; Sajiki, H. Continuous-Flow Suzuki-Miyaura and Mizoroki-Heck Reactions under Microwave Heating Conditions. *Chem. Rec.* **2019**, *19*, 3–14. [CrossRef]
21. Price, G.A.; Bogdan, A.R.; Aguirre, A.L.; Iwai, T.; Djuric, S.W.; Organ, M.G. Continuous Flow Negishi Cross-Couplings Employing Silica-Supported: Pd-PEPPSI–IPr Precatalyst. *Catal. Sci. Technol.* **2016**, *6*, 4733–4742. [CrossRef]
22. Shariatipour, M.; Heydari, A. PdII Dispersed on Magnetic Partially Reduced GO/OMWCNT Non-Covalently Modified with a Vic-Dioxime: An Efficient and Magnetically Retrievable Catalyst for Suzuki-Miyaura Coupling Reaction. *ChemistrySelect* **2021**, *6*, 1107–1117. [CrossRef]
23. Kandathil, V.; Kulkarni, B.; Siddiqa, A.; Kempasiddaiah, M.; Sasidhar, B.S.; Patil, S.A.; Patil, S.A. Immobilized N-Heterocyclic Carbene-Palladium(II) Complex on Graphene Oxide as Efficient and Recyclable Catalyst for Suzuki–Miyaura Cross-Coupling and Reduction of Nitroarenes. *Catal. Lett.* **2020**, *150*, 384–403. [CrossRef]
24. Melone, L.; Rossi, B.; Pastori, N.; Panzeri, W.; Mele, A.; Punta, C. TEMPO-Oxidized Cellulose Cross-Linked with Branched Polyethyleneimine: Nanostructured Adsorbent Sponges for Water Remediation. *Chempluschem* **2015**, *80*, 1408–1415. [CrossRef] [PubMed]
25. Fiorati, A.; Turco, G.; Travan, A.; Caneva, E.; Pastori, N.; Cametti, M.; Punta, C.; Melone, L. Mechanical and Drug Release Properties of Sponges from Cross-Linked Cellulose Nanofibers. *Chempluschem* **2017**, *82*, 848–858. [CrossRef] [PubMed]
26. Liberatori, G.; Grassi, G.; Guidi, P.; Bernardeschi, M.; Fiorati, A.; Scarcelli, V.; Genovese, M.; Faleri, C.; Protano, G.; Frenzilli, G.; et al. Effect-Based Approach to Assess Nanostructured Cellulose Sponge Removal Efficacy of Zinc Ions from Seawater to Prevent Ecological Risks. *Nanomaterials* **2020**, *10*, 1283. [CrossRef]
27. Guidi, P.; Bernardeschi, M.; Palumbo, M.; Genovese, M.; Scarcelli, V.; Fiorati, A.; Riva, L.; Punta, C.; Corsi, I.; Frenzilli, G. Suitability of a Cellulose-Based Nanomaterial for the Remediation of Heavy Metal Contaminated Freshwaters: A Case-Study Showing the Recovery of Cadmium Induced Dna Integrity Loss, Cell Proliferation Increase, Nuclear Morphology and Chromosomal Alterations. *Nanomaterials* **2020**, *10*, 1837. [CrossRef] [PubMed]
28. Guidi, P.; Bernardeschi, M.; Palumbo, M.; Scarcelli, V.; Genovese, M.; Protano, G.; Vitiello, V.; Pontorno, L.; Bonciani, L.; Buttino, I.; et al. Cellular Responses Induced by Zinc in Zebra Mussel Haemocytes. Loss of DNA Integrity as a Cellular Mechanism to Evaluate the Suitability of Nanocellulose-Based Materials in Nanoremediation. *Nanomaterials* **2021**, *11*, 2219. [CrossRef]
29. Fiorati, A.; Grassi, G.; Graziano, A.; Liberatori, G.; Pastori, N.; Melone, L.; Bonciani, L.; Pontorno, L.; Punta, C.; Corsi, I. Eco-Design of Nanostructured Cellulose Sponges for Sea-Water Decontamination from Heavy Metal Ions. *J. Clean. Prod.* **2020**, *246*, 119009. [CrossRef]
30. Bartolozzi, I.; Daddi, T.; Punta, C.; Fiorati, A.; Iraldo, F. Life Cycle Assessment of Emerging Environmental Technologies in the Early Stage of Development: A Case Study on Nanostructured Materials. *J. Ind. Ecol.* **2020**, *24*, 101–115. [CrossRef]
31. Paladini, G.; Venuti, V.; Almásy, L.; Melone, L.; Crupi, V.; Majolino, D.; Pastori, N.; Fiorati, A.; Punta, C. Cross-Linked Cellulose Nano-Sponges: A Small Angle Neutron Scattering (SANS) Study. *Cellulose* **2019**, *26*, 9005–9019. [CrossRef]
32. Paladini, G.; Venuti, V.; Crupi, V.; Majolino, D.; Fiorati, A.; Punta, C. FTIR-ATR Analysis of the H-Bond Network of Water in Branched Polyethyleneimine/TEMPO-Oxidized Cellulose Nano-Fiber Xerogels. *Cellulose* **2020**, *27*, 8605–8618. [CrossRef]
33. Paladini, G.; Venuti, V.; Crupi, V.; Majolino, D.; Fiorati, A.; Punta, C. 2D Correlation Spectroscopy (2Dcos) Analysis of Temperature-Dependent Ftir-Atr Spectra in Branched Polyethyleneimine/Tempo-Oxidized Cellulose Nano-Fiber Xerogels. *Polymers* **2021**, *13*, 528. [CrossRef] [PubMed]
34. Riva, L.; Pastori, N.; Panozzo, A.; Antonelli, M.; Punta, C. Nanostructured Cellulose-Based Sorbent Materials for Water Decontamination from Organic Dyes. *Nanomaterials* **2020**, *10*, 1570. [CrossRef] [PubMed]

35. Melone, L.; Bonafede, S.; Tushi, D.; Punta, C.; Cametti, M. Dip in Colorimetric Fluoride Sensing by a Chemically Engineered Polymeric Cellulose/ BPEI Conjugate in the Solid State. *RSC Adv.* **2015**, *5*, 83197–83205. [CrossRef]
36. Riva, L.; Fiorati, A.; Sganappa, A.; Melone, L.; Punta, C.; Cametti, M. Naked-Eye Heterogeneous Sensing of Fluoride Ions by Co-Polymeric Nanosponge Systems Comprising Aromatic-Imide-Functionalized Nanocellulose and Branched Polyethyleneimine. *Chempluschem* **2019**, *84*, 1512–1518. [CrossRef]
37. Riva, L.; Punta, C.; Sacchetti, A. Co-Polymeric Nanosponges from Cellulose Biomass as Heterogeneous Catalysts for Amine-Catalyzed Organic Reactions. *ChemCatChem* **2020**, *12*, 6214–6222. [CrossRef]
38. Riva, L.; Lotito, A.D.; Punta, C.; Sacchetti, A. Zinc-and Copper-Loaded Nanosponges from Cellulose Nanofibers Hydrogels: New Heterogeneous Catalysts for the Synthesis of Aromatic Acetals. *Gels* **2022**, *8*, 54. [CrossRef]
39. Bashar, M.M.; Zhu, H.; Yamamoto, S.; Mitsuishi, M. Highly Carboxylated and Crystalline Cellulose Nanocrystals from Jute Fiber by Facile Ammonium Persulfate Oxidation. *Cellulose* **2019**, *26*, 3671–3684. [CrossRef]
40. Johansson, L.S.; Campbell, J.M. Reproducible XPS on Biopolymers: Cellulose Studies. *Surf. Interface Anal.* **2004**, *36*, 1018–1022. [CrossRef]
41. Awada, H.; Monplaisir, D.; Daneault, C. Growth of Polyelectrolyte on Lignocellulosic Fibres: Study by ζ-Potential, FTIR, and XPS. *BioResources* **2012**, *7*, 2090–2104. [CrossRef]
42. Benkaddour, A.; Journoux-Lapp, C.; Jradi, K.; Robert, S.; Daneault, C. Study of the Hydrophobization of TEMPO-Oxidized Cellulose Gel through Two Routes: Amidation and Esterification Process. *J. Mater. Sci.* **2014**, *49*, 2832–2843. [CrossRef]
43. Naumkin, A.V.; Kraut-Vass, A.; Gaarenstroom, S.W.; Powell, C.J. *NIST X-ray Photoelectron Spectroscopy Database, Version 4.1*; National Institute of Standards and Technology: Gaithersburg, MA, USA, 2012.
44. Criado, J.J.; Fernandez, I.; Macias, B.; Salas, J.M.; Medarde, M. Novel Chelates of Pd(II) Dithiocarbamates. Spectroscopic Studies and Thermal Behaviour. *Inorg. Chim. Acta* **1990**, *174*, 67–75. [CrossRef]
45. Bertolini, J.C.; Delichere, P.; Khanra, B.C.; Massardier, J.; Noupa, C.; Tardy, B. Electronic Properties of Supported Pd Aggregates in Relation with Their Reactivity for 1,3-Butadiene Hydrogenation. *Catal. Lett.* **1990**, *6*, 215–223. [CrossRef]
46. Kishi, K.; Ikeda, S. X-ray Photoelectron Spectroscopic Study of the Reaction of Evaporated Metal Films with Chlorine Gas. *J. Phys. Chem.* **1974**, *78*, 107–112. [CrossRef]
47. Mochi, F.; Burratti, L.; Fratoddi, I.; Venditti, I.; Battocchio, C.; Carlini, L.; Iucci, G.; Casalboni, M.; De Matteis, F.; Casciardi, S.; et al. Plasmonic Sensor Based on Interaction between Silver Nanoparticles and Ni2+ or Co2+ in Water. *Nanomaterials* **2018**, *8*, 488. [CrossRef]
48. Zakharova, I.A.; Salyn, J.V.; Tatjanenko, L.V.; Mashkovsky, Y.S.; Ponticelli, G. Inhibitory Activity of Palladium(II) and Platinum(II) Complexes with Isoxazole and Its Derivatives. *J. Inorg. Biochem.* **1981**, *15*, 89–92. [CrossRef]
49. Wei, C.S.; Davies, G.H.M.; Soltani, O.; Albrecht, J.; Gao, Q.; Pathirana, C.; Hsiao, Y.; Tummala, S.; Eastgate, M.D. The Impact of Palladium(II) Reduction Pathways on the Structure and Activity of Palladium(0) Catalysts. *Angew. Chem. Int. Ed.* **2013**, *52*, 5822–5826. [CrossRef] [PubMed]
50. Isogai, A.; Saito, T.; Fukuzumi, H. TEMPO-Oxidized Cellulose Nanofibers. *Nanoscale* **2011**, *3*, 71–85. [CrossRef]
51. Pierre, G.; Punta, C.; Delattre, C.; Melone, L.; Dubessay, P.; Fiorati, A.; Pastori, N.; Galante, Y.M.; Michaud, P. TEMPO-Mediated Oxidation of Polysaccharides: An Ongoing Story. *Carbohydr. Polym.* **2017**, *165*, 71–85. [CrossRef]
52. Secchi, V.; Franchi, S.; Dettin, M.; Zamuner, A.; Beranová, K.; Vladescu, A.; Battocchio, C.; Graziani, V.; Tortora, L.; Iucci, G. Hydroxyapatite Surfaces Functionalized with a Self-Assembling Peptide: XPS, Rairs and Nexafs Study. *Nanomaterials* **2020**, *10*, 1151. [CrossRef] [PubMed]
53. Moulder, J.F.; Stickle, W.F.; Sobol, P.E.; Bomben, K.D. *Handbook of X-ray Photoelectron Spectroscopy: A Reference Book of Standard Spectra for Identification and Interpretation of XPS Data*; Physical Electronics Division, Perkin-Elmer Corporation: Boston, MA, USA, 1992; ISBN 978-0-9627026-2-4.
54. Castle, J.E. *Practical Surface Analysis by Auger and X-ray Photoelectron Spectroscopy*; Briggs, D., Seah, M.P., Eds.; John Wiley & Sons Ltd.: Chichester, UK, 1983; ISBN 047126279X.
55. Isfahani, A.L.; Mohammadpoor-Baltork, I.; Mirkhani, V.; Khosropour, A.R.; Moghadam, M.; Tangestaninejad, S.; Kia, R. Palladium Nanoparticles Immobilized on Nano-Silica Triazine Dendritic Polymer (Pdnp-NSTDP): An Efficient and Reusable Catalyst for Suzuki-Miyaura Cross-Coupling and Heck Reactions. *Adv. Synth. Catal.* **2013**, *355*, 957–972. [CrossRef]
56. Chen, W.C.; Hsu, Y.C.; Shih, W.C.; Lee, C.Y.; Chuang, W.H.; Tsai, Y.F.; Chen, P.P.Y.; Ong, T.G. Metal-Free Arylation of Benzene and Pyridine Promoted by Amino-Linked Nitrogen Heterocyclic Carbenes. *Chem. Commun.* **2012**, *48*, 6702–6704. [CrossRef] [PubMed]
57. Budén, M.E.; Guastavino, J.F.; Rossi, R.A. Room-Temperature Photoinduced Direct C-H-Arylation via Base-Promoted Homolytic Aromatic Substitution. *Org. Lett.* **2013**, *15*, 1174–1177. [CrossRef]
58. Raza, F.; Yim, D.; Park, J.H.; Kim, H.I.; Jeon, S.J.; Kim, J.H. Structuring Pd Nanoparticles on 2H-WS2 Nanosheets Induces Excellent Photocatalytic Activity for Cross-Coupling Reactions under Visible Light. *J. Am. Chem. Soc.* **2017**, *139*, 14767–14774. [CrossRef] [PubMed]
59. Lyons, D.J.M.; Dinh, A.H.; Ton, N.N.H.; Crocker, R.D.; Mai, B.K.; Nguyen, T.V. Ring Contraction of Tropylium Ions into Benzenoid Derivatives. *Org. Lett.* **2022**, *24*, 2520–2525. [CrossRef]
60. Liu, X.; Wang, Y.; Dong, L.; Chen, X.; Xin, G.; Zhang, Y.; Zang, J. One-step synthesis of shell/core structural boron and nitrogen co-doped graphitic carbon/nanodiamond as efficient electrocatalyst for the oxygen reduction reaction in alkaline media. *Electrochim. Acta* **2016**, *194*, 161–167. [CrossRef]

gels

Article

Fabrication of Poly(vinyl alcohol)/Chitosan Composite Films Strengthened with Titanium Dioxide and Polyphosphonate Additives for Packaging Applications

Tăchiță Vlad-Bubulac [1,*], Corneliu Hamciuc [1], Cristina Mihaela Rîmbu [2], Magdalena Aflori [1], Maria Butnaru [3], Alin Alexandru Enache [4] and Diana Serbezeanu [1]

1 Department of Polycondensation and Thermally Stable Polymers, "Petru Poni" Institute of Macromolecular Chemistry, 41A, Grigore Ghica Voda Alley, 700487 Iasi, Romania; chamciuc@icmpp.ro (C.H.); maflori@icmpp.ro (M.A.); diana.serbezeanu@icmpp.ro (D.S.)
2 Department of Public Health, Faculty of Veterinary Medicine "Ion Ionescu de la Brad", University of Agricultural Sciences and Veterinary Medicine, 8, Mihail Sadoveanu Alley, 707027 Iasi, Romania; crimbu@yahoo.com
3 Department of Natural Polymers, Bioactive and Biocompatible Materials, "Grigore T. Popa" University of Medicine and Pharmacy, 700115 Iasi, Romania; mariabutnaru@yahoo.com
4 S.C. Apel Laser S.R.L., 25, Vanatorilor Street, Mogosoaia, 077135 Ilfov, Romania; alin.enache@apellaser.ro
* Correspondence: tvladb@icmpp.ro

Abstract: Eco-innovation through the development of intelligent materials for food packaging is evolving, and it still has huge potential to improve food product safety, quality, and control. The design of such materials by the combination of biodegradable semi-synthetic polymers with natural ones and with some additives, which may improve certain functionalities in the targeted material, is continuing to attract attention of researchers. To fabricate composite films via casting from solution, followed by drying in atmospheric conditions, certain mass ratios of poly(vinyl alcohol) and chitosan were used as polymeric matrix, whereas TiO_2 nanoparticles and a polyphosphonate were used as reinforcing additives. The structural confirmation, surface properties, swelling behavior, and morphology of the xerogel composite films have been studied. The results confirmed the presence of all ingredients in the prepared fabrics, the contact angle of the formulation containing poly(vinyl alcohol), chitosan, and titanium dioxide in its composition exhibited the smallest value (87.67°), whereas the profilometry and scanning electron microscopy enlightened the good dispersion of the ingredients and the quality of all the composite films. Antimicrobial assay established successful antimicrobial potential of the poly(vinyl alcoohol)/chitosan-reinforced composites films against *Staphylococcus aureus*, *Methicillin-resistant Staphylococcus aureus* (MRSA), *Escherichia coli*, *Pseudomonas aeruginosa*, and *Candida albicans*. Cytotoxicity tests have revealed that the studied films are non-toxic, presented good compatibility, and they are attractive candidates for packaging applications.

Keywords: poly(vinyl alcohol); chitosan; titanium dioxide; polyphosphonate; casting from solution; xerogel composite film

Citation: Vlad-Bubulac, T.; Hamciuc, C.; Rîmbu, C.M.; Aflori, M.; Butnaru, M.; Enache, A.A.; Serbezeanu, D. Fabrication of Poly(vinyl alcohol)/Chitosan Composite Films Strengthened with Titanium Dioxide and Polyphosphonate Additives for Packaging Applications. *Gels* **2022**, *8*, 474. https://doi.org/10.3390/gels8080474

Academic Editors: Francesco Caridi, Giuseppe Paladini and Andrea Fiorati

Received: 5 July 2022
Accepted: 25 July 2022
Published: 28 July 2022

1. Introduction

Plastic packaging continues, in recent years, to be the most significant industrial use in the world (representing, on average, 30% of the total) [1–3]. Particularly, polymers, like polyethylene terephthalate, low-density polyethylene, polystyrene, polypropylene, and high-density polyethylene, were widely used as single-use packaging materials in the food and beverage industry because of their mechanical properties that could provide effective barriers to oxygen and carbon dioxide as well as their relative affordability and ease of availability [4–8]. Nevertheless, when plastic packaging based on the abovementioned polymers reaches the end life of its use, a significant amount frequently eludes formal collection and recycling processes and eventually leaks away, damaging the worldwide environment [9–11].

According to estimates, the increase in population growth will demand a 50% increase in global food supplies by the year 2050 [12]. Therefore, it is necessary to incorporate eco-innovations in primary packaging that can lessen the impact of packages on the environment and simultaneously maintain food quality and safety [13]. Additionally, more and more efforts are being put into creating bio-products for eco-innovations in food packaging, like biodegradable and compostable polymers [14,15].

The primary choice for packaging is represented by natural polymers, which should be researched and used extensively in the near future. The second-most prevalent polysaccharide in the world, chitosan, CS, is a biocompatible, biodegradable, and nontoxic substance that has proven to be a good candidate for use in packaging films. Structural versatility, attractive barrier properties, excellent film-forming and coating capabilities, as well as an innate antibacterial capability, have been proven for this biopolymer. As a result, several films made from CS have been produced and used in the food packaging sector [16–19]. Pure CS, however, is only soluble in acidic environments and has poor mechanical and thermal properties [20]. A unique bio-composite material with brand-new or improved properties could be produced by mixing biopolymers with existing polymeric matrices, natural or semi-synthetic, to meet specific needs. Such combinations of CS with starch, pectin, alginate, poly lactic acid (PLA), poly(vinyl alcohol) (PVA), gelatin, etc. have been reported to date [21–25].

Blends of PVA with CS, which combine a natural biomacromolecule and an easily biodegradable synthetic polymer, are among the most attractive bicomponent systems [26]. PVA is a thermoplastic synthetic polymer created from the hydrolysis of poly(vinyl acetate), unlike many other synthetic polymers. Due to PVA's distinct characteristics, which include high mechanical properties, chemical and thermal stability, non-toxicity, film-forming skills, and low manufacturing costs, its uses have increased over the past ten years. In addition to biodegradable goods like backing rolls, adhesives, coatings, and surfactants, PVA is utilized in a number of different industries, including those that deal with textiles, paper, and food packaging [27]. PVA, like other synthetic polymers, has applications in biology and medicine in addition to technical ones, and this has led to it being one of the main research areas for polymer scientists. Additionally, in recent years PVA have been studied for various smart applications such as shape memory hydrogels with improved viscoelasticity for printable applications [28], bio-based sensors and antimicrobial films [29], nanofibrous metallochromic sensors for colorimetric selective detection of ferric ions [30], etc.

It has also been shown that adding fillers to this PVA/CS bicomponent systems can be supplementary reinforcement, creating new composites with enhanced physical features, like water resistance, without sacrificing biodegradability. Because of their simplicity of processing, low cost, and superior synergistic characteristics, nanomaterials have been produced and employed in a wide variety of fields, including foods, medicines, and cosmetics. Incorporating nanosized compounds like nano-ZnO, nano-TiO_2, and nano-silica into the polymers has been proven in some studies to help improve the characteristics of the polymers [31–33]. In our recent paper, synergistic effects of the presence of silica nanoparticles and a polyphosphonate mixed into PVA matrix have been discussed [34].

TiO_2 is a versatile and chemically inert mineral with various pertinent uses (food, pharmaceutical, biomedical, antibacterial agent, environmental, and clean energy) [34]. Despite the fact that its use as a food additive has recently been questioned and even banned in the European Union, TiO_2's physicochemical, mechanical, and photocatalytic qualities, as well as its reactivity and thermal stability, low cost, secure manufacturing, and biocompatibility, all contribute to TiO_2's widespread application [35]. However, the capacity of nano-TiO_2 to aggregate is one of its main drawbacks. According to certain reports, the interaction of nano-TiO_2 with biopolymers including starch, gums, and chitosan can aid to lessen the spontaneous agglomeration of TiO_2, improving the functional aspects of the composite [36]. The addition of nanoparticles to the polymer matrix has two benefits:

first, it strengthens and functionalizes the matrix by improving its dispersibility, and second, it gives the mixture antibacterial capabilities [37].

In composite materials, PVA is frequently utilized as a polymer matrix, and CS is frequently employed as a reinforcing agent in an effort to broaden the application of PVA in the field of high-strength materials [26,38]. Thus, the co-existence of the hydroxyl groups in both PVA and CS, doubled by the presence of amino groups in CS, can produce intermolecular interactions, which may result in an improvement of performances of PVA composites. The use of polyphosphonate containing sulfur and phosphorus in the main chain and in the side chain in phosphaphenanthrene-type heterocycles is expected to induce thermal stability in the composite, then, to synergistically bring its own antimicrobial contribution to the composite, as was described in few studies in the recent past [39].

In the present study, optimized solutions of PVA/CS, PVA/CS/TiO_2, or PVA/CS/TiO_2/polyphosphonate have been used to obtain xerogels composite films via casting from solution method, followed by drying in ambient conditions. Schematic representation of the composites is presented in Figure 1.

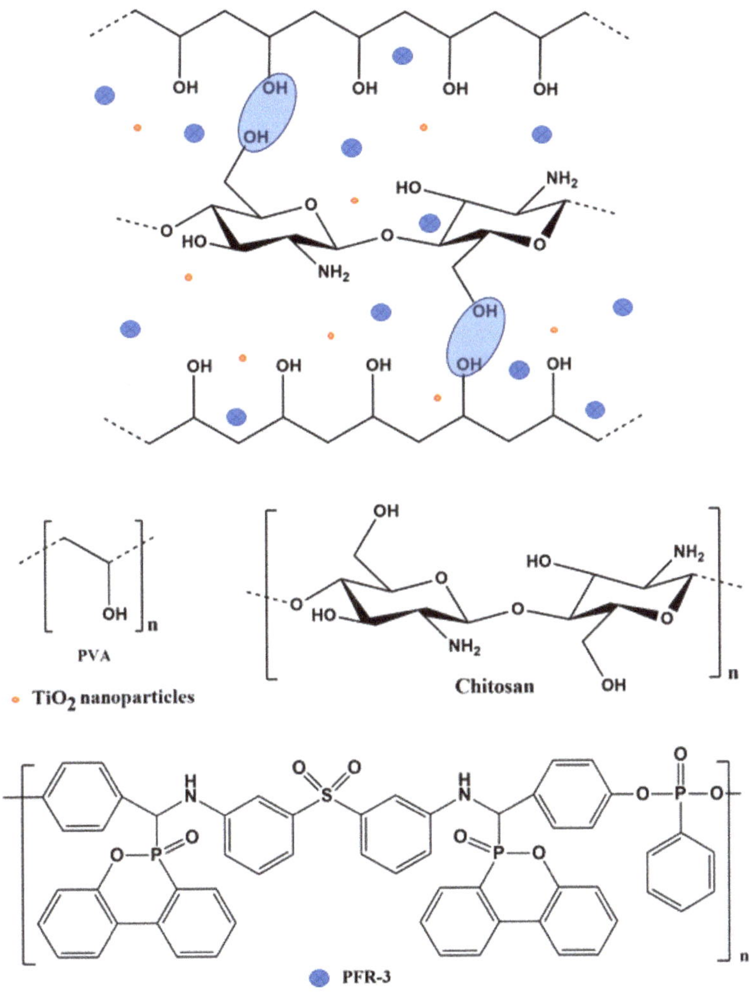

Figure 1. Schematic representation of PVA/CS/TiO_2/polyphosphonate composites.

The effect of all the ingredients on the structure-properties relation of the developed xerogel composite films was determined and discussed. Physical, chemical, morphological properties, surface properties, antimicrobial activity, and cytocompatibility of the matrices were studied by means of FTIR, SEM, swelling behavior, profilometry, contact angle measurements, antimicrobial, and cytotoxicity assays.

2. Results and Discussion

2.1. Preparation and Structural Characterization

Binary PVA/CS films and the composite PVA/CS films containing TiO_2 or TiO_2 with polyphosphonate, PFR-3, have been obtained by the casting from solution procedure. The details of this procedure are presented in the Materials and Methods section. The composition of the composite polymer matrices expressed in mass ratio for all the ingredients utilized in the preparation and the codes of the as-prepared composites are listed in Table 1.

Table 1. Preparation details of the PVA/CS composite films.

Sample	PVA (g)	Chitosan (g)	PFR-3 (g)	TiO_2 (g)
PVA/CS-0	1.5	0.5	-	-
PVA/CS-1	1.38	0.46	-	0.16
PVA/CS-2	1.2171	0.4057	0.2172	0.16
PVA/CS-3	1.0542	0.3514	0.4344	0.16

The chemical structure of the products was introspected by FTIR spectroscopy. Figure 2 presents the FTIR spectra of the binary PVA/CS-0 sample and of the PVA/CS composite films. In the spectrum of the PVA/CS-0 sample, the characteristic strong and wide band appeared at approximately 3320 cm^{-1} due to hydroxyl stretching vibration.

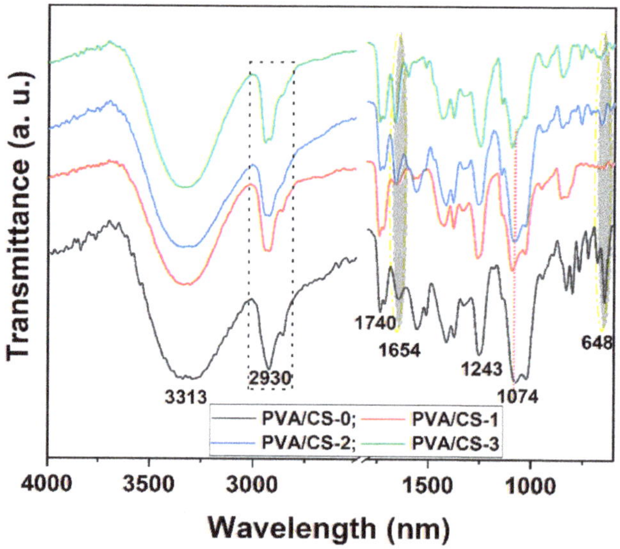

Figure 2. Fourier-transform infrared (FTIR) spectra for PVA/CS composites.

Common characteristic absorption bands were found also at about 2930 cm^{-1} with a shoulder at 2860 cm^{-1} due to asymmetric and symmetric stretching vibrations of C–H, at 1414 cm^{-1} (δ_{C-H}), 1378 cm^{-1} (ω_{C-H}), 1243 cm^{-1} (ω_{C-H}), 1074 cm^{-1}, 1023 cm^{-1} (ν_{C-O}), and 830 cm^{-1} (ρ_{CH2}) [40]. Due to C=O and C–O–C units, respectively, characteristic absorption bands for the non-hydrolyzed ester groups–O–CO–CH_3 were observed at 1735 cm^{-1} and 1245 cm^{-1}. The characteristic bands for CS could be also assigned in the FTIR spectra of the samples, at 1556 cm^{-1} (band characteristic for amide II, ν_{N-H}), 2920 cm^{-1} (ν_{C-H}), and approximately 3300 cm^{-1} (wide band with νNH vibration overlapping with νO–H of polyvinyl alcohol). The characteristic band appearing at 1557 cm^{-1} assigned for δ_{NH} (amide II) vibration of the NH_2 group, and the band observable at 1654 cm^{-1} assigned for amide I ($\nu_{C=O}$) of O=C–NHR groups, were revealed to be diminished and slightly shifted in the FTIR spectra of the ternary and quaternary composite films (samples PVA/CS-2 and PVA/CS-3), indicating some interactions between CS and PVA that could be catalyzed by the presence of the additives in the samples. Additionally, the band appearing enlarged at 846 cm^{-1} in the multi-component composites (samples PVA/CS-1, PVA/CS-2, and PVA/CS-3) is the indicative of occurrence of hydrogen bonds formed between OH groups of the ingredients present in the composites.

From the FTIR spectra of the PVA/CS-2 and PVA/CS-3 samples, characteristic absorption bands at 1470 cm^{-1}, 1210 cm^{-1} (appearing as small shoulders), and 1140 cm^{-1} (appearing as a distinctive peak), were assigned to stretching vibration of the P–Ar, P–O and P–O–C, respectively. Another distinctive band confirming the presence of the PFR-3 additive into the composition of the PVA/CS-2 and PVA/CS-3 samples could be observed as a sharp peak near 755 cm^{-1}, which is attributed to the P–O–Ph group.

2.2. Surface Characteristics of the PVA/CS Composite Films

Due to their outstanding biocompatibility, resistance to bodily fluids, mechanical qualities, anticorrosive capability, and flexibility, titanium-based composites are preferred for use in biomedical applications [41,42]. However, their properties depend on the surface, which is related with the mixing capacity of nanomaterials with titanium oxide [43]. Accordingly, it has been suggested that the interaction between titanium oxide nanoparticles (TiO_2) and biopolymers (starch, gums, and chitosan) can aid in reducing the spontaneous agglomeration of nanoparticles, thereby enhancing the functional qualities of the composite [36].

PVA/CS composite films were analyzed using a profilometer. Microscopic images of the composite films were taken (histograms generated by the profilometer) and are presented in Figure 3, and the average roughness values were calculated (Table 2). According to the histograms and the microscopic images, it has been revealed that the addition of titanium dioxide to the polymeric matrix based on chitosan and PVA resulted in the production of composite films with a homogeneous composition and a surface devoid of cracks. The average roughness, Ra, representing the arithmetic mean of the highs and lows profiles, was in the interval of 65.5–570.3 nm. The surface roughness parameters decreased when introducing the PFR-3 in the PVA/CS matrix, according to the Ra value (Table 2), suggesting its contribution to the homogeneity and quality of the multi-component composites.

Table 2. Surface tension parameters of the test liquids used in contact angle measurements performed on PVA/CS composite films.

Sample	Roughness Ra (nm)	W_a (mN/m)	γ_{SV} (mN/m)	γ^P SV (mN/m)	γ^d SV (mN/m)	γ_{SL} (mN/m)
PVA/CS-0	570.3	65.24 (W) 62.38 (EG)	22.54	2.75	19.79	30.10
PVA/CS-1	208.1	67.65 (W) 77.08 (EG)	49.84	0.014	49.82	54.98
PVA/CS-2	78.6	75.76 (W) 77.16 (EG)	38.92	1.73	37.18	35.96
PVA/CS-3	65.5	65.54 (W) 61.84 (EG)	21.49	3.14	18.55	28.45

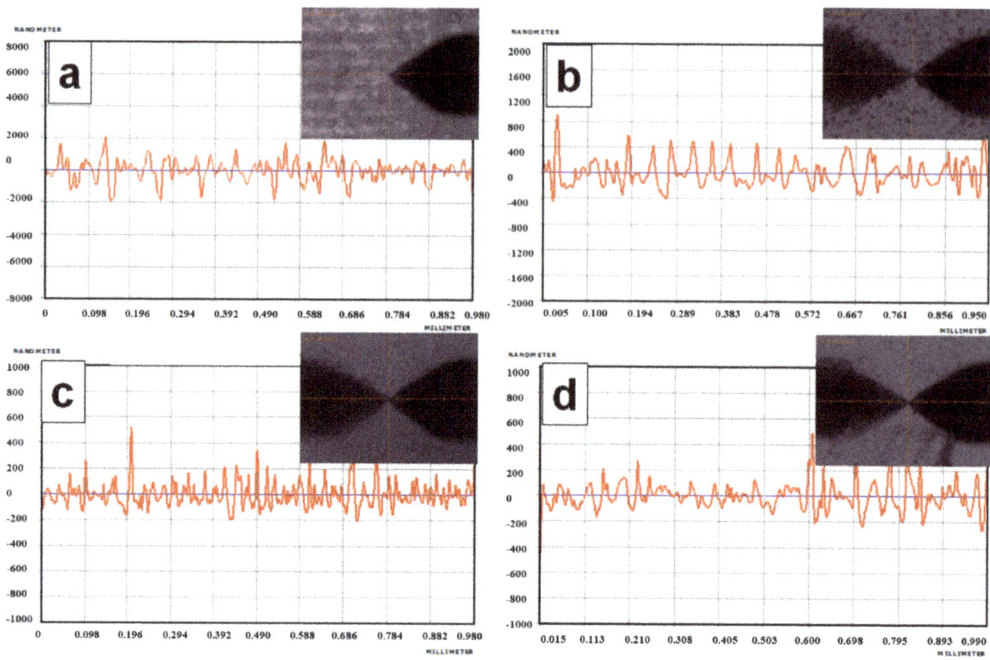

Figure 3. Microscopic images of the composite films and histograms generated by Alpha-Step D-500 Stilus profilometer (**a**) PVA/CS-0; (**b**) PVA/CS-1; (**c**) PVA/CS-2; (**d**) PVA/CS-3.

The wetting characteristics, such as work of adhesion (Wa), the total solid surface free energy (γ_{SV}), solid-liquid interfacial tension (γ_{SL}), etc., of the PVA/CS samples with respect to the W and EG were investigated. A summary of all contact angle study parameters is provided. Table 2 provides a summary of all the contact angle investigations' parameters. W and EG were used as the test liquids, whereas PVA samples were examined using contact angle measurements at room temperature. Figure 4 displays the results of measuring the contact angle for the PVA sample in W.

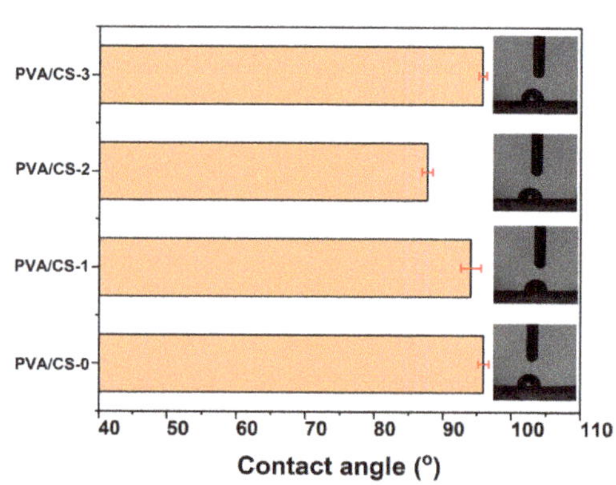

Figure 4. Contact angle of water on PVA/CS samples.

For water, the PVA/CS-0 sample exhibited a contact angle of 95.38° while by adding 0.16 g TiO_2 a slight decrease of the contact angle to 94.05° was observed in the case of the sample PVA/CS-1; a decrease was also observed in the roughness of the respective sample

(Table 2). The same behavior, in terms of contact angle and roughness decreases, has been noticed when adding 0.2172 g of polyphosphonate PFR-3, which resulted in a further decrease of the contact angle as a result of the presence of phenyl-phosphonate groups within the macromolecule of the polymeric additive that could influence the molecular hydrogen bonding; these results are consistent with the FTIR findings. Thus, in the case of PVA/CS-2, the contact angle value decreased to 87.67° due to the orientation of macromolecules and the reorganization of the polar groups from the surface of the composite sample. After continuing to add the PFR-3 additive (0.4344), the hydrophobic character was preserved as a result of the reduced concentration of OH groups, whereas the phenyl-phosphonate group's impact was disrupted by the presence of bulky, rigid phosphaphenanthrene groups.

When polyphosphonate PFR-3 was added into the PVA matrix, the work of adhesion (Wa) increased (PVA/CS-2 sample). Therefore, an increase in the work of adhesion may have resulted from the polyphosphonate's effective dispersion in the PVA/CS matrix. For the PVA/CS-0 sample, γ_{SV} was 22.54 mN/m and increased up to 38.92 mN/m in the PVA/CS-2 sample, whereas the sample containing TiO_2 nanoparticles presented the highest value of 49.84 mN/m, suggesting the different nature of the forces interacting on the surface of the different compositions. Table 2 revealed that samples PVA/CS-1 and PVA/CS-2 show lower polar surface parameters γ^P SV due to the presence of nano powders in the polymeric matrices. After continuing to add PFR-3 additive, a decrease of the Wa value and an increase of the polar surface parameters γ^P SV were achieved, probably due to the increase in the hydrophobicity of the surface of the sample PVA/CS-3. Additionally, from Table 2 it can be observed that the dispersive component values for the PVA/CS samples are greater than the polar component values, which leads to the conclusion that there are more hydrophobic groups than hydrophilic ones on the surface of the PVA/CS films.

2.3. Swelling Behavior and Morphology of PVA/CS Composite Films

The swelling behavior of the samples PVA/CS revealed that the PVA/CS-0 matrix without TiO_2 and PFR-3 additive exhibited the highest maximum swelling degree achieved at equilibrium, 834% (Figure 5). For the sample containing TiO_2, the maximum swelling degree achieved at equilibrium was 688%, whereas introducing the polyphosphonate additive into the systems further decreased the swelling behavior of the composite in the series, with the swelling degree achieved at equilibrium being 639% and 593% in the case of the PVA/CS-2 and PVA/CS-3 sample, respectively. Another observation relates to a higher initial water uptake in the first 10 min when the samples PVA/CS-0 and PVA/CS-1 reached the maximum swelling degree, whereas in the case of the samples PVA/CS-2 and PVA/CS-3, the maximum swelling degree was achieved after approximately 30 min.

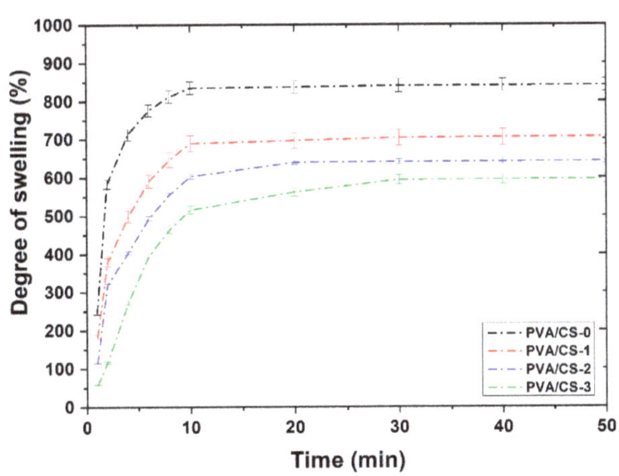

Figure 5. Swelling degree of PVA/CS composites.

The morphology of PVA/CS composite films was investigated by scanning electron microscopy. Microphotographs for PVA/CS-0 in comparison with the samples containing additive are presented in Figure 6. The morphology of the PVA/CS-0 sample appeared uniform and smooth, presenting a porous behavior. The aspect of the main polymeric matrix was also preserved in the samples containing either titanium dioxide nanoparticles or both TiO_2 and PFR-3 additive in their structure. A slight increase in the porosity could be observed in the sample PVA/CS-3 as the content of polyphosphonate PFR-3 increased.

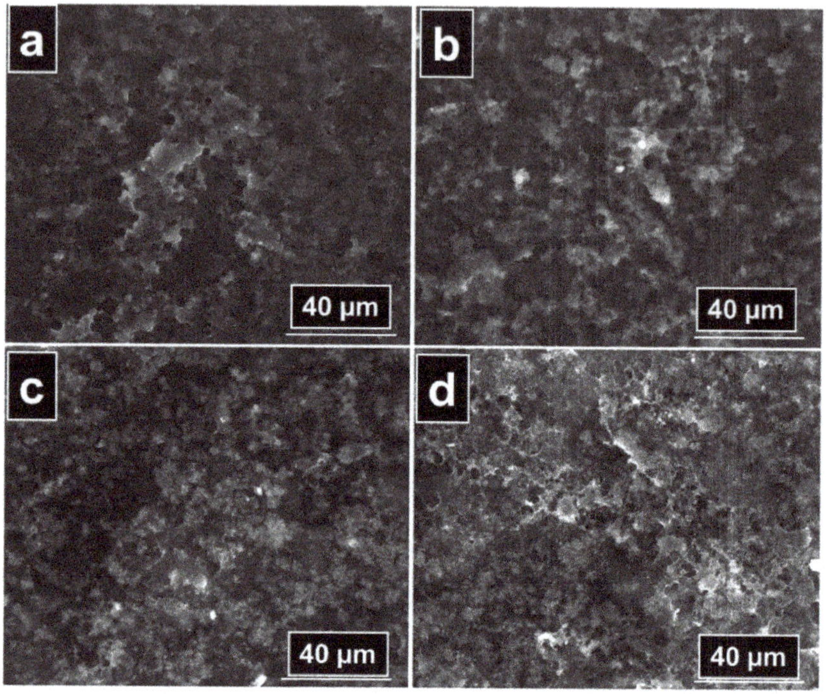

Figure 6. SEM images of PVA/CS-0 (**a**), PVA/CS-1 (**b**), PVA/CS-2 (**c**), and PVA/CS-3 (**d**) composite films.

2.4. *In Vitro Evaluation of Antimicrobial Potential of the PVA/CS Composite Films*

Due to the chitosan and the specific conditions of the culture medium and incubation parameters (37 °C/24 h), the tested matrices changed their disc-like shape and/or lost contact with the microbial culture, leading to a limitation in the interpretation of the antimicrobial activity.

However, qualitative evaluation of the antimicrobial potential of the PVA/CS composites films against Staphylococcus aureus ATCC 25923, *Methicillin-resistant Staphylococcus aureus* (MRSA) ATCC 43300, Escherichia coli ATCC 25922, Pseudomonas aeruginosa ATCC 27853, and Candida albicans ATCC 90028 (Table 3) was successful. Samples PVA/CS-2 and PVA/CS-3 contain both TiO_2 and PFR-3 additive in their structure. The presence of TiO_2 at a similar concentration as in PVA/CS-1 reflects the antimicrobial nature of the matrix, which was highlighted in PVA/CS-2 against Candida albicans and in PVA/CS-3 against Staphylococcus aureus, Pseudomonas aeruginosa, and Candida albicans. At the same time, inhibition of antimicrobial activity was observed against MRSA (matrices PVA/CS-2 and PVA/CS-3), Escherichia coli (matrices PVA/CS-2 and PVA/CS-3), Pseudomonas aeruginosa (PVA/CS-2), and Staphylococcus aureus (PVA/CS-2).

Table 3. Qualitative evaluation of the antimicrobial activity of APV-chitosan-PFR-TiO_2 matrices.

Sample	*Staphylococcus aureus*	*MRSA*	*Escherichia coli*	*Pseudomonas aeruginosa*	*Candida albicans*
PVA/CS-0	+	+	+	+	+
PVA/CS-1	+++	++	++	+	++
PVA/CS-2	+	+	+	+	+
PVA/CS-3	+	+	+	+	+

Legend: [+] = inhibition zone present; [++] = clear inhibition zone the size of the matrix disc, [+++] = clear zone of inhibition went beyond the imprint area of the matrix disc.

Comparing the results obtained for the PVA/CS-0 film (the sample containing only PVA and chitosan) and the functionalized composite films (PVA/CS-1, PVA/CS-2, PVA/CS-3), it can be seen that the PVA/CS-1 sample (containing PVA, CS, and TiO_2 nanopowder) has the best antimicrobial activity against both Gram-positive bacteria (*Staphylococcus aureus*, MRSA) and Gram-negative bacteria (*Escherichia coli, Pseudomonas aeruginosa*). The yeast *Candida albicans* also proved to be sensitive to the action of the compounds in the PVA/CS-3 sample.

In the case of the PVA/CS-2 and PVA/CS-3 composite films containing both TiO_2 nanoparticles and PFR-3 additive in their structure, the presence of TiO_2 at a similar concentration as in PVA/CS-1 highlights the antimicrobial effect of the sample on all strains tested. The inclusion of PRF in their composition did not significantly alter the antimicrobial potential of the tested materials.

The tests performed in our study confirmed the well-recognized antimicrobial capability of TiO_2. The antimicrobial potential and mechanisms of action of TiO_2 on bacterial cells are complex and diverse, exhibiting a broad spectrum of antimicrobial activity against bacteria (Gram-positive and Gram-negative), yeasts, and implicitly antibiotic-resistant microorganisms [44]. The mechanism thought to be responsible for the antimicrobial effect of TiO_2 is commonly associated with reactive oxygen species (ROS), which, when irradiated with bandgaps, produce photoinduced charges in the presence of O_2 [45]. Thus, the oxidative effect alters several cellular structures, but the first to be affected are the cell wall and cell membranes, leading to cell lysis and loss of cell integrity [46]. Specific studies on antimicrobial mechanisms have shown that microorganisms exposed to photocatalytic TiO_2 NPs exhibited cell inactivation at the level of the regulatory network and signal transduction, a significant reduction in respiratory chain activity, and inhibition of the ability to assimilate and transport iron and phosphorus. These processes, with the extensive cell wall and membrane changes, were the main factors explaining the biocidal activity of TiO_2 NPs [44].

2.5. *In Vitro Citotoxicity and Proliferation Assay*

The cytotoxicity tests undertaken in this step of the study (MTT and cell morphology analysis) have a screening value and are performed and analyzed in accordance with ISO-10993-5 standards recommendations for cytotoxicity as a first-intended and eliminatory requirement for biocompatibility.

PVA/CS composite films have been subjected to cytocompatibility tests that were performed by culturing MCF 7 in the physical presence of material samples (Figure 7). The MCF 7 cells are a standard line for cytocompatibility testing, as recommended by ASTM.

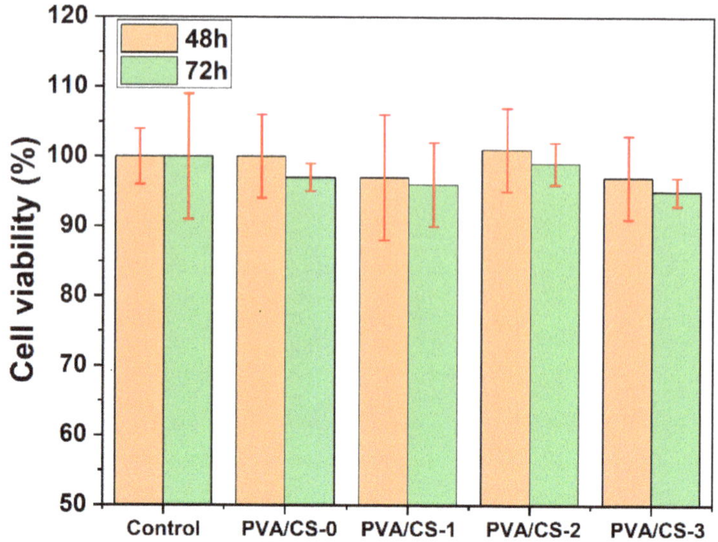

Figure 7. MTT test results for all PVA/CS composites.

Figure 7 shows the results of the MTT test, in which cell viability was expressed as an average and standard variation, obtained from the processing of absorbance values. Cell viability in control cultures was considered 100% ± SD and experimental values—percentage of control value. Figure 7 shows the cell viability after 48 and 72 h of direct cell interaction with the material samples.

The test revealed that all analyzed material samples decrease the cell viability by about 20% compared to control cultures. On the other hand, cell viability in the experimental cultures is approximatively the same, regardless of the TiO_2 ratio, which means that TiO_2 does not influence the biocompatibility of the materials. The decreasing of the cell viability in experimental cultures with the same value could be explained through the mechanical action of the direct contact of the membranes on the cells. The in vitro experimental condition for cell viability testing leads to the conclusion that polyvinyl alcohol/chitosan matrices with TiO_2 nanoparticles and/or polyphosphonate additive PFR-3 express a cytocompatibility, according to ISO 10993-5 recommendations.

2.6. Analysis of Cell Morphology

The effect of direct exposure of cultured cells to the sample material was assessed microscopically by phase contrast, overlapped to fluorescence images of DAPI-stained nuclear DNA. The images of the fixed and stained cells at 20× objective magnification are shown in Figure 8.

From the images showing the experimental cultures, no differences of cell morphology were observed between the different compositions of the samples. All analyzed cultures contain epithelial cells with a shape typical of the MCF 7 cell line. It is observed that at 72 h of culture, the cells grow in confluent monolayers, with a cell density without significant differences between the binary PVA/CS sample and multi-component PVA/CS films containing TiO_2 and/or PFR-3 additives.

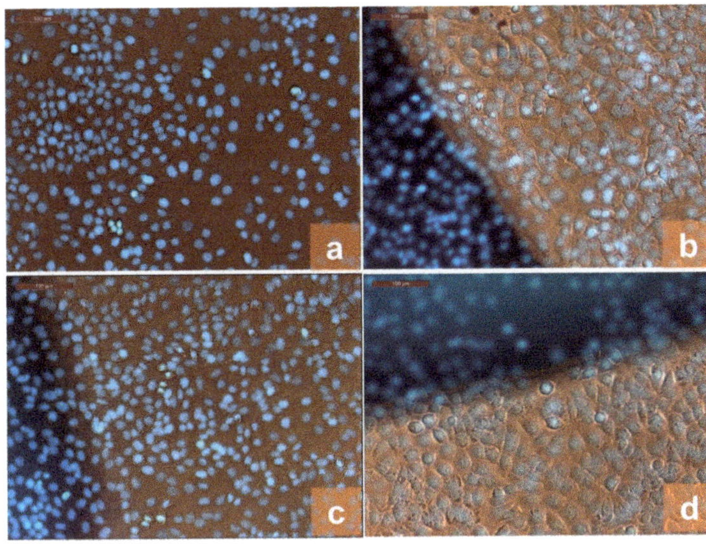

Figure 8. Microscopic images of DAPI-stained cells at 72 h of culturing in contact with PVA/CS-0 (**a**), PVA/CS-1 (**b**), PVA/CS-2 (**c**), and PVA/CS-0 (**d**) at 20× objective magnification.

3. Conclusions

PVA/CS composite films were obtained by the casting from solution technique, starting from appropriate amounts of PVA and CS, as polymer matrices, and TiO_2 and PFR-3 polyphosphonate were utilized as reinforcing additives. Infrared spectroscopy enabled structural validation and the identification of distinctive bands for certain functionalities in the prepared composite films. According to the Ra value, the surface roughness characteristics decreased when the PFR-3 was introduced into the PVA/CH matrix, indicating its contribution to the homogeneity and quality of the four-component composites. Scanning electron microscopy demonstrated a small increase in porosity in the PVA/CS-3 sample as the polyphosphonate PFR-3 level increased. In vitro evaluation of antimicrobial potential in the series confirmed the antimicrobial role of TiO_2 in the PVA/CS-TiO_2 matrices. Cell viability testing revealed that polyvinyl alcohol/chitosan matrices containing TiO_2 nanoparticles or TiO_2/polyphosphonate additive PFR-3 exhibit cytocompatibility in accordance with ISO 10993-5 guidelines.

4. Materials and Methods

4.1. Materials

Partially hydrolyzed PVA (Mowiol 26–88 (KURARAY POVAL 26–88)) with a molecular weight of 160,000 g/mol and an 87.7 percent degree of hydrolysis was supplied by ZAUBA (Munchen, Germany) and used as received. Chitosan hydrochloride (Mw = 302.11 kDa, DAC degree = 82%), titanium (IV) oxide nanopowder (21 nm primary particle size), and glacial acetic acid were purchased from Sigma-Aldrich (St. Louis, MO, USA). 3,3′-Diaminodiphenyl sulfone, 4-hydroxybenzaldehyde, tetrahydrofuran anhydrous (THF), *N*,*N*-dimethylformamide (DMF), and phenylphosphonic dichloride were purchased from Sigma-Aldrich (Taufkirchen, Germany).

9,10-Dihydro-9-oxa-10-phosphaphenanthrene-10-oxide (DOPO) was purchased from Chemos GmbH, Germany and dehydrated before use. Polyphosphonate additive (PFR-3) has been prepared in our laboratory according to an adopted procedure from a previous work [47], starting from 3,3′-diaminodiphenyl sulfone and 4-hydroxybenzaldehyde. All other reagents were used as received from commercial sources.

4.2. Preparation of PFR-3 Polyphophonate Additive

In the first step, a DOPO-containing bisphenol was obtained via condensation reaction of 20 mmol 4-hydroxybenzaldehyde (2.44 g) with 10 mmol 3,3′-diaminodiphenyl sulfone, in 20 mL THF, followed by the in-situ addition of DOPO (40 mmol, 8.64 g solved in 15 mL THF) to the newly formed azomethine groups. PFR-3 have been prepared through polycondensation reaction of equimolecular amounts of phenylphosphonic dichloride and DOPO-containing bisphenol, in DMF.

PFR-3, Yield: 95%.

FTIR (KBr, cm^{-1}): 3063 (=C–H), 1589 (C–C Aromatic), 1475 (P–Ar), 1353 and 1149 (O=S=O), 1200 (–P=O), 924 (P–O–Ar).

^{1}H NMR (400 MHz, DMSO-d6, δ, ppm): 8.18–8.10 (m, 4H), 7.49 (m, 2H), 7.47–7.39 (m, 5H), 7.33–7.05 (m, 20H), 6.95–6.66 (m, 6H), 6.65–6.62 (m, 3H), 5.60–5.50 (m, 1H), 5.30–5.00 (m, 1H).

4.3. Preparation of PVA/CS Films following the Casting Procedure

The first step in preparing these materials as composite films involved the treating of polyvinyl alcohol with oligophosphonate, a reaction which takes place in polar organic solvent, in this case DMF, working with concentrations of 5% PVA, by heating the mixtures at 95 °C for about 4–5 h. After the dissolution of the components is complete, the mixture is allowed to cool down to about 40 °C; then, the solution is poured into crystallizers, and the solvent is allowed to evaporate. The second stage consists in transferring the film obtained after drying in the crystallizer over a previously obtained chitosan solution (1% solution in water treated with glacial acetic acid 2%). After complete dissolution of the reaction mixture, the mixture was treated with an aqueous suspension of a predetermined amount of titanium dioxide nanoparticles, homogenized, and previously dispersed by ultrasound for 15 min. The resulting homogeneous composite mixture was poured into 10 × 10 cm^2 Teflon plates to slowly evaporate the solvent in ambient conditions. Afterwards, they were placed in the oven for an advanced drying treatment by heating to approximately 40–50 °C, under a vacuum, for a period of 8 h. The obtained films are stored and used for the characterization and testing of new materials. The complete data on the synthesis of composites in this series are presented in Table 1, whereas the schematic diagram of the preparation is given in Figure 9.

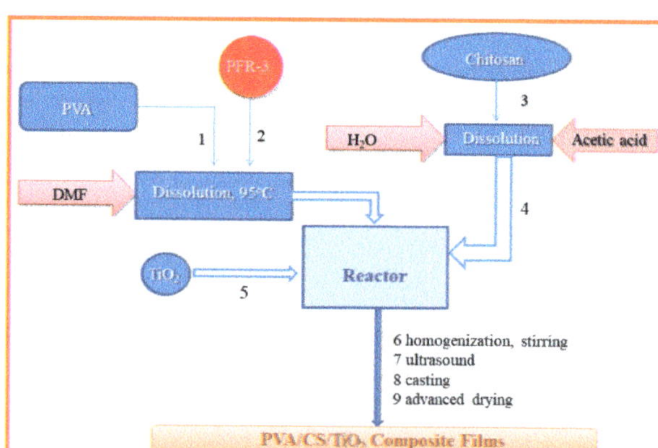

Figure 9. General scheme of the process for the preparation of films based on PVA and chitosan.

4.4. Composite Films Structure and Performances Characterization

4.4.1. Chemical Structure of PVA/CS/TiO_2 Composite Films

The chemical structure of the PVA/CS/TiO_2 films was analyzed using Bio-Rad 'FTS 135′ FTIR spectrometer equipped with a Specac "Golden Gate" ATR accessory. To record scans between 4000 and 500 cm^{-1} at a resolution of 4 cm^{-1}, a LUMOS Microscope Fourier Transform Infrared (FTIR) spectrophotometer (Bruker Optik GmbH, Ettlingen, Germany), equipped with an attenuated total reflection (ATR) device was used.

4.4.2. Roughness and Contact Angle Measurements

The roughness of the composite films was analyzed at a recording speed of 0.10 mm/s using the Tencor Alpha-Step D-500 High Sensitivity Stylus Profiler (KLA-Tencor Corporation, Milpitas, CA, USA). By applying a stylus force of 15 mg and a long-range cutoff filter of 60 µm, arithmetic mean roughness, Ra, was obtained, which is the average of the absolute value along the sampling length.

Two test liquids—double-distilled water (W) and ethylene glycol (EG)—were used to test their static contact angles on the composite film surfaces. The measurements were performed in triplicate, at room temperature, on a CAM 101 system (KSV Instruments, Helsinki, Finland) equipped with a liquid dispenser, video camera, and drop shape analysis software. The geometric mean approach was employed to calculate the surface tension parameters. This measurement allows for the calculation of parameters that describe the composite film's surface and its capacity for absorption, such as free surface energy (SV), solid-liquid interfacial tension (SL), and work of adhesion (Wa). Detailed experimental setup and calculations were described in a previous publication [34].

4.4.3. Degree of Swelling

Swelling behavior of the PVA/CS composite films prepared via the solution casting method was studied. Thus, dried samples of 0.5 × 0.5 cm^2 were weighted and submerged in 10 mL Millipore water in a closed bottle that was set in a thermostatic bath at 37 °C. The composite film samples were taken at predetermined intervals, and, after the extra water was removed with filter paper, the films weights were measured. The following equation was used to compute the swelling ratio, which represents the water absorption of each sample:

$$SD(\%) = \frac{W_t - W_d}{W_d} \times 100 \quad (1)$$

where SD (%) represents the amount of absorbed Millipore water; W_d—weight of the dry composite film; W_t—weight of the hydrated sample.

4.4.4. Microscopic Morphology

Microscopic examinations have been performed on Environmental Scanning Electron Microscope Type Quanta 200, operating at 10 kV with secondary electrons in low vacuum mode (LFD detector). The composite samples were fractured, and their cross-section surfaces were analyzed using scanning electron microscopy (SEM). An Energy Dispersive X-ray (EDX) system is a feature of the Quanta 200 microscope that allows for qualitative, quantitative analysis, and elemental mapping.

4.4.5. Evaluation of Antimicrobial Activity

Diffusimetric determinations were performed to highlight the antimicrobial activity of the PVA/CS samples containing PFR-3 and/or TiO_2 additives. This technique is common for such tests and is based on the principle of contact of the test matrices with the surface of a culture medium inoculated with different microbial species. The antimicrobial tests were performed with standardized strains: *Staphylococcus aureus* ATCC 25923, *MRSA* ATCC 43300, *Escherichia coli* ATCC 25922, *Pseudomonas aeruginosa* ATCC 27853, and *Candida albicans* ATCC 90028. Standard microbial suspensions with a density of 0.5 McFarland, prepared with a spectrophotometer, were used. An aliquot (500 µL) of the bacterial suspension was applied to the surface of the Mueller-Hinton aggregate culture medium (BioRad, Taufkirchen, Germany) using an exudate swab. After drying for 10 min in a thermostat with the lid open, the test matrices were spread on the surface of the medium. The results were evaluated after incubation at 37 °C for 24 h. The antimicrobial activity was highlighted by identifying the areas of microbial inhibition formed upon contact with the tested matrices.

4.4.6. Evaluation of Cytotoxicity and Cell Morphology

The PVA/CS composite films were cut into 4 mm Ø discs, decontaminated with 70% aqueous ethyl alcohol solution for 20 min and then repeatedly washed in sterile ultrapure water and HBSS (Hank's Balanced Salt Solution) to remove any remaining contaminants. Following washing, the samples were equilibrated for 24 h in culture media (DMEM Ham/F12, supplemented with 10% fetal bovine serum and 1% antibiotic mixture). For equilibration, 0.5 mL of culture media were utilized for each piece of 4 mm Ø material. Using the MTT assay, the cytotoxicity of the growth medium and the material sample were both evaluated. Additionally, phase-contrast microscopy and fluorescence microscopy were used to examine the direct effects of the material samples on cell morphology. For biocompatibility study, the human epithelial cell line MCF 7, at the density of 20×10^4/well cells, was plated in 48-well culture plates and DMEM Ham/F12 culture medium supplemented with 10% bovine fetal serum and 1% mixture of Penicillin-Streptomycin-Neomycin antibiotics (all for in vitro use). Thus, material samples or the media of their extraction were put over the cell monolayers in the 48-well culture plates in order to assess the viability, density, and morphology of the cells.

The MTT technique uses tetrazolium salts as an oxidized substrate for mitochondrial dehydrogenases to measure the activity of cellular metabolism. The basic idea behind the procedure is that the yellow MTT compound is reduced into an insoluble purple (formazan) product. The formazan salts are solubilized with isopropyl alcohol to obtain a blue-violet solution, the intensity of which is directly proportional to the number of living cells in the culture. Briefly, for MTT experiments, the culture medium of the cell cultures was replaced with work MTT solution of 0.25 mg/mL and incubated with cells for 3 h. The MTT solution was replaced with an equal volume of isopropyl alcohol to solubilize the formazan crystals. The absorbance of the formazan solution was measured spectrophotometrically at 570 nm, using a TECAN UV/VIS plate reader. An equivalent amount of isopropanol was used as the blank reference sample. The cell viability was calculate applying the equation:

$$Cell\ viability\ (\%) = \frac{A_{Sample}}{A_{control}} \times 100 \quad (2)$$

Each membrane sample was evaluated in triplicate, at 48 and 72 h of cell incubation with materials. The cultures developed in the absence of the material were used as growth control.

The cells were fixed in 4% paraformaldehyde solution and permeabilized for 30 min at +40 °C, in 0.05% Triton X 100 in PBS. After 72 h of interaction between the cells and material, the DAPI (2-(4-amidinophenyl)-1H-indole-6-carboxamidines) staining protocol was carried out for the cell morphology examination. PBS was used to properly clean the permeabilized cells before they were incubated with DAPI 10 μg/mL work solution for 30 min at room temperature. After being washed in PBS, stained cells were examined using fluorescence microscopy at λex 340 nm and λem 488 nm. To emphasize the cell nucleus and form, phase contrast was overlapped with the fluorescence.

Author Contributions: Conceptualization, C.H., D.S. and T.V.-B.; methodology, C.H., T.V.-B., C.M.R. and M.B.; validation, D.S., C.H., T.V.-B., M.A., C.M.R. and A.A.E.; formal analysis, C.H. and T.V.-B.; investigation, C.M.R., D.S., M.B., T.V.-B. and C.H.; resources, M.A. and A.A.E.; writing—original draft preparation, T.V.-B., D.S., C.M.R. and C.H.; writing—review and editing, T.V.-B., D.S., C.M.R. and C.H.; project administration, A.A.E. and M.A.; funding acquisition, A.A.E. and M.A. All authors have read and agreed to the published version of the manuscript.

Funding: This research received no external funding.

Institutional Review Board Statement: Not applicable.

Informed Consent Statement: Not applicable.

Data Availability Statement: The data that support the findings of the current study are listed within the article.

Acknowledgments: The authors acknowledge the financial support of this research through the project "Partnerships for knowledge transfer in the field of polymer materials used in biomedical engineering" ID P_40_443, Contract no. 86/8.09.2016, SMIS 105689, co-financed by the European Regional Development Fund by the Competitiveness Operational Program 2014–2020, Axis 1 Research, Technological Development and Innovation in support of economic competitiveness and business development, Action 1.2.3 Knowledge Transfer Partnerships. The authors Tăchiță Vlad-Bubulac and Diana Serbezeanu acknowledge the financial support through CNCSIS– UEFISCSU, Project Number PN-III-P1-1.1-TE2019-0639 nr. 89/03.09.2020.

Conflicts of Interest: The authors declare no conflict of interest.

References

1. Phelan, A.; Meissner, K.; Humphrey, J.; Ross, H. Plastic pollution and packaging: Corporate commitments and actions from the food and beverage sector. *J. Clean. Prod.* **2022**, *331*, 129827. [CrossRef]
2. Geyer, R.; Jambeck Jenna, R.; Law Kara, L. Production, use, and fate of all plastics ever made. *Sci. Adv.* **2017**, *3*, e1700782. [CrossRef]
3. Walker, T.R.; McGuinty, E.; Charlebois, S.; Music, J. Single-use plastic packaging in the Canadian food industry: Consumer behavior and perceptions. *Humanit. Soc. Sci. Commun.* **2021**, *8*, 80. [CrossRef]
4. Ncube, L.K.; Ude, A.U.; Ogunmuyiwa, E.N.; Zulkifli, R.; Beas, I.N. Environmental Impact of Food Packaging Materials: A Review of Contemporary Development from Conventional Plastics to Polylactic Acid Based Materials. *Materials* **2020**, *13*, 4994. [CrossRef] [PubMed]
5. Cruz, R.M.S.; Alves, V.; Khmelinskii, I.; Vieira, M.C. Chapter 2—New food packaging systems. In *Food Packaging and Preservation*; Grumezescu, A.M., Holban, A.M., Eds.; Academic Press: Cambridge, MA, USA, 2018; pp. 63–85.
6. Da Rocha, M.; de Souza, M.M.; Prentice, C. Chapter 9—Biodegradable films: An alternative food packaging. In *Food Packaging and Preservation*; Grumezescu, A.M., Holban, A.M., Eds.; Academic Press: Cambridge, MA, USA, 2018; pp. 307–342.
7. Madera-Santana, T.J.; Freile-Pelegrín, Y.; Azamar-Barrios, J.A. Physicochemical and morphological properties of plasticized poly(vinyl alcohol)-agar biodegradable films. *Int. J. Biol. Macromol.* **2014**, *69*, 176–184. [CrossRef]
8. Siracusa, V.; Rocculi, P.; Romani, S.; Rosa, M.D. Biodegradable polymers for food packaging: A review. *Trends Food Sci. Technol.* **2008**, *19*, 634–643. [CrossRef]
9. Lombardi, M.; Rana, R.; Fellner, J. Material flow analysis and sustainability of the Italian plastic packaging management. *J. Clean. Prod.* **2021**, *287*, 125573. [CrossRef]
10. Van Eygen, E.; Laner, D.; Fellner, J. Circular economy of plastic packaging: Current practice and perspectives in Austria. *Waste Manag.* **2018**, *72*, 55–64. [CrossRef] [PubMed]
11. MacArthur, E. Beyond plastic waste. *Science* **2017**, *358*, 843. [CrossRef]
12. Guillard, V.; Gaucel, S.; Fornaciari, C.; Angellier-Coussy, H.; Buche, P.; Gontard, N. The Next Generation of Sustainable Food Packaging to Preserve Our Environment in a Circular Economy Context. *Front. Nutr.* **2018**, *5*, 121. [CrossRef] [PubMed]
13. Girotto, F.; Alibardi, L.; Cossu, R. Food waste generation and industrial uses: A review. *Waste Manag.* **2015**, *45*, 32–41. [CrossRef] [PubMed]
14. Philp, J.C.; Bartsev, A.; Ritchie, R.J.; Baucher, M.-A.; Guy, K. Bioplastics science from a policy vantage point. *New Biotechnol.* **2013**, *30*, 635–646. [CrossRef]
15. Cammarelle, A.; Bimbo, F.; Lombardi, M.; Viscecchia, R. Health and eco-innovations in food packaging. In *Sustainable Packaging*; Muthu, S.S., Ed.; Springer: Singapore, 2021; pp. 31–70.
16. Flórez, M.; Guerra-Rodríguez, E.; Cazón, P.; Vázquez, M. Chitosan for food packaging: Recent advances in active and intelligent films. *Food Hydrocoll.* **2022**, *124*, 107328. [CrossRef]
17. Souza, V.G.L.; Pires, J.R.A.; Rodrigues, C.; Coelhoso, I.M.; Fernando, A.L. Chitosan Composites in Packaging Industry-Current Trends and Future Challenges. *Polymers* **2020**, *12*, 417. [CrossRef]
18. Shahbaz, U.; Basharat, S.; Javed, U.; Bibi, A.; Yu, X.B. Chitosan: A multipurpose polymer in food industry. *Polym. Bull.* **2022**. [CrossRef]
19. Babaei-Ghazvini, A.; Acharya, B.; Korber, D.R. Antimicrobial Biodegradable Food Packaging Based on Chitosan and Metal/Metal-Oxide Bio-Nanocomposites: A Review. *Polymers* **2021**, *13*, 2790. [CrossRef]
20. Wang, K.; Du, L.; Zhang, C.; Lu, Z.; Lu, F.; Zhao, H. Preparation of chitosan/curdlan/carboxymethyl cellulose blended film and its characterization. *J. Food Sci. Technol.* **2019**, *56*, 5396–5404. [CrossRef]
21. Jridi, M.; Hajji, S.; Ayed, H.B.; Lassoued, I.; Mbarek, A.; Kammoun, M.; Souissi, N.; Nasri, M. Physical, structural, antioxidant and antimicrobial properties of gelatin–chitosan composite edible films. *Int. J. Biol. Macromol.* **2014**, *67*, 373–379. [CrossRef]

22. Liu, Y.; Wang, S.; Zhang, R. Composite poly(lactic acid)/chitosan nanofibrous scaffolds for cardiac tissue engineering. *Int. J. Biol. Macromol.* **2017**, *103*, 1130–1137. [CrossRef] [PubMed]
23. Wang, H.; Zhang, R.; Zhang, H.; Jiang, S.; Liu, H.; Sun, M.; Jiang, S. Kinetics and functional effectiveness of nisin loaded antimicrobial packaging film based on chitosan/poly(vinyl alcohol). *Carbohydr. Polym.* **2015**, *127*, 64–71. [CrossRef] [PubMed]
24. Serbezeanu, D.; Bercea, M.; Butnaru, M.; Enache, A.A.; Rîmbu, C.M.; Vlad-Bubulac, T. Development of histamine reinforced poly(vinyl alcohol)/chitosan blended films for potential biomedical applications. *J. Appl. Polym. Sci.* **2022**, *139*, 51912. [CrossRef]
25. Bonilla, J.; Fortunati, E.; Atarés, L.; Chiralt, A.; Kenny, J.M. Physical, structural and antimicrobial properties of poly vinyl alcohol–chitosan biodegradable films. *Food Hydrocoll.* **2014**, *35*, 463–470. [CrossRef]
26. Choo, K.; Ching, Y.C.; Chuah, C.H.; Julai, S.; Liou, N.-S.J.M. Preparation and characterization of polyvinyl alcohol-chitosan composite films reinforced with cellulose nanofiber. *Materials* **2016**, *9*, 644. [CrossRef] [PubMed]
27. Suganthi, S.; Vignesh, S.; Kalyana Sundar, J.; Raj, V. Fabrication of PVA polymer films with improved antibacterial activity by fine-tuning via organic acids for food packaging applications. *Appl. Water Sci.* **2020**, *10*, 100. [CrossRef]
28. Nurly, H.; Yan, Q.; Song, B.; Shi, Y. Effect of carbon nanotubes reinforcement on the polyvinyl alcohol—Polyethylene glycol double-network hydrogel composites: A general approach to shape memory and printability. *Eur. Polym. J.* **2019**, *110*, 114–122. [CrossRef]
29. Halonen, N.; Pálvölgyi, P.S.; Bassani, A.; Fiorentini, C.; Nair, R.; Spigno, G.; Kordas, K. Bio-Based Smart Materials for Food Packaging and Sensors—A Review. *Front. Mater.* **2020**, *7*, 82. [CrossRef]
30. El-Naggar, M.E.; El-Newehy, M.H.; Aldalbahi, A.; Salem, W.M.; Khattab, T.A. Immobilization of anthocyanin extract from red-cabbage into electrospun polyvinyl alcohol nanofibers for colorimetric selective detection of ferric ions. *J. Environ. Chem. Eng.* **2021**, *9*, 105072. [CrossRef]
31. Emamhadi, M.A.; Sarafraz, M.; Akbari, M.; Thai, V.N.; Fakhri, Y.; Linh, N.T.T.; Mousavi Khaneghah, A. Nanomaterials for food packaging applications: A systematic review. *Food Chem. Toxicol.* **2020**, *146*, 111825. [CrossRef]
32. Chang, X.; Hou, Y.; Liu, Q.; Hu, Z.; Xie, Q.; Shan, Y.; Li, G.; Ding, S. Physicochemical and antimicrobial properties of chitosan composite films incorporated with glycerol monolaurate and nano-TiO_2. *Food Hydrocoll.* **2021**, *119*, 106846. [CrossRef]
33. Paduraru, A.; Ghitulica, C.; Trusca, R.; Surdu, V.A.; Neacsu, I.A.; Holban, A.M.; Birca, A.C.; Iordache, F.; Vasile, B.S. Antimicrobial Wound Dressings as Potential Materials for Skin Tissue Regeneration. *Materials* **2019**, *12*, 1859. [CrossRef]
34. Hamciuc, C.; Vlad-Bubulac, T.; Serbezeanu, D.; Hamciuc, E.; Aflori, M.; Lisa, G.; Anghel, I.; Şofran, I.-E.; Trofin, A. Tailoring thermal and flame retardant properties via synergistic effect in polyvinyl alcohol nanocomposites based on polyphosphonate and/or SiO_2 nanoparticles. *Compos. Part C* **2020**, *3*, 100063. [CrossRef]
35. Ali, H.M.; Babar, H.; Shah, T.R.; Sajid, M.U.; Qasim, M.A.; Javed, S. Preparation Techniques of TiO_2 Nanofluids and Challenges: A Review. *Appl. Sci.* **2018**, *8*, 587. [CrossRef]
36. Díaz-Visurraga, J.; Meléndrez, M.F.; García, A.; Paulraj, M.; Cárdenas, G. Semitransparent chitosan-TiO_2 nanotubes composite film for food package applications. *J. Appl. Polym. Sci.* **2010**, *116*, 3503–3515. [CrossRef]
37. Shahrousvand, M.; Hoseinian, M.S.; Ghollasi, M.; Karbalaeimahdi, A.; Salimi, A.; Tabar, F.A. Flexible magnetic polyurethane/Fe_2O_3 nanoparticles as organic-inorganic nanocomposites for biomedical applications: Properties and cell behavior. *Mater. Sci. Eng. C* **2017**, *74*, 556–567. [CrossRef]
38. Yan, J.; Li, M.; Wang, H.; Lian, X.; Fan, Y.; Xie, Z.; Niu, B.; Li, W. Preparation and property studies of chitosan-PVA biodegradable antibacterial multilayer films doped with Cu_2O and nano-chitosan composites. *Food Control* **2021**, *126*, 108049. [CrossRef]
39. Guazzelli, E.; Lusiani, N.; Monni, G.; Oliva, M.; Pelosi, C.; Wurm, F.R.; Pretti, C.; Martinelli, E. Amphiphilic Polyphosphonate Copolymers as New Additives for PDMS-Based Antifouling Coatings. *Polymers* **2021**, *13*, 3414. [CrossRef]
40. Mansur, H.S.; Sadahira, C.M.; Souza, A.N.; Mansur, A.A.P. FTIR spectroscopy characterization of poly (vinyl alcohol) hydrogel with different hydrolysis degree and chemically crosslinked with glutaraldehyde. *Mater. Sci. Eng. C* **2008**, *28*, 539–548. [CrossRef]
41. Gerhardt, L.C.; Jell, G.M.; Boccaccini, A.R. Titanium dioxide (TiO(2)) nanoparticles filled poly(D,L lactid acid) (PDLLA) matrix composites for bone tissue engineering. *J. Mater. Sci. Mater. Med.* **2007**, *18*, 1287–1298. [CrossRef]
42. Fei Yin, Z.; Wu, L.; Gui Yang, H.; Hua Su, Y. Recent progress in biomedical applications of titanium dioxide. *PCCP* **2013**, *15*, 4844–4858. [CrossRef]
43. Marszewski, M.; Jaroniec, M. Scaffold-assisted synthesis of crystalline mesoporous titania materials. *RSC Adv.* **2015**, *5*, 61960–61972. [CrossRef]
44. De Dicastillo, C.L.; Correa, M.G.; Martínez, F.B.; Streitt, C.; Galotto, M.J. Antimicrobial effect of titanium dioxide nanoparticles. In *Antimicrobial Resistance*; Mareș, M., Lim, S.H.E., Lai, K.-S., Cristina, R.-T., Eds.; InTechOpen: London, UK, 2020.
45. Verdier, T.; Coutand, M.; Bertron, A.; Roques, C. Antibacterial Activity of TiO2 Photocatalyst Alone or in Coatings on E. coli: The Influence of Methodological Aspects. *Coatings* **2014**, *4*, 670–686. [CrossRef]
46. Khezerlou, A.; Alizadeh-Sani, M.; Azizi-Lalabadi, M.; Ehsani, A. Nanoparticles and their antimicrobial properties against pathogens including bacteria, fungi, parasites and viruses. *Microb. Pathog.* **2018**, *123*, 505–526. [CrossRef] [PubMed]
47. Hamciuc, C.; Vlad-Bubulac, T.; Serbezeanu, D.; Carja, I.-D.; Hamciuc, E.; Lisa, G.; Pérez, V.F. Environmentally friendly fire-resistant epoxy resins based on a new oligophosphonate with high flame retardant efficiency. *RSC Adv.* **2016**, *6*, 22764–22776. [CrossRef]

gels

Article

Metallic Strontium as a Precursor of the $Al_2O_3/SrCO_3$ Xerogels Obtained by the One-Pot Sol–Gel Method

Eliza Romanczuk-Ruszuk [1], Bogna Sztorch [2], Zbigniew Oksiuta [1] and Robert E. Przekop [2,*]

[1] Institute of Biomedical Engineering, Faculty of Mechanical Engineering, Bialystok University of Technology, Wiejska 45C Street, 15-351 Bialystok, Poland; e.romanczuk@pb.edu.pl (E.R.-R.); z.oksiuta@pb.edu.pl (Z.O.)
[2] Centre for Advanced Technologies, Adam Mickiewicz University in Poznan, Uniwersytetu Poznanskiego 10 Street, 61-614 Poznan, Poland; bogna.sztorch@amu.edu.pl
* Correspondence: rprzekop@amu.edu.pl or r.przekop@gmail.com

Abstract: Two series of binary xerogel systems of Sr/Al with molar ratios of 0.1, 0.25, 0.5, and 1.0 were synthesized by the sol–gel technique with metallic strontium component as a precursor. The influence of the metallic precursor on the properties of the final xerogel was determined. The properties of the gels were determined on the basis of X-ray powder diffraction (XRD), thermogravimetric analysis (TGA), low temperature nitrogen adsorption, transmission, and scanning electron microscopy with Energy Dispersive X-ray Spectroscopy (TEM, SEM, and SEM/EDS). The $Al_2O_3/SrCO_3$ xerogels were tested as supports for platinum catalysts. Hydrogen chemisorption was used to determine the platinum dispersion of the Pt/Al_2O_3-$SrCO_3$ systems. The original method of synthesis allows to obtain highly dispersed and stable strontium carbonate phases that allow for obtaining a high (42–50%) dispersion of platinum nanoparticles.

Keywords: sol–gel; metallic precursor; $SrCO_3$; xerogels; alumina; binary gels; one-pot

Citation: Romanczuk-Ruszuk, E.; Sztorch, B.; Oksiuta, Z.; Przekop, R.E. Metallic Strontium as a Precursor of the $Al_2O_3/SrCO_3$ Xerogels Obtained by the One-Pot Sol–Gel Method. *Gels* **2022**, *8*, 473. https://doi.org/10.3390/gels8080473

Academic Editors: Francesco Caridi, Giuseppe Paladini and Andrea Fiorati

Received: 29 June 2022
Accepted: 25 July 2022
Published: 27 July 2022

1. Introduction

Xerogels are described as porous systems obtained by drying wet gels and retaining their porous structure after drying [1]. The advantages of xerogels are thermal stability, large surface area, and porosity. Xerogels are biocompatible and nontoxic and can be easily modified [2]. Silica xerogels are the most popular xerogels and can be used as fillers for polymer composites. They are characterized by low density, high thermal stability, low thermal conductivity, and good hydrophobic properties [3].

It is well-known that metal oxides characterized by large specific surfaces and thermal stability, such as Al_2O_3, SiO_2, TiO_2, and CeO_2, can be used as catalyst supports. In addition to the well-recognized oxides, oxides of the alkaline earth metal groups are also used for catalytic processes. Strontium oxide (SrO) is an example of an oxide of the alkaline earth metal group. It can catalyze numerous synthetic reactions, like nitroaldol reactions, selective oxidation of propane, and oxidative coupling of methane [4–6]. Strontium has a lower electronegativity among metals from Group II of the Periodic Table. Therefore, strontium oxide has a higher basic strength compared to other group II oxides. The electronegativity increases in the order MgO < CaO < SrO < BaO [4]. Nevertheless, there is limited research on using SrO as a catalyst. The problem is related to the preparation and use of SrO as the base catalyst, as there are difficulties in the preparation of SrO [7].

Strontium carbonate ($SrCO_3$) takes advantage of the production of X-ray tubes, hard magnets, ceramics, and special glasses [8]. Before or during the high-temperature production processes, strontium carbonate ($SrCO_3$) is decomposed into strontium oxide (SrO) and carbon dioxide (CO_2). Strontium carbonate decomposes into strontium oxide and carbon dioxide during calcination at a temperature above 1000 °C (1273 K) in atmospheric conditions [9,10]. Recently, much work has focused on the use of carbonate as a catalyst support. It is worth noting that carbonate is not a typical catalyst support. In the work of

Omat et al. [11] it was found that cobalt deposited on $SrCO_3$ showed exceptional activity for the dry reforming of methane. Another study investigated the selectivity and reactivity of cobalt deposited on an alkaline earth metal carbonate for the catalytic preferential oxidation of CO [12,13]. The catalytic properties of $Co/SrCO_3$ were greater compared to cobalt catalysts supported on popular metal oxides. Moreover, $Co/SrCO_3$ showed the best productivity. In Iida et al. [14], the reforming of the catalytic activity of toluene of $Ru/SrCO_3$-Al_2O_3 and Ru/Al_2O_3 catalysts in steam was compared. The research shows that the carbonate catalyst shows higher activity.

The aim of this study was to determine the properties of $Al_2O_3/SrCO_3$ xerogels obtained from metallic strontium. Therefore, a new synthesis approach using an $Al_2O_3/SrCO_3$ using the sol–gel method is presented. In the proposed method, the precursors of one of the matrix components were used in metallic form. Our previous work has indicated that the introduction of a metallic precursor changes the properties of the final product [15].

The advantage of using the sol–gel synthesis is the purity of the materials obtained without inorganic admixtures and residual ion content [16].

2. Results and Discussion

2.1. X-ray Powder Diffraction

Figure 1 presents X-ray diffraction patterns of $Al_2O_3/SrCO_3$ systems annealed in 500 °C (773 K) with different strontium (Sr) content and Sr1.0 annealed in different temperatures. The XRD results of $Al_2O_3/SrCO_3$ samples with a different amount of strontium indicate only the presence of $SrCO_3$ in the structure. The appearance of $SrCO_3$ in the XRD results can be explained by the strontium acetate decomposition, which takes place at 400–480 °C (673–753 K) [17]. As the Sr content increases, there is a change in the broadening and intensity of the peaks (Figure 1a). As expected, the highest intensity of the peaks in the sample with an Al/Sr molar ratio of 1.0 was detected. The XRD plot of the Sr1.0 annealed at different temperatures varies with the temperature operated (Figure 1b). XRD analysis of Sr1.0 materials annealed at 1000 °C (1273 K), 1150 °C (1423 K), and 1300 °C (1573 K) shows the same phases, but the peaks differ in intensity. The difference in the intensity of the peaks after different annealing times may be due to the amount of $SrCO_3$ crystallized [18]. In each of these materials one or more of the following compounds was identified: $SrCO_3$, Al_2O_3, SrO, or $SrAl_2O_4$. The appearance of $SrAl_2O_4$ can be explained by the interfacial reaction between $SrCO_3$ and Al_2O_3 at a temperature above 500 °C (773 K) and the diffusion of Al^{3+} ions in the $SrCO_3$ lattice, which causes the formation of $SrAl_2O_4$ [19]. Moreover, the presence of $SrAl_2O_4$ may be related to the decomposition of strontium carbonate to SrO at higher temperatures and to the reaction with Al_2O_3, which is represented as follows [20]:

$$SrCO_3 \rightarrow SrO + CO_2 \quad (1)$$

$$SrO + Al_2O_3 \rightarrow SrAl_2O_4 \quad (2)$$

The presence of SrO in samples at temperatures above 1000 °C (1273 K) can be explained by the fact that in the temperature range 900–1175 °C (1173–1448 K) the equilibrium state moves toward the carbonate decomposition [21]. The SrO-CO_2-$SrCO_3$ equilibrium diagram by Rhodes et al. [22] shows that at temperatures below 900–1000 °C (1173–1272 K) the equilibrium moves toward the formation of $SrCO_3$; furthermore, at higher temperatures the possibility of $SrCO_3$ decomposition increases [21–23].

2.2. SEM, TEM, and EDS Analysis

Figure 2 shows the surface of the samples with different molar ratios of Sr before annealing. Note that the surfaces of the tested materials vary depending on the amount of Sr. The SEM image of the Sr0.1 sample surface presents a granular structure. The Sr1.0 sample has a completely different structure compared to the other tested materials. The SEM EDS analysis (Figure 2e) shows that the elements in the examined xerogels systems decompose regularly. Oxygen clusters and a higher strontium content are observed in the Sr1.0 sample compared to the other tested xerogels.

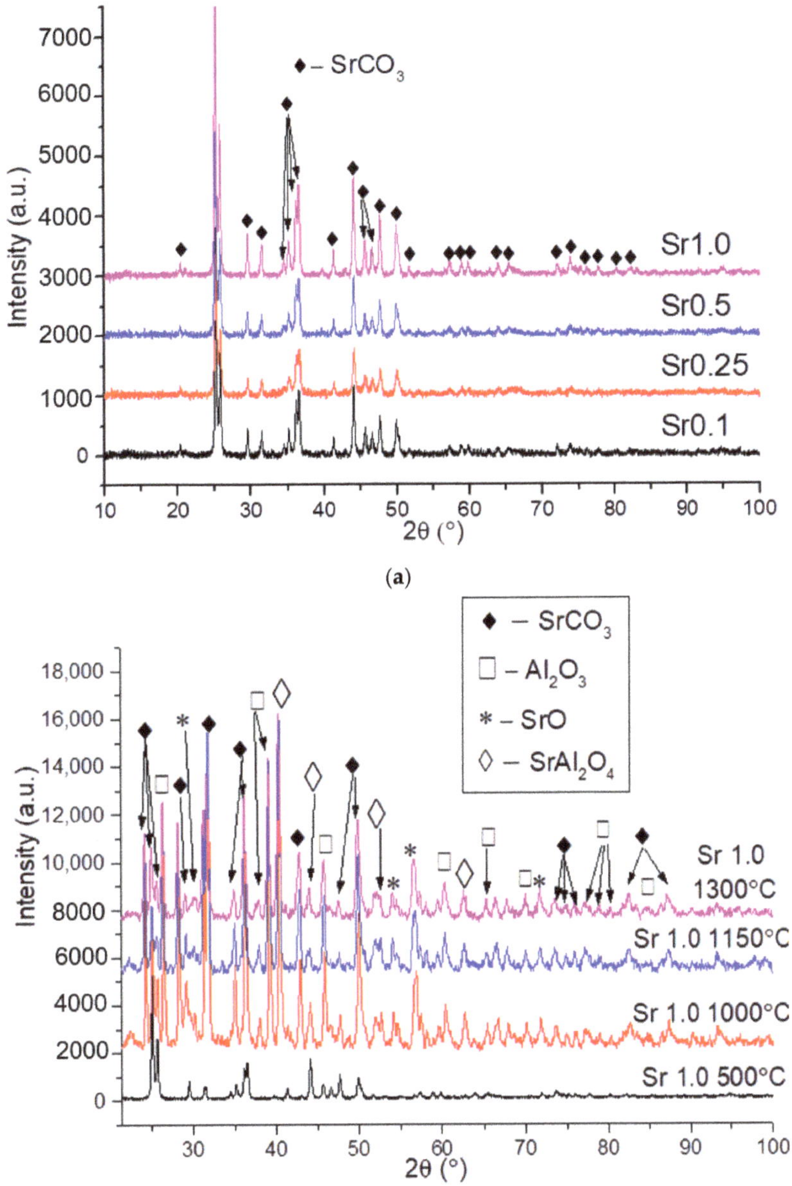

(**a**)

(**b**)

Figure 1. XRD patterns of $Al_2O_3/SrCO_3$ systems: (**a**) annealed at 500 °C (773 K) with Sr molar ratios of 0.1, 0.25, 0.5 and 1.0; (**b**) Sr 1.0 molar ratio at different temperature.

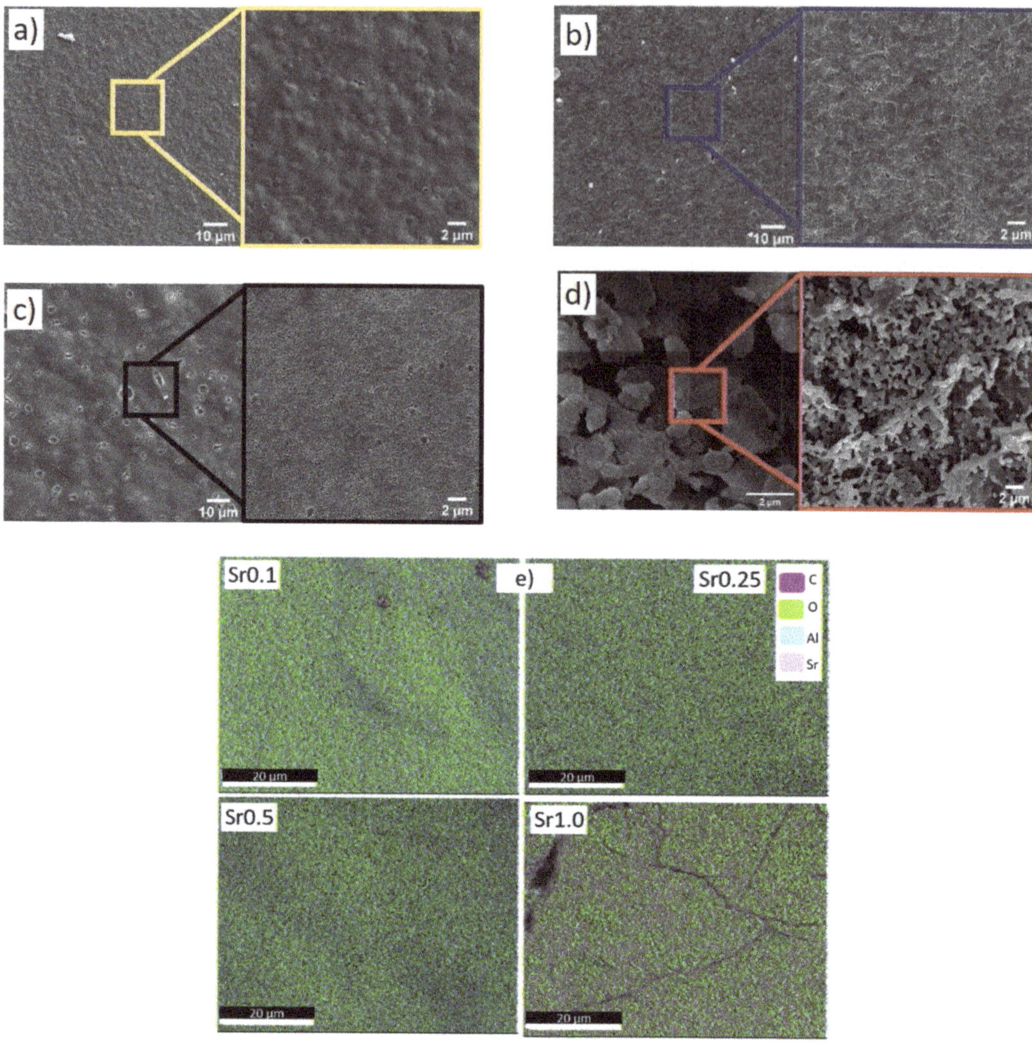

Figure 2. SEM micrographs of the $Al_2O_3/SrCO_3$ xerogels with various molar ratios: (**a**) Sr0.1, (**b**) Sr0.25, (**c**) Sr0.5, (**d**) Sr1.0, and (**e**) SEM-EDS analysis.

Figure 3 presents a comparison of the SEM micrographs of the $Al_2O_3/SrCO_3$ samples with Sr to an Al ratio of 1.0 annealed at different temperatures. The structure of the gel after annealing at 1000 °C (1273 K) (Figure 3a) is non-homogeneous. The surface of the gels after annealing at 1150 °C (1423 K) and 1300 °C (1573 K) is similar (Figure 3b,c, respectively). In the material after annealing at 1150 °C (1423 K) and 1300 °C (1573 K), a monolithic structure with a uniform composition distribution without any traces of crystallization of the strontium phases is observed. In the xerogel annealed at 1000 °C (1273 K), a hierarchical structure with the effect of phase aggregation can be observed.

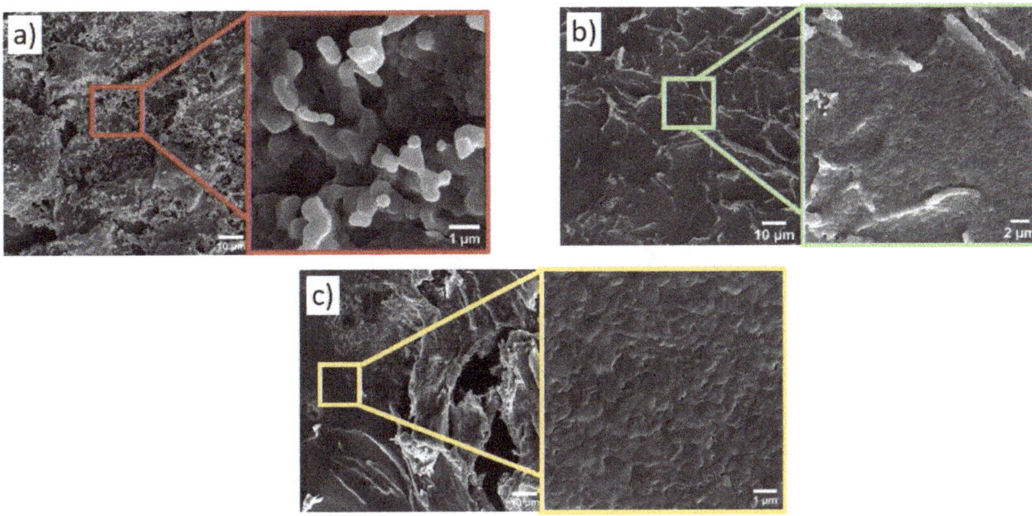

Figure 3. SEM micrographs of surface $Al_2O_3/SrCO_3$ xerogels with the 1.0 Sr molar ratio annealed at: (**a**) 1000 °C (1273 K), (**b**) 1150 °C (1423 K), and (**c**) 1300 °C (1573 K).

Figure 4 presents a comparison of the TEM micrographs of the $Al_2O_3/SrCO_3$ samples with Sr to an Al ratio of 1.0 annealed at different temperatures. The black areas represent a strontium-rich phase (strontium carbonate), as confirmed by the XRD data presented in Figure 1a. The structures shown are typical for sol–gel systems [15].

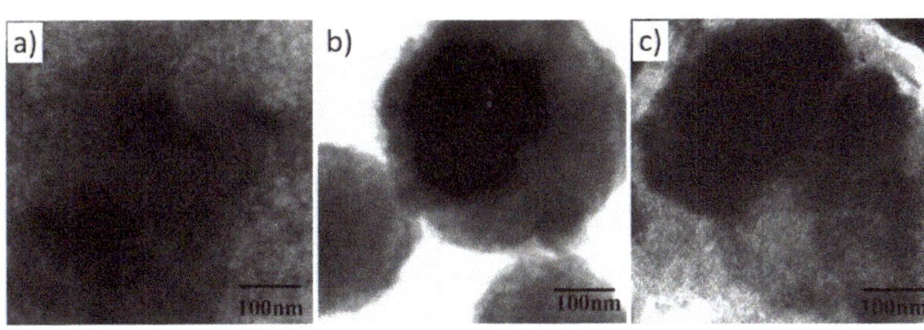

Figure 4. TEM images of $Al_2O_3/SrCO_3$ xerogels after annealing at 500 °C with various Sr/Al molar ratios: (**a**) 0.1, (**b**) 0.5, and (**c**) 1.0.

Figure 5 shows TEM images of $Al_2O_3/SrCO_3$ sol–gel with Pt after annealing at 500 °C of samples with a molar ratio of Al/Sr 0.1, 0.25, and 0.5. The presented structures differ depending on the Al/Sr molar ratio. The Sr0.5Pt sample contains rod-shaped crystals, which are not observed in the other tested materials. The amorphous alumina structure is visible. The large dark fields show agglomerated strontium carbonate, and the smallest fields show platinum areas (marked with a red arrow).

2.3. Porous Structure—Low Temperature Nitrogen Adsorption–Desorption

Table 1 shows surface area, pore equivalent diameter, and volume. Figure 6 shows plots of isotherms, pore volume distribution, and pore area distribution. The adsorption isotherms for samples with different Sr/Al ratios are type IV with the hysteresis loop (IUPAC) present in the range of the relative pressure p/p_0 0.5–0.8, which is characteristic of mesoporous structures [24]. Additionally, all the analyzed samples are characterized by this isotherm. The shape of the hysteresis loops is similar for the samples with an Sr/Al ratio higher than 0.1. The adsorption isotherms for platinum samples differ from those with different Sr contents. Type III with the H_2 hysteresis loop (except for the Sr1.0Pt system) is in the relative pressure range p/p_0 0.5–1.

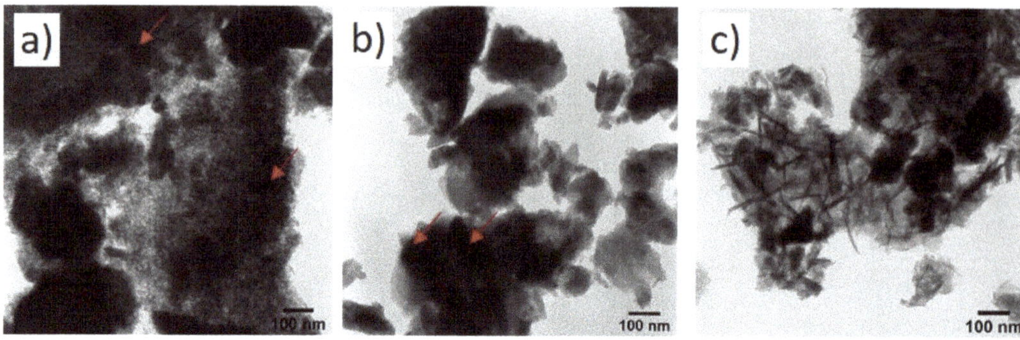

Figure 5. TEM images of $Al_2O_3/SrCO_3$ sol–gel with Pt after annealing at 500 °C with molar ratios of: (**a**) Sr0.1 Pt, (**b**) Sr0.25 Pt, and (**c**) Sr0.5 Pt.

Table 1. Textural properties of Sr systems.

System Composition	Surface Area S_{BET} [m^2/g]	Average Pore Diameter D_{BJH} [nm]	Average Pore Volume D_{BJH} [cm^3/g]
Sr0.1	135	7	0.30
Sr0.25	204	6	0.34
Sr0.5	132	6	0.25
Sr1.0	69	6	0.13
Sr0.1 Pt	154	7	0.29
Sr0.25 Pt	209	6	0.31
Sr0.5 Pt	159	7	0.25
Sr1.0 Pt	75	6	0.12

The pore distribution curves depend on the desorptive branch of the BJH isotherm. In catalytic systems, the presence of scattered platinum on the surface leads to a slight increase in surface area (Table 1). In every system tested, the surface area of the Sr0.1 sample is twice that of a sample with an Sr/Al molar ratio of 1.0. Thus, the prepared $Al_20_3/SrCO_3$ xerogel samples exhibited specific surface areas above 100 m^2/g (except for the Sr1.0 and Sr1.0 Pt). The surface area of the samples in which $SrCO_3$ was used for the synthesis is greater compared to the samples with metal strontium. The average pore diameter is smaller for $Al_2O_3/SrCO_3$ systems compared to the other samples. The average pore diameter and average pore volume are similar for the metallic strontium and platinum strontium systems. However, for systems with Sr/Al 0.5 and 1.0 molar ratios, a change in the geometry (a slight increase in the diameter) of the pores is visible, which may indicate the location of platinum crystals in the pores of the xerogel of a smaller size. For the systems with Pt, a slight increase in the specific surface area was also observed. This phenomenon may be related to the influence of platinum on the oxidative decomposition of carbon deposit residues that may occur in xerogel systems.

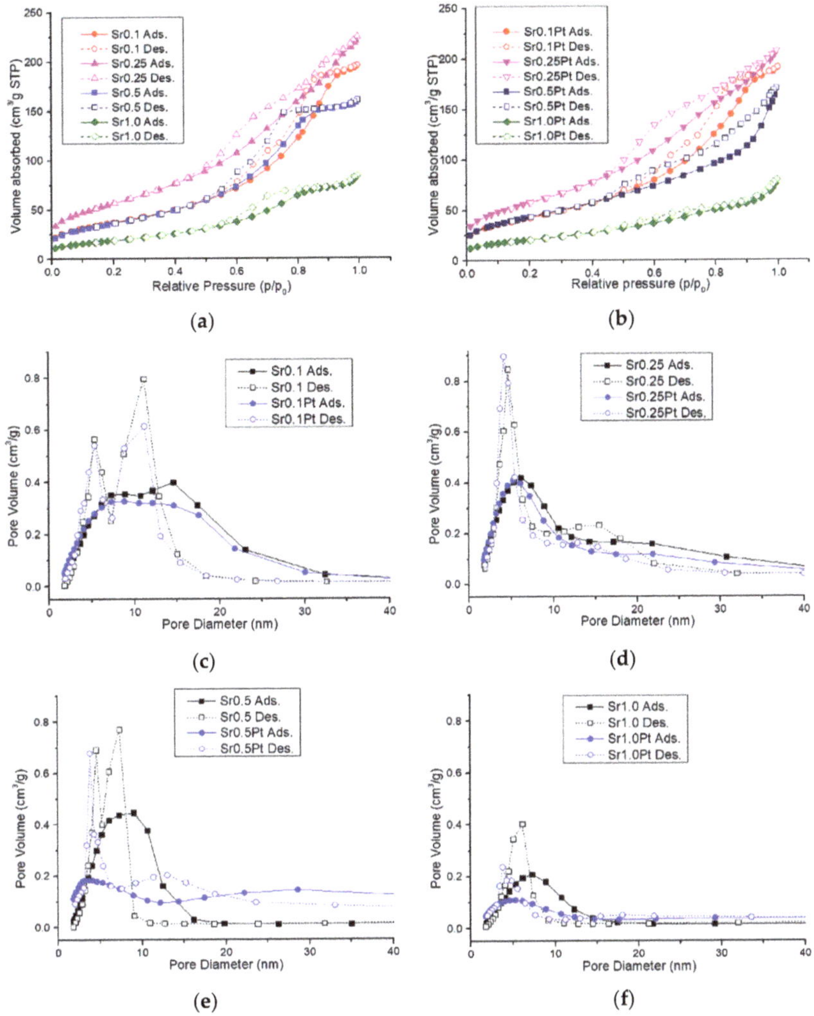

Figure 6. (**a**,**b**) Isotherm of $Al_2O_3/SrCO_3$ and Pt/Al_2O_3-$SrCO_3$, (**c**–**f**) pore volume distribution of $Al_2O_3/SrCO_3$ and Pt/Al_2O_3-$SrCO_3$.

2.4. Chemisorption of Hydrogen on Pt-Al-Sr Catalysts

Table 2 shows hydrogen chemisorption results on Pt/Al_2O_3-$SrCO_3$ systems with 1% metal content loading. Based on the results of hydrogen chemisorption, the platinum dispersion, the metallic surface area values, and the volume of the adsorbed hydrogen were determined. The results of hydrogen chemisorption for samples with a metallic strontium precursor showed no significant differences. The lowest metal dispersion occurred in the sample with an Sr/Al molar ratio equal to 1. Note that the $Al_2O_3/SrCO_3$ system is alkaline. In such system (alkaline), dispersion is significantly lower, because the chemical nature of the surface of materials with large pores and low surface area is important. In our study, the alkaline nature of the system reduces platinum dispersion, but the presence of nanopores stabilizes the platinum nanocrystallites and, as a result, dispersion is beneficial.

Table 2. Platinum dispersion, surface area, and volume of adsorbed hydrogen of the Pt/Al_2O_3-$SrCO_3$ systems.

System Composition	Metal (Pt) Dispersion [%]	Metallic Surface [m^2/g_{metal}]	Volume of Adsorbed Hydrogen [cm^3/g]
Sr0.1 Pt	47	116.94	0.27 ± 0.006
Sr0.25 Pt	49	122.55	0.29 ± 0.004
Sr0.5 Pt	48	119.15	0.28 ± 0.001
Sr1.0 Pt	42	102.94	0.24 ± 0.003

2.5. Thermal Analysis

For thermogravimetric testing, xerogel dried for one week at room temperature was used. Thermograms of the analyzed systems are shown in Figure 7a and the DTG curves in Figure 7b. Four visible areas of thermal changes in the xerogel were observed. The first step at temperatures up to 100 °C (373 K) is to remove the water adsorbed by the system. The second step of mass loss between 180 and 260 °C (453 to 533 K) was attributed to a correspondence to the removal of internally absorbed and trapped solvent residues and to the water of hydration in the gel. The third step of mass loss between 350 and 480 °C (623 to 753 K) can be attributed to the decomposition of anhydrous strontium acetate to $SrCO_3$, as confirmed by the XRD results (Figure 1). In the sample Sr0.5 and Sr1.0, there is a fourth stage of weight loss above 900 °C (1173 K) with the decomposition of the strontium carbonate.

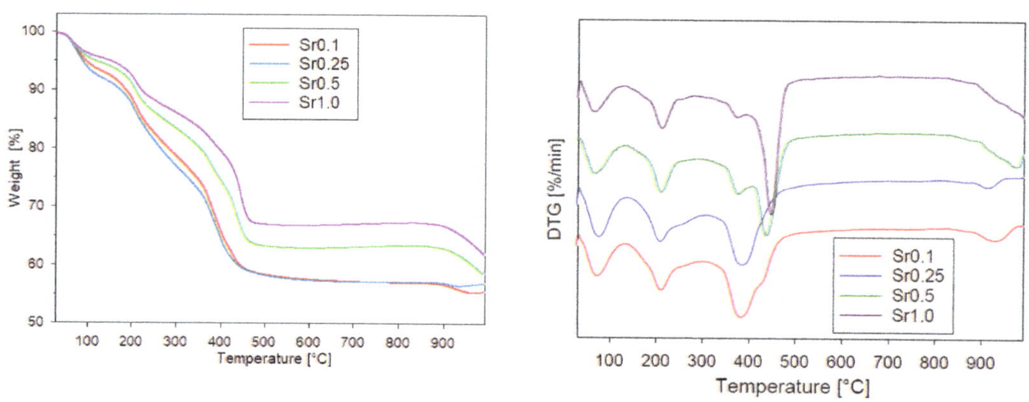

(a) (b)

Figure 7. (**a**) TGA weight loss, and (**b**) DTG curves of Al_2O_3-$SrCO_3$ xerogels obtained with a metallic strontium precursor.

Furthermore, the decomposition processes can be described in the following steps [17–25]:

$$Sr(CH_3COO)_2 \rightarrow SrCO_{3\,(s)} + CH_3COCH_{3\,(g)} + CO_{2\,(g)} \quad (3)$$

$$SrCO_{3\,(s)} \rightarrow SrO_{\,(s)} + CO_{2\,(g)} \quad (4)$$

$$Al_2O_3 + SrO \xrightarrow{>1150\ ^\circ C} SrAl_2O_4 \quad (5)$$

The DTG curves shown in Figure 7b indicate a multi-stage distribution. The decomposition of anhydrous strontium acetate to $SrCO_3$ is shifted to the right as the strontium molar ratio increases.

3. Conclusions

In this study, a new method of obtaining binary Al_2O_3-$SrCO_3$ xerogels systems is presented. The obtained xerogels are characterized by the presence of a stable carbonate phase for the full range of concentrations, which is confirmed by the XRD results. The effectiveness of the synthesis method using a reactive metallic precursor (metallic strontium) was confirmed. The obtained $Al_2O_3/SrCO_3$ xerogels are characterized by a high dispersion of the carbonate phase and a large specific surface area for alkaline systems. Carbonate xerogels with an alkaline element are characterized by a similar dispersion of the metallic phase (42–50%) in all the tested systems, which is a very good result for alkaline systems. Changes in the nanoporosity system may confirm the theory of stabilization of platinum nanoclusters in the structure of nanopore carriers obtained by the sol–gel method. Xerogels obtained by the described method are also an attractive precursor for high-temperature ceramics with a strictly defined microstructure.

Carbonate xerogels with an element of basic nature are characterized by a similar dispersion of the metallic phase (42–50%) in all the tested systems, which is a very good result for alkaline systems.

Some strontium aluminates (such as $SrAl_2O_4$) are used as phosphors. Phosphors based on strontium aluminate are characterized by good luminescent properties such as long-lasting afterglow and high quantum efficiency in comparison to classic sulfide phosphors [26,27].

4. Materials and Methods

4.1. Materials

Strontium carbonate, aluminum isopropoxide, acetic acid, and toluene were purchased from Sigma-Aldrich (Saint Louis, MO, USA) and used as received.

4.2. Preparation

Al_2O_3-$SrCO_3$ mixed systems with different molar ratios of Sr to Al: 0.1; 0.25; 0.5; and 1.0 were synthesized by aqueous sol–gel methods. Alumina gel was prepared according to our previous study [15]. Reactions were executed in a 1L glass reactor equipped with a mixer and a thermostat. Aluminum isopropoxide (500 g of fine powder) was added and hydrolyzed in 440mL of water at 75 °C. Then the obtained suspension was stirred for 2 h, and 175 g (167.5 cm^3) of 98% acetic acid was peptized. The sol was heated under reflux for 24 h at 95 °C (368 K), followed by metallic strontium addition in small portions as the second component. The next step was refluxing the resulting mixture for 72 h with vigorous stirring. The obtained product was a homogeneous liquid gel. To obtain a monolithic xerogel, part of the obtained gel was spilled into Petri dishes and dried for 5 days at room temperature. Next, the dry gel was annealed in a tube furnace at 500 °C (773 K) for 6 h under air flow. Part of the sample was air dried for 3 h, then annealed in 1000 °C (1273 K). Next, a portion of the sample was taken and air dried for 3 h and annealed in 1150 °C (1423 K). Finally, the sample batch was air dried for 3 h and annealed in 1300 °C (1573 K). The sample obtained after the annealing was crushed and sieved. Two grain fractions were collected: 0.1–0.2 mm and <0.1 mm. The particle size fraction with a diameter of 0.1 to 0.2 mm was used to determine the porous structure.

To prepare the Pt catalysts, 1.98 g of the support annealed at 500 °C (773 K) was weighed and placed in a 100 mL round-bottom flask. The samples were wetted with 2 mL of distilled water, then 1 mL of aqueous H_2PtCl_6 solution (platinum content: 20 mg Pt/mL) was added. The round-bottom flask was placed on a vacuum evaporator, and the solvent was evaporated. The dried system was heat treated at a temperature of 500 degrees for 2 h in air atmosphere.

Figure 8 shows the scheme of the synthesis of $Al_2O_3/SrCO_3$ xerogels.

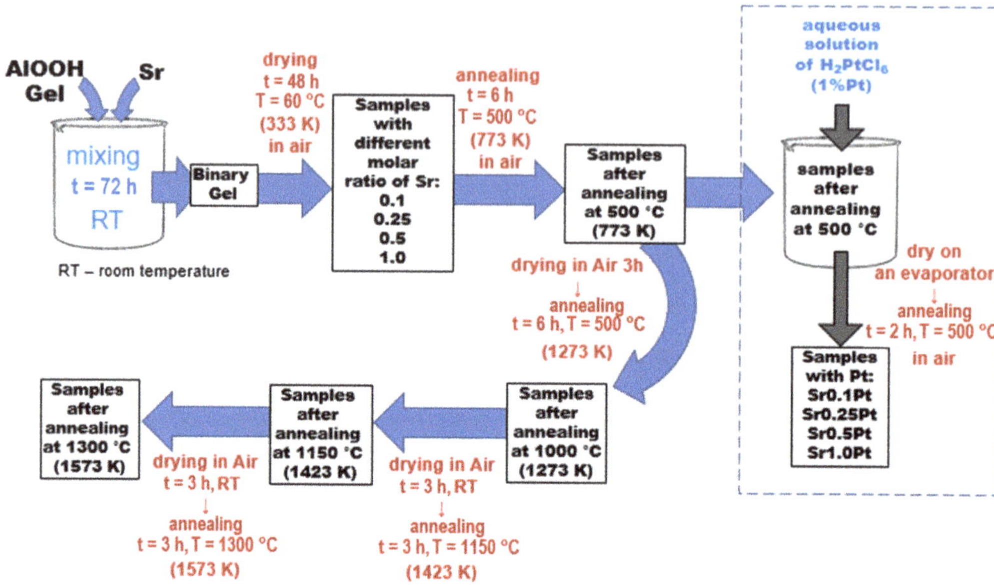

Figure 8. Schematic procedure of $Al_2O_3/SrCO_3$ xerogels synthesis.

4.3. Characterization

The samples obtained in this work were characterized by the following techniques.

4.3.1. X-ray Diffraction Analysis

The X-ray diffraction (XRD) analysis was performed using the Philips PW1050 diffractometer (Almelo, the Netherlands) working in the θ–2θ geometry with Cu-K_α (λ = 0.15406 nm) radiation of 35 kV and 20 mA. For all the samples, an angular range (2θ) of 10° to 100° with a step width of 0.01° and a step time of 3 s was used [28].

4.3.2. SEM, TEM, and EDS (Energy Dispersive X-ray Spectroscopy) Analysis

The surface morphology of the oxide xerogels was depicted by a Scanning Electron Microscope (SEM, Hitachi 3000N, Tokyo, Japan), which was operated in high-vacuum conditions at 15 kV acceleration voltage and low-vacuum conditions at 20 kV acceleration voltage. Chemical composition was performed using the Energy Dispersive Spectroscopy (EDS). TEM observations were performed using a JEOL 200 CX (Tokyo, Japan) transmission electron microscope worked at 80 kV.

4.3.3. Porous Structure

In order to determine the porosity of the structure, measurements of low-temperature nitrogen adsorption were carried out using the Autosorb iQ Station 2 (Quantachrome Instruments, Boynton Beach, Florida, United States) in the standard analysis mode; 200–300 mg of material with a particle size fraction from 0.1 to 0.2 mm were tested. Prior to testing, the samples were degassed for 10 h at 350 °C (623 K) and 0.4 Pa to constant weight. The adsorption and desorption isotherm branches were assumed in the $p/p0$ 0–1 range. Quantachrome ASiQwin software (version 2.0) was used. The Boer t-method and the BJH method were used to determine the distribution of the pore surface and pore volume. The volume and diameter of the pores were determined by the BJH method from the adsorption branch of the isotherm.

The BET multipoint linear regression method was used to calculate the surface area using the p/p0 0.1–0.3 window and the available seven degrees of freedom (nine data points) [25].

4.3.4. Thermal Analysis

On the NETZSCH TG 209 F1 Libra thermogravimeter (Selb, Germany), the thermal conversion of unprocessed (air-dried) samples was carried out. Five mg of the sample was placed in an alumina vessel (volume 85 μL) and heated from 20 °C·min^{-1} to 1000 °C (1273 K). For analysis, the fraction with a grain size <0.1 mm was used. TG traces were recorded with air flow (20 cm^3·min^{-1}) with a resolution of 0.1 μg. Drying under vacuum or at elevated temperature was not applied [28].

4.3.5. Chemisorption of Hydrogen on Pt-Al-Sr Catalysts

Hydrogen chemisorption was carried out by means of an ASAP 2010C sorptometer (Micromeritics, Norcross, GA, USA). Samples had previously been reduced with H_2 at 400 °C (673 K) during 2 h. Then the samples were evacuated in a sorptometer at room temperature for 0.25 h, then at 350 °C (623 K) for 1 h. After 1 h, the samples were reduced with a hydrogen flow (2.4 L/h) at 350 °C (623 K) and degassed for 2 h at 350 °C (623 K). Hydrogen chemisorption studies were performed at 35 °C (308 K). The platinum dispersion was determined from the total amount of chemically adsorbed hydrogen. The following equation was used to calculate the metal surface area S [29]:

$$S = \frac{v_m \cdot N_A \cdot n \cdot a_m \cdot 100}{22414 \cdot m \cdot wt} \left[m^2 \cdot g_{Pt}^{-1} \right] \quad (6)$$

where v_m—volume of adsorbed hydrogen (cm^3), N_A—Avogadro's number (6.022 × 1023 mol^{-1}), $n = 1$ is the chemisorption stoichiometry, wt (%)—the metal loading, a_m—the surface area (m^2), and m—the sample mass (g). The following formula was used to calculate the dispersion of the active phase:

$$D = \frac{S \cdot M}{a_m \cdot N_A} \quad (7)$$

where S is the metal surface area, M is the platinum atomic weight, N_A is Avogadro's number, and a_m is the surface occupied by one platinum atom.

Author Contributions: Conceptualization, R.E.P. and E.R.-R.; methodology, R.E.P. and E.R.-R.; formal analysis, R.E.P. and E.R.-R.; investigation, R.E.P., E.R.-R. and B.S. data curation, B.S. and E.R.-R.; writing—original draft preparation, E.R.-R., R.E.P., B.S. and Z.O.; writing—review and editing, R.E.P., E.R.-R. and Z.O.; visualization, E.R.-R. and B.S.; supervision, R.E.P. and Z.O.; project administration, B.S.; funding acquisition, R.E.P. All authors have read and agreed to the published version of the manuscript.

Funding: Research funded by the statutory funds of the Center for Advanced Technology AdamMickiewicz University.

Institutional Review Board Statement: Not applicable.

Informed Consent Statement: Not applicable.

Data Availability Statement: Not applicable.

Conflicts of Interest: The authors declare no conflict of interest.

References

1. Tuysuz, H.; Schuth, F. Chapter Two—Ordered Mesoporous Materials as Catalysts. *Adv. Catal.* **2012**, *55*, 127–239.
2. Ray, S.; Gusain, R.; Kumar, N. Chapter four—Adsorption in the context of water purification. In *Carbon Nanomaterial-Based Adsorbents for Water Purification*; Elsevier: Amsterdam, The Netherlands, 2020; pp. 67–100.
3. Kaya, G.; Deveci, H. Synergistic effects of silica aerogels/xerogels on properties of polymer composites: A review. *J. Ind. Eng. Chem.* **2020**, *89*, 13–27. [CrossRef]
4. Matsuhashi, H.; Iwamoto, A.; Sasaki, M.; Yoshida, K.; Aritani, H. Synthesis of SrO–Al_2O_3 Solid Base Catalysts from Strontium Hydroxide and Aluminum Alkoxide by a Solid-liquid Interface Reaction. *J. Jpn. Pet. Inst.* **2021**, *64*, 103–111. [CrossRef]

5. Akutu, K.; Kabashima, H.; Seki, T.; Hattori, H. Nitroaldol reaction over solid base catalysts. *Appl. Catal. A Gen.* **2001**, *247*, 65–74. [CrossRef]
6. Lee, H.; Wu, W.; Chen, B.H.; Liao, J.D. Heterogeneous Catalysts Using Strontium Oxide Agglomerates Depositing upon Titanium Plate for Enhancing Biodiesel Production. *Catalysts* **2021**, *11*, 30. [CrossRef]
7. Busca, G. Base and Basic Materials in Chemical and Environmental Processes. Liquid Versus Solid Basicity. *Chem. Rev.* **2010**, *110*, 2217–2249. [CrossRef]
8. Hibbins, S. Strontium and strontium compounds. In *Kirk-Othmer Encyclopedia of Chemical Technology*; Ley, C., Ed.; Wiley: New York, NY, USA, 2000.
9. Sezer, R.; Yilmaz, E.; Ertürk, S.; Cüneyt, A. Calcination of Strontium Carbonate in Rotary Kiln Furnace. In *10th International Symposium on High-Temperature Metallurgical Processing*; Springer: Cham, Switzerland; Berlin/Heidelberg, Germany, 2019.
10. Feyzi, M.; Shahbazi, Z. Preparation, kinetic and thermodynamic studies of Al–Sr nanocatalysts for biodiesel production. *J. Taiwan Inst. Chem. Eng.* **2017**, *71*, 145–155. [CrossRef]
11. Omata, K.; Nukui, N.; Hottai, T.; Showa, Y.; Yamada, M. Strontium carbonate supported cobalt catalyst for dry reforming of methane under pressure. *Catal. Commun.* **2004**, *5*, 755–758. [CrossRef]
12. Kobayashi, Y.; Omata, K.; Yamada, M. Screening of Additives to a Co/$SrCO_3$ Catalyst by Artificial Neural Network for Preferential Oxidation of CO in Excess H_2. *Ind. Eng. Chem. Res.* **2010**, *49*, 1541–1549. [CrossRef]
13. Omata, K.; Kobayashi, Y.; Yamada, M. Artificial neural network-aided development of supported Co catalyst for preferential oxidation of CO in excess hydrogen. *Catal. Commun.* **2005**, *6*, 563–567. [CrossRef]
14. Iida, H.; Deguchi, S.; Torigai, M.; Osawa, Y. Steam reforming of toluene over Ru/$SrCO_3$-Al_2O_3 catalyst under extremely low steam-to-carbon ratio conditions. *Fuel* **2020**, *272*, 117703. [CrossRef]
15. Przekop, R.; Marciniak, P.; Sztorch, B.; Czapik, A.; Stodolny, M.; Martyla, A. One-pot synthesis of Al_2O_3-$La_2O_2CO_3$ systems obtained from the metallic precursor by the sol–gel method. *J. Non-Cryst. Solids* **2018**, *479*, 105–112. [CrossRef]
16. Marciniak, P.; Sztorch, B.; Martyła, A.; Czapik, A.; Stodolny, M.; Przekop, R.E. Metallic Calcium as a Precursor for Sol–Gel Synthesis of $CaCO_3$-SiO_2 and CaO-SiO_2 Systems. *Ceramics* **2021**, *4*, 278–290. [CrossRef]
17. Duan, Y.; Li, J.; Yang, X.; Cao, X.; Hu, L.; Wang, Z.; Liu, Y.; Wang, C. Thermal investigation of strontium acetate hemihydrate in nitrogen gas. *J. Therm. Anal. Calorim.* **2008**, *91*, 169–174. [CrossRef]
18. Kwon, J.O.; Seok, S.I.; Jung, D. Annealing effect on photoluminescence properties of Er doped Al_2O_3–SiO_2 sol–gel films. *J. Non-Cryst. Solids* **2006**, *352*, 2841–2845. [CrossRef]
19. Chang, Y.; Hsiang, H. Phase Evolution During Formation of $SrAl_2O_4$ from $SrCO_3$ and Al_2O_3/AlOOH. *J. Am. Ceram. Soc.* **2007**, *90*, 2759–2765. [CrossRef]
20. Garcés, R.S.; Torres, J.; Valdes, A.F. Synthesis of $SrAl_2O_4$ and $Sr_3Al_2O_6$ at high temperature, starting from mechanically activated $SrCO_3$ and Al_2O_3 in blends of 3:1 molar ratio. *Ceram. Int.* **2012**, *38*, 889–894. [CrossRef]
21. Mizera; Kowalczyk, A.; Chmielarz, L.; Drożdż, E. Catalysts Based on Strontium Titanate Doped with Ni/Co/Cu for Dry Reforming of Methane. *Materials* **2021**, *14*, 7227. [CrossRef]
22. Rhodes, N.; Barde, A.; Randih, K.; Li, L.; Hahn, D.; Mei, R.; Klausner, J.; AuYeyng, N. Inside Back Cover: Solar Thermochemical Energy Storage Through Carbonation Cycles of $SrCO_3$/SrO Supported on $SrZrO_3$. *ChemSusChem* **2015**, *8*, 3913. [CrossRef]
23. Bagherisereshki, E.; Tran, J.; Lei, F.; AuYeung, N. Investigation into SrO/$SrCO_3$ for High Temperature Thermochemical Energy Storage. *Sol. Energy* **2018**, *160*, 85–93. [CrossRef]
24. Du, J.; Liu, Z.; Li, Z.; Han, B.; Huang, Y.; Zhang, J. Synthesis of mesoporous $SrCO_3$ spheres and hollow $CaCO_3$ spheres in room-temperature ionic liquid. *Microporous Mesoporous Mater.* **2005**, *83*, 145–149. [CrossRef]
25. Przekop, R.E.; Marciniak, P.; Sztorch, B.; Czapik, A.; Stodolny, M.; Martyła, A. New method for the synthesis of Al_2O_3–CaO and Al_2O_3–CaO–$CaCO_3$ systems from a metallic precursor by the sol–gel route. *J. Aust. Ceram. Soc.* **2018**, *54*, 679–690. [CrossRef]
26. Kaya, S.; Karacaoglu, E.; Karas, B. Effect of Al/Sr ratio on the luminescence properties of $SrAl_2O_4$:Eu_2^+, Dy_3^+ phosphors. *Ceram. Int.* **2012**, *38*, 3701–3706. [CrossRef]
27. Liepina, V.; Smits, K.; Millers, D.; Grigorjeva, L.; Monty, C. IOP Conference Series: Materials Science and Engineering. In *the Luminescent Properties of Persistent Strontium Aluminate Phosphor Prepared By Solar Induced Solid State Synthesis*; IOP Publishing Ltd.: Bristol, UK, 2012.
28. Przekop, R.; Marciniak, P.; Sztorch, B.; Czapik, A.; Stodolny, M.; Martyła, A. One-pot synthesis method of SiO_2-$La_2O_2CO_3$ andSiO_2-La_2O_3 systems using metallic lanthanum as a precursor. *J. Non-Cryst. Solids* **2019**, *520*, 119444.
29. Cunha, D.; Cruz, G. Hydrogenation of benzene and toluene over Ir particles supported on γ-Al_2O_3. *Appl. Catal. A Gen.* **2002**, *236*, 55–56. [CrossRef]

gels

MDPI

Article

Ordered Mesoporous Silica Prepared in Different Solvent Conditions: Application for Cu(II) and Pb(II) Adsorption

Ana-Maria Putz [1], Oleksandr I. Ivankov [2], Alexander I. Kuklin [2], Vasyl Ryukhtin [3], Cătălin Ianăşi [1], Mihaela Ciopec [4], Adina Negrea [4], László Trif [5], Zsolt Endre Horváth [6] and László Almásy [7,*]

1 "Coriolan Drăgulescu" Institute of Chemistry, Bv. Mihai Viteazul, No. 24, 300223 Timisoara, Romania; putzanamaria@acad-icht.tm.edu.ro (A.-M.P.); cianasic@yahoo.com (C.I.)
2 Frank Laboratory of Neutron Physics, Joint Institute for Nuclear Research, Joliot-Curie 6, 141980 Dubna, Russia; ivankov@jinr.ru (O.I.I.); kuklin@jinr.ru (A.I.K.)
3 Nuclear Physics Institute, ASCR, Husinec—Řež 130, 250 68 Řež, Czech Republic; ryukhtin@ujf.cas.cz
4 Faculty of Industrial Chemistry and Environmental Engineering, Politehnica University Timişoara, 6th Vasile Pârvan Bvd., 300223 Timisoara, Romania; mihaela.ciopec@upt.ro (M.C.); adina.negrea@chim.upt.ro (A.N.)
5 Institute of Materials and Environmental Chemistry, Research Centre for Natural Sciences, Magyar Tudósok Körútja 2, 1117 Budapest, Hungary; trif.laszlo@ttk.hu
6 Institute for Technical Physics and Material Science, Centre for Energy Research, Konkoly-Thege Miklós út 29-33, 1121 Budapest, Hungary; horvath.zsolt.endre@ek-cer.hu
7 Institute for Energy Security and Environmental Safety, Centre for Energy Research, Konkoly-Thege Miklós út 29-33, 1121 Budapest, Hungary
* Correspondence: almasy.laszlo@ek-cer.hu

Abstract: In this work, the synthesis of ordered mesoporous silica of MCM-41 type was investigated aimed at improving its morphology by varying the synthesis conditions in a one-pot process, employing different temperatures and solvent conditions. 2-methoxyethanol was used as co-solvent to ethanol. The co-solvent ratio and the synthesis temperature were varied. The pore morphology of the materials was characterized by nitrogen porosimetry and small angle neutron scattering (SANS), and the particle morphology by transmission electron microscopy (TEM) and ultra-small angle neutron scattering (USANS). The thermal behavior was investigated by simultaneous thermogravimetry-differential scanning calorimetry (TG-DSC) measurements. The SANS and N_2 sorption results demonstrated that a well-ordered mesoporous structure was obtained at all conditions in the synthesis at room temperature. Addition of methoxyethanol led to an increase of the pore wall thickness. Simultaneously, an increase of methoxyethanol content led to lowering of the mean particle size from 300 to 230 nm, according to the ultra-small angle scattering data. The ordered porosity and high specific surfaces make these materials suitable for applications such as adsorbents in environmental remediation. Batch adsorption measurements of metal ion removal from aqueous solutions of Cu(II) and Pb(II) showed that the materials exhibit dominantly monolayer surface adsorption characteristics. The adsorption capacities were 9.7 mg/g for Cu(II) and 18.8 mg/g for Pb(II) at pH 5, making these materials competitive in performance to various composite materials.

Keywords: MCM-41; SANS; USANS; SAXS; Stöber method; 2-methoxyethanol; Langmuir isotherm

Citation: Putz, A.-M.; Ivankov, O.I.; Kuklin, A.I.; Ryukhtin, V.; Ianăşi, C.; Ciopec, M.; Negrea, A.; Trif, L.; Horváth, Z.E.; Almásy, L. Ordered Mesoporous Silica Prepared in Different Solvent Conditions: Application for Cu(II) and Pb(II) Adsorption. *Gels* **2022**, *8*, 443. https://doi.org/10.3390/gels8070443

Academic Editors: Francesco Caridi, Giuseppe Paladini and Andrea Fiorati

Received: 23 June 2022
Accepted: 13 July 2022
Published: 15 July 2022

1. Introduction

Numerous research has dealt with the problem of obtaining suitable adsorbents in order to remove the hazardous pollutants. Adsorption technology is the most promising and frequently used approach due to its simplicity, high efficiency and low cost [1]. Various kinds of sorbent materials, both natural and synthetic, have been used to remove heavy metal ions from aqueous solution. Low cost industrial and agricultural by-products and waste materials usually have low adsorption efficiency because of their low pore volumes and poor pore structure [2,3]. Novel nanomaterials, including carbon-based nanomaterials, metal-oxide based nanomaterials, nanoparticles and various nanocomposites have been systematically applied for the removal of heavy metal ions from wastewater [4–11].

Due to their simplicity and abundance, silica and silica-based sorbents obtained from natural bio or waste sources represent an emerging approach in applications for removal of industrial pollution [12–17].

Synthetic silica based mesoporous materials, in spite of their higher costs, have also received wide attention and have been widely used for the adsorption of heavy metal ions due to their exceptionally large specific surface area, regular pore structure and suitability for surface modifications [18–21].

Silica materials with ordered porosity and open pore network, such as MCM-41 and SBA-15, are especially suitable for capturing species from aqueous solutions. In the classical synthesis of MCM-41-type materials, the silica precursors and the surfactant molecules used for pore templates are dissolved in water-ethanol mixtures. Full or partial replacement of ethanol with another co-solvent can modify the reaction speed of the sol-gel process, and change the morphology of the resulting silica gel [22,23]. The more hydrophobic 2-methoxyethanol has been used previously as a co-solvent in sol-gel synthesis of uniform spherical silica nanoparticles [24]. Hydrophobic organic co-solvents can affect the pore size by mingling with the surfactant molecules and acting as spacers inside the micelles [25,26]. For example, it was proven that ethoxyethanol acts not only as solvent but also as co-surfactant, controlling the morphology and pore structure. In alkali-catalyzed reactions, the co-surfactants promote the water solubility of the silica precursor tetraethoxysilane (TEOS), facilitating the hydrolysis reactions [27,28]. In previous works, 2-methoxyethanol was used as a protic polar solvent playing different specific roles: as solvent and complexing agent in the synthesis of mixed silica–titania [29], or as a solvent as well as a stabilizer of alkoxide in the hydrolysis-precipitation reaction, allowing one to control the reactivity of the precursors by adjustment of the quantity of 2-methoxyethanol [30].

In the present study, we applied a simple sol-gel synthesis procedure to obtain sorbent materials suitable, among other applications, for water pollutant removal. Two series of ordered mesoporous silica were synthetized, one at room temperature and the other one at 50 °C, and different amounts of 2-methoxyethanol were used in the range of 0:1 to 3:1 relative to the ethanol content. Cetyltrimethylammonium bromide (CTAB) was used as a pore forming agent. The influence of the mixed solvent composition and synthesis temperature on the particle morphology and the pore network was studied by small angle neutron scattering and nitrogen porosimetry.

The applicability of the obtained materials as adsorbents for heavy metals ions from water was studied with the example of Pb(II) and Cu(II).

2. Results and Discussion

2.1. FT-IR Analysis

All samples (xerogels and thermally treated materials) show the specific vibration bands for the silica skeleton at 1055–1085 cm^{-1}, 785–805 cm^{-1} and 450–455 cm^{-1} (Figure 1). This is assigned to the asymmetric stretching [31,32], symmetric stretching [33,34] and bending rocking mode vibration of the Si-O-Si network [32], respectively. For all calcined samples, the asymmetric stretching vibration bands are shifted to higher values by approximately 20–30 cm^{-1}. This effect is related to the densification of the silica network [34]. The 1640 cm^{-1} vibration band belongs to molecular water [35]. The presence of the silanol groups was confirmed by the presence of the band centered about 960 cm^{-1}, which is associated with the stretching mode of the Si-OH groups [35]. The broad band centered at around 3440–3405 cm^{-1} corresponds to the O-H stretching bands of hydrogen-bonded water molecules (H-O-H$\cdots$H) and the Si-O-H stretching of surface silanols hydrogen-bonded to water (-SiOH$\cdots H_2O$) [33]. They become more intense after calcination due to the increased hygroscopicity of the calcined samples, because the removal of the surfactant molecules allows some water to enter the pores [36].

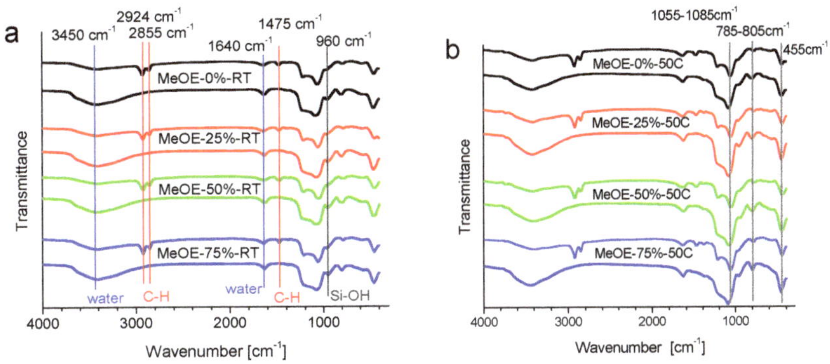

Figure 1. FT-IR spectra of mesoporous silica synthesized at room temperature (**a**) and at 50 °C (**b**), before and after calcination. Vertical lines in panel (**a**) show the characteristic bands of CTAB that disappear after thermal treatment (red); the vibrations related to water (blue) and Si-OH vibrations (dark grey). The vertical lines in panel (**b**) show the characteristic bands of the silica network.

Another group of characteristic vibrations is due to surfactant molecules. In the non-calcined samples, the bands at 2924 and 2855 cm^{-1} are the asymmetric and symmetric stretching vibrations of the alkyl chain, respectively, and the band around 1475 cm^{-1}, is the bending vibration of the C–H groups [35]. These bands completely disappear after calcination for all samples (Figure 1).

2.2. Thermal Analysis

Thermogravimetric and scanning calorimetry analyses in synthetic air were performed on the RT series of xerogels not subjected to calcination. The weight loss curves for the four samples are shown in Figure 2a. On the DTG curves (Figure 2b), five distinct weight loss regions can be identified, from which the second and the third steps are overlapped. In the first temperature range between 25 and 150 °C, physisorbed water evaporates, which is accompanied by a small and broad endotherm (Figure 2c). Between 150 and 350 °C, the surfactant decomposes and burns out. The two overlapped peaks on the DTG curve can be attributed to the separate oxidation and decomposition of the tetramethyl ammonium headgroups and the long alkyl chains of the CTAB. The oxidative burning of the organic matter is accompanied by a sharp and large exothermic signal on the heat flow curves around 340 °C. Between 400 and 700 °C, the small weight loss, accompanied by a small and broad exotherm, can be attributed to the complete oxidation and elimination of the decomposed organic compounds from the pores.

The spike-like exothermic peaks in the DSC curves at around 720–740 °C indicate phase transformations in the silica matrix.

In their thermal behavior, no systematic differences between the four samples were observed, which can be explained by the small differences in the morphology, and indicate that not much of methoxyethanol remained entrapped in the pores after the washing procedure during synthesis. The total weight loss of about 50% in the whole temperature range corresponds to the typical content of CTAB surfactant in the MCM-41 type materials [36].

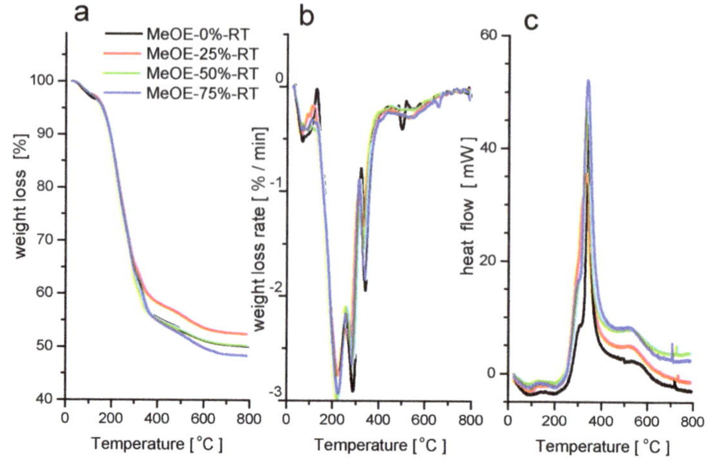

Figure 2. TG (**a**), DTG (**b**) and DSC (**c**) curves for samples prepared at room temperature.

2.3. Nitrogen Porosimetry

The nitrogen physisorption isotherms for all materials are shown in Figure 3. The materials present type IVb isotherm according to IUPAC classification. Type IVb is specific for mesoporous materials having conical or cylindrical pores, closed at one end [37,38]. The narrow hysteresis in the interval 0.2–0.4 P/P_0 suggests the presence of some wide pores that could have more access to the external surface [38].

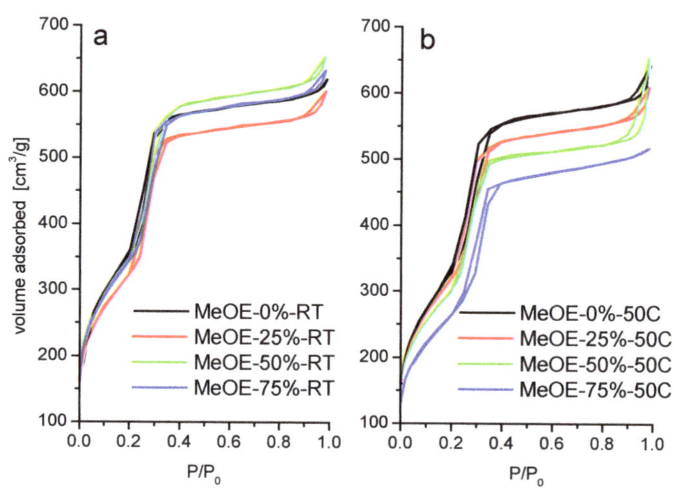

Figure 3. N_2 adsorption-desorption isotherms of calcined samples prepared at room temperature (**a**) and at 50 °C (**b**).

The pore size distribution (Figure 4) is narrow and centered around 3.5 nm for all samples. The BJH mean pore size shows a small increase from 3.0 to 3.3 nm, with the increase of methoxyethanol concentration in the solvent. The total pore volume does not change appreciably, except for the samples prepared with the highest methoxyethanol content at 50 °C when it drops by about 40%. This is related to the partial collapse of the pores, as shown later by the SANS results.

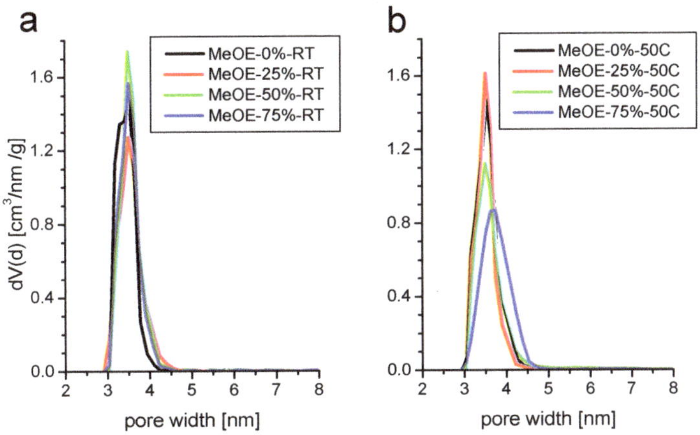

Figure 4. Pore size distribution of samples synthetized at room temperature (**a**) and at 50 °C (**b**).

The different texture of the materials was also observed in the surface fractal dimensions evaluated by Frenkel-Halsey-Hill (FHH) method [39]. The FHH model is used to determine the fractal geometry and calculate surface structure and irregularities. Value of Df 2 indicates smooth surface and 3 is characteristic for a porous surface [39]. The FHH data (Table 1) indicate that using methoxyethanol as co-solvent a smoother surface is obtained compared to the materials prepared only with ethanol.

Table 1. Textural parameters of the calcined MCM-41 samples.

Sample	BET Surface Area (m^2/g)	BET Constant	Pore Width (DFT) (nm)	Mean Pore Size (BJH) (nm)	Mean Pore Size (BJH) (nm)	Total Pore Volume (cm^3/g)	Df (Adsorpt.)	Df (Desorpt.)
MeOE-0%-RT	1793	10	3.53	3.30	3.07	0.961	2.846	2.876
MeOE-25%-RT	1540	12	3.53	3.39	3.06	0.933	2.762	2.830
MeOE-50%-RT	1620	13	3.53	3.41	3.34	1.013	2.761	2.828
MeOE-75%-RT	1547	15	3.53	3.41	3.31	0.981	2.764	2.839
MeOE-0%-50C	1568	13	3.53	3.39	3.07	0.998	2.750	2.864
MeOE-25%-50C	1446	15	3.53	3.40	3.04	0.946	2.757	2.865
MeOE-50%-50C	1428	12	3.53	3.06	3.06	1.016	2.750	2.864
MeOE-75%-50C	1074	24	3.78	3.39	3.31	0.804	2.738	2.822

Materials synthetized at 50 °C present almost the same characteristics. The isotherms are of type IVb with a narrow hysteresis (Figure 3b). The surface area decreases with increasing methoxyethanol content. For this series, the pore size distribution is somewhat broader, especially for the highest methoxyethanol content (Figure 4b).

All textural parameters for surface area, total pore volume and mean pore diameter are shown in Table 1. The highest BET surface area of 1793 m^2/g was obtained for sample synthesized at room temperature, and all samples had surface areas above 1000 m^2/g. With increasing 2-methoxyethanol content in the solvent mixture, the surface area decreased in both series. The changes in the surface area and roughness can be related to the lowering of the speed of hydrolysis and condensation with addition of methoxyethanol having hydrophobic methyl and methylene groups.

2.4. Electron Microscopy

TEM micrographs of the two investigated samples showed spherical and slightly elongated particles with sizes of 200–400 nm, typical for silica obtained in the Stöber synthesis process (Figure 5). Well-developed parallel channels with hexagonal ordering confirmed the structure of the MCM-41 materials. There were no noticeable differences between the samples MeOE-0%-RT and MeOE-75%-RT, in the selected areas of the TEM images.

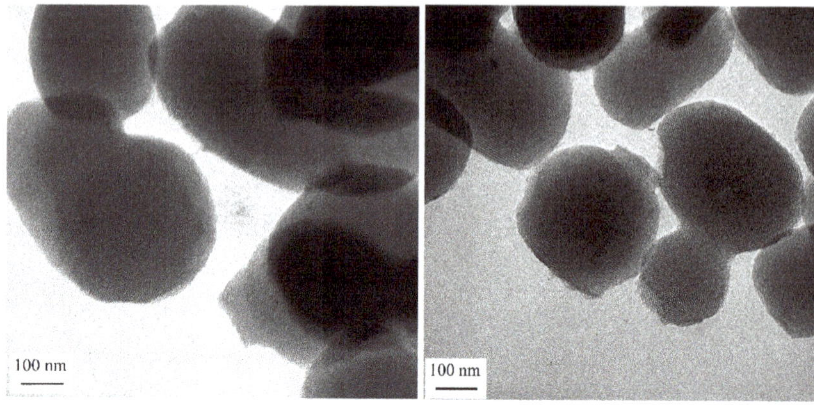

(a) (b)

Figure 5. Characteristic TEM images of mesoporous silica prepared at room temperature without 2-methoxyethanol (**a**) and in solvent with 3:1 methoxyethanol/ethanol ratio (**b**).

2.5. Small Angle Neutron Scattering

SANS curves for the two series taken on spectrometer YuMO are shown in Figure 6a,b. All samples show the typical organized structure of MCM-41 type, exhibiting the (100), (110), (200) reflections of the hexagonal pore structure [40–42]. The (110) and (200) peaks are smeared because of the moderate instrumental resolution related to the relaxed collimation and finite detector ring size, characteristic for small-angle neutron instruments. The sample prepared at 50 °C using 75% methoxyethanol is disordered (Figure 6b) and the diffraction peaks are nearly invisible. No well-ordered structure formed in the synthesis at these conditions, and the xerogel filled with CTAB was weaker and the pore structure was destroyed during calcination. Another characteristic difference between the samples prepared at the two temperatures is the width of the first diffraction peak, which is broader for the 50 °C synthesis, indicating a weaker long-range ordering. For samples prepared at room temperature, a significant difference was seen in the (100) peak position, between the material prepared in pure ethanol solvent, and the three samples prepared in mixed solvents. The peak position for MeOE-0%-RT sample was smaller by 3%, showing larger interpore distance, which indicates a larger wall thickness (Figure 7). This effect can be related to the slower condensation rate of the silica the presence of 2-methoxyethanol as cosolvent.

Measurements of the RT sample series in the very low q range, performed on the MAUD instrument, allowed us to assess the overall dimensions of the nanoparticles. The scattering curves measured with the medium and low-resolution instrument settings (solid and open symbols) are shown in Figure 8. The scattering data were fitted to a model of polydisperse spherical particles with log-normal size distribution. The resulting mean diameters were in the range of 200–300 nm, which is typical for Stöber synthesis. Interestingly, a well distinguished monotonous change of the particle size in the function of methoxyethanol content was observed: the larger (300 nm) particles were obtained using pure ethanol-water solvent mixture, and the gradual replacement of ethanol by 2-methoxyethanol resulted in smaller (230 nm) particles.

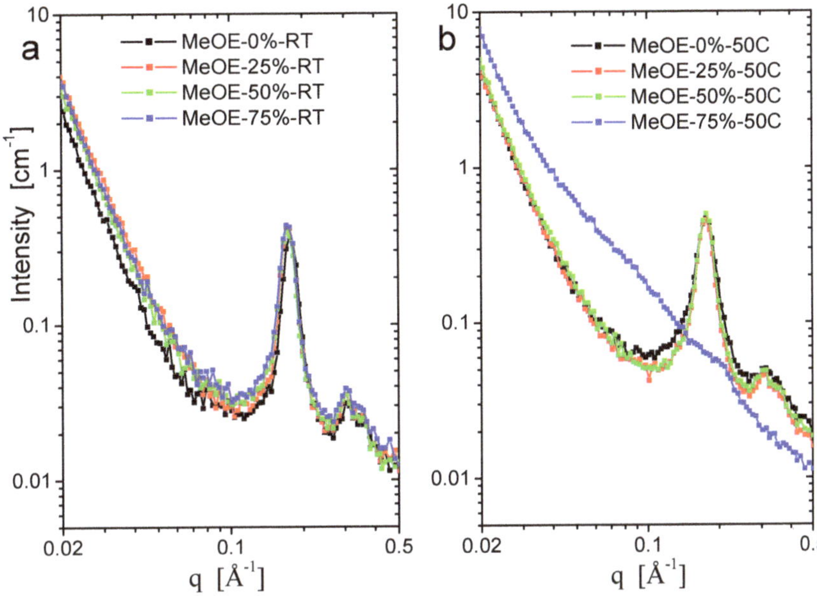

Figure 6. SANS scattering curves of samples prepared at room temperature (**a**) and at 50 °C (**b**).

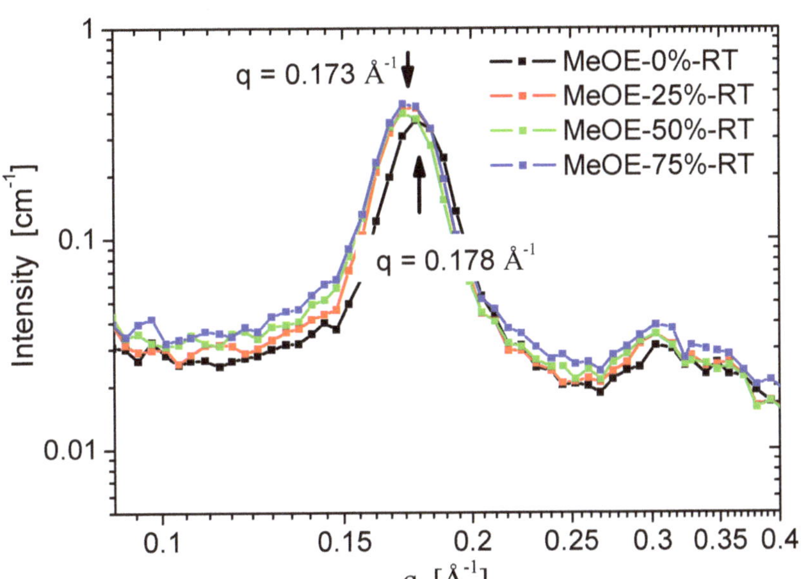

Figure 7. Variation of the first diffraction peak position with the content of 2-methoxyethanol in the solvent mixture.

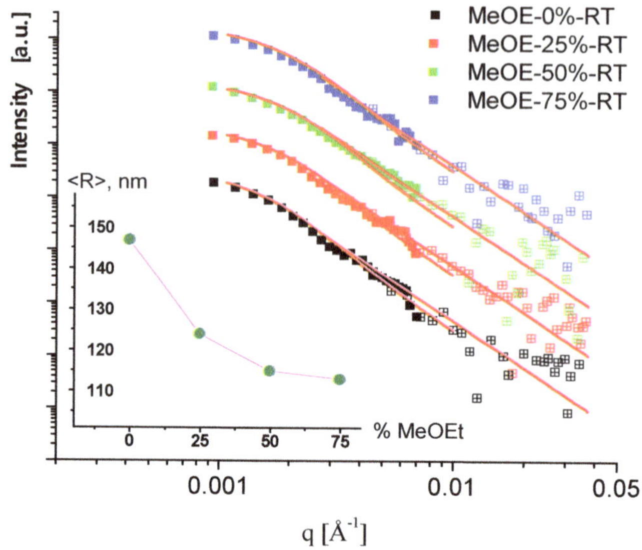

Figure 8. USANS scattering curves of samples prepared at room temperature. Symbols are the measured data, and red solid lines are the fitted model of polydisperse spheres. Data are shifted vertically. In the inset, the mean radius of the particles is shown in the function of 2-methoxyethanol content in the sol.

2.6. Metal Ion Adsorption

Adsorption isotherms are powerful tools for the analysis of adsorption processes. They establish the relationship between the equilibrium concentration and the amount adsorbed by unit mass of the adsorbent at a constant temperature. Adsorption studies were performed at pH 5 because it was shown previously for these kind of materials that the maximum adsorption capacities for Cu^{2+} and Pb^{2+} were observed at pH 5.0 [19]. First, all eight samples were tested for adsorption of Cu(II) and Pb(II), and then the material with the best adsorption capacity for Cu(II), (sample MeOE-75%-RT), and the material with the best adsorption capacity for Pb(II), (sample MeOE-0%-RT) were selected for construction of the adsorption isotherms. The equilibrium time for adsorption was determined varying the time in the range 15–180 min for a Me(II) ions concentration of 10 mg/L. After 2 h, the adsorption capacity has a constant value. Two hours of contact time was used in constructing the adsorption isotherms.

The adsorption capacity q (mg/g), was determined using the following equation:

$$q = (C_0 - C_f)\frac{V}{m} \quad (1)$$

where C_0 and C_f are the initial and final concentrations of metallic ions in solution in the beginning and the end of the adsorption test, respectively (mg/L), V is the volume of the solution (L) and m is the mass of the adsorbent (g).

The adsorption isotherms of Me(II) ions are presented in Figure 9. The Langmuir, Freundlich and Sips isotherms were used to model experimental data in order to establish the adsorption mechanism and the maximum adsorption capacity [43,44].

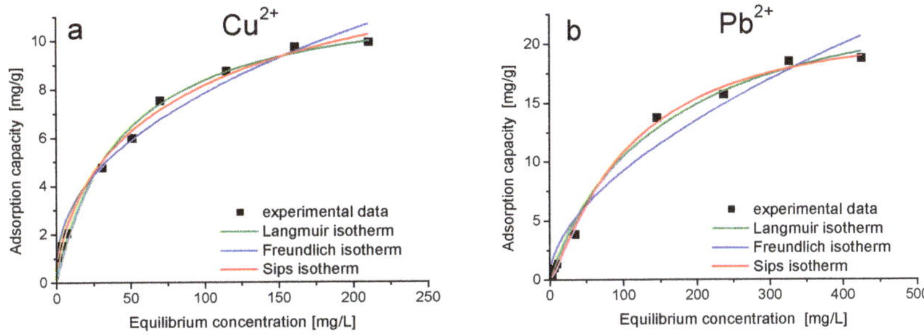

Figure 9. Adsorption isotherms of (**a**) Cu(II) on MeOE-75%-RT and (**b**) Pb(II) on MeOE-0%-RT.

The non-linear form of the Langmuir isotherm equation [45] can be expressed as follows:

$$q_e = \frac{q_L\, K_L\, C_f}{1 + K_L\, C_f} \quad (2)$$

where q_e is the equilibrium absorption capacity (mg/g); C_f is the equilibrium ion concentration in solution (mg/L); q_L is the Langmuir maximum adsorption capacity (mg/g) and K_L is the Langmuir constant.

The Freundlich isotherm is applicable to heterogeneous adsorption surfaces. The non-linear form of the Freundlich isotherm equation [46] is:

$$q_e = K_F\, C_f^{1/n_f} \quad (3)$$

where K_F and n_f are the characteristic constants related to the adsorption capacity of the adsorbent and the intensity of adsorption.

The Sips or Langmuir-Freundlich isotherm (Equation (4)) derives from the Freundlich and Langmuir models. It reduces to a Freundlich isotherm at low adsorbate concentrations. At high adsorbate concentrations, the Sips isotherm has the characteristics of the Langmuir isotherm and therefore can be used to calculate the monolayer adsorption capacity [47].

$$q_e = \frac{q_S\, K_S\, C_e^{1/n_S}}{1 + K_S\, C_e^{1/n_S}} \quad (4)$$

where K_S (mg/g) is a constant related to the adsorption capacity of the adsorbent and n_s is the heterogeneity factor. As the surface heterogeneity is higher, the deviation of $1/n_s$ value from 1 will be higher. The parameters of Langmuir, Freundlich and Sips isotherms determined using non-linear regression are listed in Table 2.

Table 2. Langmuir, Freundlich and Sips isotherm parameters for Cu(II) ions adsorption on sample MeOE-75%-RT and Pb(II) on MeOE-0%-RT.

Adsorbent /Metal Ion	$q_{m,exp}$ (mg/g)	Langmuir Isotherm			Freundlich Isotherm			Sips Isotherm			
		q_L (mg/g)	K_L	R^2	K_F (mg/g)	$1/n_F$	R^2	K_S	q_S (mg/g)	$1/n_s$	R^2
MeOE-75%-RT /Cu^{2+}	9.7	7.3	0.022	0.988	1.14	0.41	0.978	0.0026	10.2	0.3	0.990
MeOE-0%-RT /Pb^{2+}	18.8	26.3	0.065	0.993	0.7	0.5	0.963	0.0061	22.2	0.6	0.995

The adsorption capacity increased with increasing equilibrium concentration of Me(II) ions. The experimental isotherms show the approach of the adsorption capacity to constant value and, correspondingly, the large deviation of the fitted Freundlich model curves from the data. The maximum adsorption capacity of Cu(II) on sample MeOE-75%-RT for an initial Cu(II) concentration of 250 mg/L is $q_{m,exp}$ = 9.7 mg/g, and the maximum adsorption capacity of Pb(II) on sample MeOE-0%-RT is $q_{m,exp}$ = 18.8 mg/g for an initial Pb(II) concentration of 500 mg/L.

For both materials, the adsorption is better described by the Sips isotherm. The maximum adsorption capacity obtained by modelling the experimental data with Sips isotherm is q_s= 10.2 mg Cu(II)/g sorbent, a value very close to the experimental value $q_{m,exp}$ = 9.7 mg Cu(II)/g sorbent for sample MeOE-75%-RT and q_s = 22.2 mg/g, a value very close to the experimental value $q_{m,exp}$ = 18.8 mg Pb(II)/g for sample MeOE-0%-RT at the highest salt concentration. Parameter $1/n$ in the Sips model is a measure of the adsorption intensity or surface heterogeneity. For $1/n = 1$, the partition between the two phases is independent of the concentration. The situation $1/n < 1$ is the most commonly encountered and corresponds to Langmuir isotherm, while $1/n > 1$ indicates cooperative adsorption involving strong interactions between the molecules of the adsorbate [48,49]. The value of the heterogeneity factor $1/n_s$ was obtained as 0.3 for sample MeOE-75%-RT and 0.6 for sample MeOE-0%-RT, indicating that the heterogeneity of the surface was very low. This suggests that the adsorption mechanism approaches the monolayer adsorption and the adsorption of Me(II) ions onto adsorbent materials was a favourable process.

The maximum adsorption capacities of these materials were compared to some other adsorbents from recent literature (Table 3). Concerning the Pb(II) adsorption, the obtained value of 18.8 mg/g proved that our mesoporous silica has better adsorption properties compared with kaolin, activated carbon and zeolite, but worse than the functionalized mesoporous silica and the composite materials of iron oxide nanoparticles with functionalized porous silica or cellulose (Table 3). Concerning the Cu(II) adsorption, the obtained value of 9.7 mg/g for the adsorption capacity is higher than that of zeolites, similar to magnetite nanoparticles, but worse than the adsorption capacities of other metal oxide nanoparticles or mesoporous silica having different functional groups. Therefore, the functionalization of mesoporous silica materials is necessary to improve their adsorption capacity for metal ions removal. On the other hand, the ease of preparation and low cost of the nonfunctionalized mesoporous silica may compensate for their lower performance in certain applications.

Table 3. Comparison of the Me(II) adsorption capacity of the prepared mesoporous silica samples to other adsorbent materials.

Sorbent	Metal Ion	Adsorption Capacity (mg/g)	Reference
Kaolin	Pb(II)	4.5	[50]
activated carbon		6.7	[50]
magnetic chlorapatite nanoparticles		238	[51]
zeolite		9.9	[52]
thiol functionalized iron-oxide loaded FDU-12 mesoporous silica		287	[53]
Fe_3O_4@carboxymethyl-cellulose		152	[54]
Fe_3O_4@SiO_2@DMSA		50.5	[55]
Fe_3O_4@SiO_2@TSD		417	[56]
Fe_3O_4@SiO_2 -NH_2		76.6	[57]
pretreated *Aspergillus niger*		32.6	[58]
maghemite nanoparticle		68.9	[59]
magnetite nanoparticles		37.3	[60]
TiO_2 nanoparticles		21.7	[61]
Al_2O_3 nanoparticles		41.2	[61]
MgO nanoparticles		148	[61]
Chitosan/graphene oxide		461	[62]
silica@ketoenol-pyrazole		41.8	[63]
$ZnCl_2$-MCM-41		479	[64]

Table 3. *Cont.*

Sorbent	Metal Ion	Adsorption Capacity (mg/g)	Reference
EDTA/SBA-15		273	[65]
waste silica coated by iron oxide		8.2	[66]
silica-magnetite composite		14.9	[67]
citrate coated SPION		58.9	[68]
gelatin-siloxane hybrid		3.75	[11]
chitosane-alginate hydrogel		85	[9]
thiol functionalized silica/magnetite		0.8	[69]
magnetic nano-zeolite		476.1	[70]
nano-silica made of *Saccharum officinarum*		148	[71]
Calcined MCM-41		18.8	**Present paper**
zeolite	Cu(II)	8.5	[52]
$Fe_3O_4@SiO_2$ -NH		29.9	[57]
pretreated *Aspergillus niger*		28.7	[58]
maghemite nanoparticle		34.0	[59]
magnetite nanoparticles		10.8	[60]
TiO_2 nanoparticles		50.2	[61]
Al_2O_3 nanoparticles		47.9	[61]
MgO nanoparticles		149.1	[61]
Chitosan/graphene oxide		423.8	[62]
silica@ketoenol-pyrazole		76.9	[63]
waste silica coated by iron oxide		3.4	[66]
magnetic nano-zeolite		59.9	[70]
steel slag/CNT composite		132.8	[72]
bifunctional silica nanospheres		139.8	[73]
nanosilica/nanopolyaniline		108	[74]
nanosilica/crosslinked nanopolyaniline		105	[74]
gelatin-siloxane hybrid		1.76	[11]
core-shell magnetite-silica NP		41	[75]
Calcined MCM-41		9.7	**Present paper**

3. Conclusions

In this paper we studied morphology evolution of ordered mesoporous silica nanoparticles of MCM-41 type, prepared at different solvent conditions using 2-methoxyethanol as a co-solvent and CTAB as a pore forming molecule, by conventional pin-hole and slit-smeared ultra-small-angle neutron scattering. The use of mixed solvent resulted in larger wall thickness and interpore distance, and the hexagonal mesoporous structure was maintained up to 75/25 methoxyethanol/ethanol ratio. This effect can be related to the slower condensation rate of the silica network in the presence of 2-methoxyethanol as cosolvent. These structural changes were accompanied by a 10% decrease of the surface area, as measured by nitrogen sorption. A weaker long-range ordering was obtained for the samples synthesized at elevated temperature of 50 °C. These textural and structural observations show that varying the solvent conditions and temperature allows one to tune and optimize the porosity and intrapore surface of the materials, which is of primary importance for their applications as sorbent materials. In future work the effect of other synthesis conditions should be explored aiming to achieve the highest porosity, high specific surface and surface quality modification by functional groups, for achieving efficient binding of metal and organic pollutants.

Selected samples were tested for adsorption of metal pollutants from aqueous phase. The adsorption equilibrium data of Pb(II) and Cu(II) followed the Sips and the Langmuir isotherms, while the Freundlich model did not fit the experimental data. The modeling showed that the heterogeneity of the adsorbent materials surface was very low, suggesting that the adsorption mechanism approached monolayer adsorption. Comparison of the adsorption performance of the prepared materials to various adsorbents from the literature revealed that the adsorption capacity of the present mesoporous silica is comparable to some simple materials, but lower than many of the much more expensive composite materials.

This study showed that while the modification of the structure, overall dimension and pore size can be tuned easily within narrow limits by changing the solvent composition, the resulting effect on the adsorption performance is not particularly strong within the investigated synthesis conditions. This is attributed to the relatively small change of the pore size and internal surface by the simple solvent variation method. Replacement of the traditional CTAB with longer or shorter alkyl chain surfactants can efficiently vary the porosity within a certain range. Furthermore, using more eco-friendly biosurfactants may be beneficial in long term, after resolving the most prohibitive factors such as their high costs and low productivity [76].

4. Materials and Methods

4.1. Sample Preparation

The following chemicals were used: tetraethoxysilane (TEOS), (99%, for analysis, Fluka Chemie, Buchs, Switzerland); hexadecyltrimethyl ammonium bromide (CTAB, Sigma-Aldrich, Darmstadt, Germany); ethanol (Chimopar, Bucharest, Romania); 2-methoxyethanol (99%, for analysis, Sigma-Aldrich); ammonia solution 25% (Adra Chim SRL, Bucharest, Romania).

Ordered mesoporous silica materials were prepared by sol-gel synthesis in alkaline conditions, using TEOS as silica precursor and CTAB as pore forming agent, adopting traditional methods [22,77] for using solvent mixtures. As solvent, ethanol and 2-methoxyethanol were used in different proportions. A quantity of 1 g CTAB was added to 192.49 mL of distilled H_2O under stirring. After the solution turned clear, 68 mL of alcohol was added and then 23.2 mL of aqueous ammonia solution (25%) was added, and the samples were stirred for 30 min. Ethanol was used for the first sample from each series and it was progressively substituted with 2-methoxyethanol for the next three samples in each series. After that, 4 mL of TEOS was poured into the solution slowly under stirring. Stirring continued for 3 h. After about 20 h, the solid product was recovered by filtration and washed several times with distilled water with repeated filtrations until the pH of the washing water approached the pH value of the distilled water. The samples were dried at room temperature until the next day, then further dried for several hours at 90 °C. CTAB was removed by calcination at 550 °C during 6 h, after a heating ramp with 1 °C/min. The sample names and synthesis parameters are shown in Table 4.

Table 4. Synopsis of the synthesized samples.

Sample Name	CTAB (g)	TEOS (mL)	Synthesis Temperature	H_2O (mL)	Ethanol (mL)	2-Methoxyethanol (mL)
MeOET-0%-RT	1	4	r.t.	192.5	68	0
MeOET-25%-RT	1	4	r.t.	192.5	51	17
MeOET-50%-RT	1	4	r.t.	192.5	34	34
MeOET-75%-RT	1	4	r.t.	192.5	17	51
MeOET-0%-50C	1	4	50 °C	192.5	68	0
MeOET-25%-50C	1	4	50 °C	192.5	51	17
MeOET-50%-50C	1	4	50 °C	192.5	34	34
MeOET-75%-50C	1	4	50 °C	192.5	17	51

4.2. Characterization Methods

FTIR spectra were taken on KBr pellets with a JASCO–FT/IR-4200 apparatus. Samples after drying (the xerogel samples) as well as after calcination were studied. N_2 physisorption measurements were done at 77 K using a Quantachrome Nova 1200e apparatus. Prior to the analysis the samples were dried and degassed in vacuum at 80 °C for 4 h.

The surface area was determined by the BET (Brunauer–Emmet–Teller) equation in the relative pressure range (P/P_0) 0.01–0.25. Pore size distribution was evaluated with BJH (Barrett-Joyner-Halenda) and DFT (Density Functional Theory) methods. Total pore volumes were determined using the relative pressure point closest to 1.

Thermal measurements were performed on a Setaram LabsysEvo (Lyon, France) TG-DSC system, in flowing (90 mL/min) high purity (99.999%) synthetic air (20% O_2 in N_2) atmosphere. Samples were weighed into 100 µL Al_2O_3 crucibles (the reference cell was empty) and were heated from 25 °C to 800 °C with a heating rate of 10 °C/min. The obtained data was blank corrected and further processed with the software Calisto Processing, ver. 2.08. The temperature scale and calorimetric sensitivity were calibrated by a multipoint calibration method in which seven different certified reference materials were used to cover the thermal analyzer's entire operating temperature range.

Transmission electron microscopy (TEM) investigation was carried out on a Philips CM20 transmission electron microscope equipped with LaB_6 electron gun operating at 200 kV. The samples for TEM were prepared by drop-drying suspensions on holey carbon foil coated copper grids.

Small-angle neutron scattering measurements were performed on the YuMO time-of-flight instrument operating at the IBR-2 pulsed reactor source in Dubna, Russia [78,79]. The scattered neutrons were detected using the time-of-flight method by a two-detector set-up with ring wire detectors [80]. Measurements were performed on dry powders at room temperature. Corrections for transmission, background and empty aluminium container scattering were performed using SAS software [81].

Ultrasmall-angle neutron scattering measurements were performed on a double-bent crystal instrument MAUD operating at the thermal channel of the LVR15 10 MW research reactor in Řež, Czech Republic [82].

4.3. Metal Ion Adsorption Measurements

For the metallic ion adsorption experiments, 1000 mg/L metallic ion solution in HNO_3 0.5 mol/L (Merck standard solution) was used. All other metallic ion solutions were prepared from this solution with appropriate dilution. The pH of the sample solutions was pH 5 and was adjusted by using NaOH buffer solutions with the concentration in the 0.05–0.2 M range, measured using a Mettler Toledo Seven Compact S210 pH meter. Distilled water was used in all experiments.

Equilibrium isotherms were constructed using 0.1 g adsorbent materials placed in 25 mL Cu(II) solution at different initial concentrations (5–250 mg/L), and 25 mL Pb(II) solution at different initial concentrations (5–500 mg/L), at room temperature (25 ± 1 °C) and with a stirring time 2 h, using a thermostated shaker bath Julabo SW23, and shaking speed 200 rpm. The equilibrium time for isotherms adsorption was determined varying the time in the range 15–180 min for a Me(II) ions concentration of 10mg/L. After determination of the residual concentration of the metallic ions and the adsorption capacity, it was found that after 2 h the adsorption capacity has a constant value. After stirring, the samples were separated trough filtration. The residual concentration of the metallic ions in filtrate was analysed by atomic absorption spectrometry using a Varian SpectrAA 280 Fast Sequential Atomic Absorption Spectrometer with an air-acetylene flame.

Author Contributions: Conceptualization, A.-M.P. and L.A.; methodology, A.-M.P., L.T., A.N., M.C. and L.A.; investigation, A.-M.P., C.I., L.T., A.N., M.C., O.I.I., A.I.K., V.R. and Z.E.H.; resources, A.-M.P. and O.I.I.; data curation, A.-M.P., A.N., M.C., L.A. and V.R.; writing—original draft preparation, A.-M.P., A.N., M.C. and L.A.; writing—review and editing, A.-M.P. and L.A. All authors have read and agreed to the published version of the manuscript.

Funding: This research received no external funding.

Data Availability Statement: Experimental data are available from the authors.

Acknowledgments: A.-M.P. and L.A. thank the Inter-Academic Exchange Program between Romanian Academy and the Hungarian Academy of Sciences. A.-M.P. and O.I.I. thank the scientific cooperation program between Romania and Joint Institute for Nuclear Research, Dubna, Russia. USANS measurements were carried out at the CANAM infrastructure of NPI CAS Řež with the use of reactor LVR-15 supported by the project LM2018120. V.R. acknowledges support from the Czech

Academy of Sciences in the frame of the program "Strategie AV21, No. 23". L.A. is grateful to A. Elbakyan for facilitating access to scientific literature.

Conflicts of Interest: The authors declare no conflict of interest. The funders had no role in the design of the study; in the collection, analyses, or interpretation of data; in the writing of the manuscript; or in the decision to publish the results.

References

1. Yuan, Q.; Li, N.; Chi, Y.; Geng, W.; Yan, W.; Zhao, Y.; Li, X.; Dong, B. Effect of Large Pore Size of Multifunctional Mesoporous Microsphere on Removal of Heavy Metal Ions. *J. Hazard. Mater.* **2013**, *254–255*, 157–165. [CrossRef] [PubMed]
2. De Gisi, S.; Lofrano, G.; Grassi, M.; Notarnicola, M. Characteristics and Adsorption Capacities of Low-Cost Sorbents for Wastewater Treatment: A Review. *Sustain. Mater. Technol.* **2016**, *9*, 10–40. [CrossRef]
3. Beveridge, A.; Pickering, W.F. The Influence of Surfactants on the Adsorption of Heavy Metal Ions by Clays. *Water Res.* **1983**, *17*, 215–225. [CrossRef]
4. Yang, J.; Hou, B.; Wang, J.; Tian, B.; Bi, J.; Wang, N.; Li, X.; Huang, X. Nanomaterials for the Removal of Heavy Metals from Wastewater. *Nanomaterials* **2019**, *9*, 424. [CrossRef]
5. Hu, H.; Zhao, D.; Wu, C.; Xie, R. Sulfidized Nanoscale Zerovalent Iron Supported by Oyster Powder for Efficient Removal of Cr (VI): Characterization, Performance, and Mechanisms. *Materials* **2022**, *15*, 3898. [CrossRef] [PubMed]
6. Shen, J.; Wang, N.; Wang, Y.G.; Yu, D.; Ouyang, X. Efficient Adsorption of Pb(II) from Aqueous Solutions by Metal Organic Framework (Zn-BDC) Coated Magnetic Montmorillonite. *Polymers* **2018**, *10*, 1383. [CrossRef] [PubMed]
7. Herman, P.; Kiss, A.; Fábián, I.; Kalmár, J.; Nagy, G. In situ remediation efficacy of hybrid aerogel adsorbent in model aquatic culture of *Paramecium caudatum* exposed to Hg(II). *Chemosphere* **2021**, *275*, 130019. [CrossRef] [PubMed]
8. Herman, P.; Fábián, I.; Kalmár, J. Mesoporous Silica–Gelatin Aerogels for the Selective Adsorption of Aqueous Hg(II). *ACS Appl. Nano Mater.* **2020**, *3*, 195. [CrossRef]
9. Hamza, M.F.; Hamad, N.A.; Hamad, D.M.; Khalafalla, M.S.; Abdel-Rahman, A.A.; Zeid, I.F.; Wei, Y.; Hessien, M.M.; Fouda, A.; Salem, W.M. Synthesis of Eco-Friendly Biopolymer, Alginate-Chitosan Composite to Adsorb the Heavy Metals, Cd(II) and Pb(II) from Contaminated Effluents. *Materials* **2021**, *14*, 2189. [CrossRef]
10. Zhang, J.; He, H.; Owusu, A.; Henderson, M.J.; Yan, M.; Tian, Q.; Almásy, L. Efficient and selective sorption of uranyl by hydrated Ti_3C_2TX-Li agglomerates. *Environ. Nanotechnol. Monit. Manag.* **2021**, *16*, 100528. [CrossRef]
11. Wojciechowska, P.; Cierpiszewski, R.; Maciejewski, H. Gelatin-Siloxane Hybrid Monoliths as Novel Heavy Metal Adsorbents. *Appl. Sci.* **2022**, *12*, 1258. [CrossRef]
12. Kaliannan, D.; Palaninaicker, S.; Palanivel, V.; Dhamodaran, K.; Kadirvelu, K.; Kumaran, S. Sol-Gel Mediated Synthesis of Silica Nanoparticle from *Bambusa Vulgaris* Leaves and Its Environmental Applications: Kinetics and Isotherms Studies. *J. Sol-Gel Sci. Technol.* **2019**, *90*, 653–664. [CrossRef]
13. Li, Y.; Wang, R.; Luo, X.; Chen, Z.; Wang, L.; Zhou, Y.; Liu, W.; Cheng, M.; Zhang, C. Synthesis of Rice Husk-Based MCM-41 for Removal of Aflatoxin B1 from Peanut Oil. *Toxins* **2022**, *14*, 87. [CrossRef] [PubMed]
14. Sarti, E.; Chenet, T.; Stevanin, C.; Costa, V.; Cavazzini, A.; Catani, M.; Martucci, A.; Precisvalle, N.; Beltrami, G.; Pasti, L. High-Silica Zeolites as Sorbent Media for Adsorption and Pre-Concentration of Pharmaceuticals in Aqueous Solutions. *Molecules* **2020**, *25*, 3331. [CrossRef] [PubMed]
15. Medeiros de Paula, G.; do Nascimento Rocha de Paula, L.; Freire Rodrigues, M.G. Production of MCM-41 and SBA-15 Hybrid Silicas from Industrial Waste. *Silicon* **2022**, *14*, 439–447. [CrossRef]
16. Razak, N.A.A.; Othman, N.H.; Shayuti, M.S.M.; Jumahat, A.; Sapiai, N.; Lau, W.J. Agricultural and industrial waste-derived mesoporous silica nanoparticles: A review on chemical synthesis route. *J. Env. Chem. Eng.* **2022**, *10*, 107322. [CrossRef]
17. Miao, C.; Liang, L.; Zhang, F.; Chen, S.; Shang, K.; Jiang, J.; Zhang, Y.; Ouyang, J. Review of the fabrication and application of porous materials from silicon-rich industrial solid waste. *Int. J. Miner. Metall. Mater.* **2022**, *29*, 424–438. [CrossRef]
18. Wu, S.; Li, F.; Xu, R.; Wei, S.; Li, G. Synthesis of Thiol-Functionalized MCM-41 Mesoporous Silicas and Its Application in Cu(II), Pb(II), Ag(I), and Cr(III) Removal. *J. Nanopart. Res.* **2010**, *12*, 2111–2124. [CrossRef]
19. Zhu, W.; Wang, J.; Wu, D.; Li, X.; Luo, Y.; Han, C.; Ma, W.; He, S. Investigating the Heavy Metal Adsorption of Mesoporous Silica Materials Prepared by Microwave Synthesis. *Nanoscale Res. Lett.* **2017**, *12*, 323. [CrossRef]
20. Putz, A.-M.; Ciopec, M.; Negrea, A.; Grad, O.; Ianăşi, C.; Ivankov, O.; Milanović, M.; Stijepović, I.; Almásy, L. Comparison of Structure and Adsorption Properties of Mesoporous Silica Functionalized with Aminopropyl Groups by the Co-Condensation and the Post Grafting Methods. *Materials* **2021**, *14*, 628. [CrossRef]
21. Gervas, C.; Mubofu, E.B.; Mdoe, J.E.G.; Revaprasadu, N. Functionalized mesoporous organo-silica nanosorbents for removal of chromium (III) ions from tanneries wastewater. *J. Porous Mater.* **2016**, *23*, 83–93. [CrossRef]
22. Liu, S.; Lu, L.; Yang, Z.; Cool, P.; Vansant, E.F. Further Investigations on the Modified Stöber Method for Spherical MCM-41. *Mater. Chem. Phys.* **2006**, *97*, 203–206. [CrossRef]
23. Liu, S.; Cool, P.; Collart, O.; Van Der Voort, P.; Vansant, E.F.; Lebedev, O.I.; Van Tendeloo, G.; Jiang, M. The Influence of the Alcohol Concentration on the Structural Ordering of Mesoporous Silica: Cosurfactant versus Cosolvent. *J. Phys. Chem. B* **2003**, *107*, 10405–10411. [CrossRef]

24. Ngoc Thi Le, H.; Jeong, H.K. Synthesis and Characterization of Uniform Silica Nanoparticles on Nickel Substrate by Spin Coating and Sol–Gel Method. *Chem. Phys. Lett.* **2014**, *592*, 349–354. [CrossRef]
25. El-Safty, S.A. Instant Synthesis of Mesoporous Monolithic Materials with Controllable Geometry, Dimension and Stability: A Review. *J. Porous Mater.* **2011**, *18*, 259–287. [CrossRef]
26. Courtois, J.; Byström, E.; Irgum, K. Novel Monolithic Materials Using Poly(Ethylene Glycol) as Porogen for Protein Separation. *Polymer* **2006**, *47*, 2603–2611. [CrossRef]
27. Tan, S.; Wu, Q.; Wang, J.; Wang, Y.; Liu, X.; Sui, K.; Deng, X.; Wang, H.; Wu, M. Dynamic Self-Assembly Synthesis and Controlled Release as Drug Vehicles of Porous Hollow Silica Nanoparticles. *Microporous Mesoporous Mater.* **2011**, *142*, 601–608. [CrossRef]
28. Zhang, Y.B.; Qian, X.F.; Li, Z.K.; Yin, J.; Zhu, Z.K. Synthesis of Novel Mesoporous Silica Spheres with Starburst Pore Canal Structure. *J. Solid State Chem.* **2004**, *177*, 844–848. [CrossRef]
29. Pabón, E.; Retuert, J.; Quijada, R. Synthesis of Mixed Silica–Titania by the Sol–Gel Method Using Polyethylenimine: Porosity and Catalytic Properties. *J. Porous Mater.* **2007**, *14*, 151–158. [CrossRef]
30. Pirson, A.; Mohsine, A.; Marchot, P.; Michaux, B.; van Cantfort, O.; Pirard, J.P.; Lecloux, A.J. Synthesis of SiO_2-TiO_2 Xerogels by Sol-Gel Process. *J. Sol-Gel Sci. Technol.* **1995**, *4*, 179–185. [CrossRef]
31. Dudás, Z.; Len, A.; Ianăşi, C.; Paladini, G. Structural Modifications Caused by the Increasing MTES Amount in Hybrid MTES/TEOS-based Silica Xerogels. *Mater. Charact.* **2020**, *167*, 110519. [CrossRef]
32. Al-Oweini, R.; El-Rassy, H. Synthesis and Characterization by FTIR Spectroscopy of Silica Aerogels Prepared Using Several $Si(OR)_4$ and $R''Si(OR')_3$ Precursors. *J. Mol. Struct.* **2009**, *919*, 140–145. [CrossRef]
33. Brinker, C.J.; Scherer, G.W. *Sol-Gel Science: The Physics and Chemistry of Sol-Gel Processing*; Academic Press: New York, NY, USA, 1990; pp. 581–585.
34. Gallardo, J.; Galliano, P.; Durán, A. Thermal Evolution of Hybrid Sol-Gel Silica Coatings: A Structural Analysis. *J. Sol-Gel Sci. Technol.* **2000**, *19*, 393–397. [CrossRef]
35. Lenza, R.F.S.; Vasconcelos, W.L. Preparation of Silica by Sol-Gel Method Using Formamide. *Mat. Res.* **2001**, *4*, 175–179. [CrossRef]
36. Putz, A.-M.; Cecilia, S.; Ianăşi, C.; Dudás, Z.; Székely, K.N.; Plocek, J.; Sfârloagă, P.; Săcărescu, L.; Almásy, L. Pore Ordering in Mesoporous Matrices Induced by Different Directing Agents. *J. Porous Mater.* **2015**, *22*, 321–331. [CrossRef]
37. Cychosz, K.A.; Thommes, M. Progress in the Physisorption Characterization of Nanoporous Gas Storage Materials. *Engineering* **2018**, *4*, 559–566. [CrossRef]
38. Thommes, M.; Kaneko, K.; Neimark, A.V.; Olivier, J.P.; Rodriguez-Reinoso, F.; Rouquerol, J.; Sing, K.S.W. Physisorption of Gases, with Special Reference to the Evaluation of Surface Area and Pore Size Distribution (IUPAC Technical Report). *Pure Appl. Chem.* **2015**, *87*, 1051–1069. [CrossRef]
39. Ahmad, A.L.; Mustafa, N.N.N. Pore Surface Fractal Analysis of Palladium-Alumina Ceramic Membrane Using Frenkel-Halsey-Hill (FHH) Model. *J. Colloid Interface Sci.* **2006**, *301*, 575–584. [CrossRef]
40. Lelong, G.; Bhattacharyya, S.; Kline, S.; Cacciaguerra, T.; Gonzalez, M.A.; Saboungi, M.-L. Effect of Surfactant Concentration on the Morphology and Texture of MCM-41 Materials. *J. Phys. Chem. C* **2008**, *112*, 10674–10680. [CrossRef]
41. Soper, A.K. Radical Re-Appraisal of Water Structure in Hydrophilic Confinement. *Chem. Phys. Lett.* **2013**, *590*, 1–15. [CrossRef]
42. Goworek, J.; Kierys, A.; Gac, W.; Borówka, A.; Kusak, R. Thermal Degradation of CTAB in As-Synthesized MCM-41. *J. Therm. Anal. Calorim.* **2009**, *96*, 375–382. [CrossRef]
43. Foo, K.Y.; Hameed, B.H. Insights into the Modeling of Adsorption Isotherm Systems. *Chem. Eng. J.* **2010**, *156*, 2–10. [CrossRef]
44. Alberti, G.; Amendola, V.; Pesavento, M.; Biesuz, R. Beyond the Synthesis of Novel Solid Phases: Review on Modelling of Sorption Phenomena. *Coord. Chem. Rev.* **2012**, *256*, 28–45. [CrossRef]
45. Langmuir, I. The Adsorption of Gases on Plane Surfaces of Glass, Mica and Platinum. *J. Am. Chem. Soc.* **1918**, *40*, 1361–1403. [CrossRef]
46. Freundlich, H. Über die Adsorption in Lösungen. *Z. Phys. Chem.* **1906**, *57*, 385–470. [CrossRef]
47. Sips, R. On the Structure of a Catalyst Surface. *J. Chem. Phys.* **1948**, *16*, 490–495. [CrossRef]
48. Ciopec, M.; Davidescu, C.; Negrea, A.; Lupa, L.; Negrea, P.; Popa, A.; Muntean, C. Use of D2EHPA impregnated XAD7 resin for the removal of Cd(II) and Zn(II) from aqueous solutions. *Environ. Eng. Manag. J.* **2011**, *10*, 1597–1608. [CrossRef]
49. Anirudhan, T.S.; Senan, P.; Suchithra, P.S. Evaluation of Iron(III)-Coordinated Amino-Functionalized Poly(Glycidyl Methacrylate)-Grafted Cellulose for Arsenic(V) Adsorption from Aqueous Solutions. *Water Air Soil Pollut.* **2011**, *220*, 101–116. [CrossRef]
50. Mishra, P.C.; Patel, R.K. Removal of Lead and Zinc Ions from Water by Low Cost Adsorbents. *J. Hazard. Mater.* **2009**, *168*, 319–325. [CrossRef]
51. Keochaiyom, B.; Wan, J.; Zeng, G.; Huang, D.; Xue, W.; Hu, L.; Huang, C.; Zhang, C.; Cheng, M. Synthesis and Application of Magnetic Chlorapatite Nanoparticles for Zinc (II), Cadmium (II) and Lead (II) Removal from Water Solutions. *J. Colloid Interface Sci.* **2017**, *505*, 824–835. [CrossRef]
52. Nguyen, T.C.; Loganathan, P.; Nguyen, T.V.; Vigneswaran, S.; Kandasamy, J.; Naidu, R. Simultaneous Adsorption of Cd, Cr, Cu, Pb, and Zn by an Iron-Coated Australian Zeolite in Batch and Fixed-Bed Column Studies. *Chem. Eng. J.* **2015**, *270*, 393–404. [CrossRef]
53. Hubert, B.; Setiawan, Y.; Soetaredjo, F.E.; Hartono, S.B. Adsorption of Pb on Thiol Modified Magnetic Mesoporous Silica. *ARPN J. Eng. Appl. Sci.* **2018**, *13*, 859–863.

54. Fan, H.; Ma, X.; Zhou, S.; Huang, J.; Liu, Y.; Liu, Y. Highly Efficient Removal of Heavy Metal Ions by Carboxymethyl Cellulose-Immobilized Fe_3O_4 Nanoparticles Prepared via High-Gravity Technology. *Carbohydr. Polym.* **2019**, *213*, 39–49. [CrossRef] [PubMed]
55. Tian, Q.; Wang, X.; Mao, F.; Guo, X. Absorption Performance of DMSA Modified Fe_3O_4@SiO_2 Core/Shell Magnetic Nanocomposite for Pb^{2+} Removal. *J. Cent. South Univ.* **2018**, *25*, 709–718. [CrossRef]
56. Ahmad, N.; Sereshti, H.; Mousazadeh, M.; Rashidi Nodeh, H.; Kamboh, M.A.; Mohamad, S. New Magnetic Silica-Based Hybrid Organic-Inorganic Nanocomposite for the Removal of Lead(II) and Nickel(II) Ions from Aqueous Solutions. *Mater. Chem. Phys.* **2019**, *226*, 73–81. [CrossRef]
57. Wang, J.; Zheng, S.; Shao, Y.; Liu, J.; Xu, Z.; Zhu, D. Amino-Functionalized Fe_3O_4@SiO_2 Core-Shell Magnetic Nanomaterial as a Novel Adsorbent for Aqueous Heavy Metals Removal. *J. Colloid Interface Sci.* **2010**, *349*, 293–299. [CrossRef]
58. Dursun, A.Y. A Comparative Study on Determination of the Equilibrium, Kinetic and Thermodynamic Parameters of Biosorption of Copper(II) and Lead(II) Ions onto Pretreated *Aspergillus niger*. *Biochem. Eng. J.* **2006**, *28*, 187–195. [CrossRef]
59. Rajput, S.; Singh, L.P.; Pittman, C.U.; Mohan, D. Lead (Pb^{2+}) and Copper (Cu^{2+}) Remediation from Water Using Superparamagnetic Maghemite (γ-Fe_2O_3) Nanoparticles Synthesized by Flame Spray Pyrolysis (FSP). *J. Colloid Interface Sci.* **2017**, *492*, 176–190. [CrossRef]
60. Giraldo, L.; Erto, A.; Moreno-Piraján, J.C. Magnetite Nanoparticles for Removal of Heavy Metals from Aqueous Solutions: Synthesis and Characterization. *Adsorption* **2013**, *19*, 465–474. [CrossRef]
61. Mahdavi, S.; Jalali, M.; Afkhami, A. Heavy Metals Removal from Aqueous Solutions Using TiO_2, MgO, and Al_2O_3 Nanoparticles. *Chem. Eng. Commun.* **2013**, *200*, 448–470. [CrossRef]
62. Hadi Najafabadi, H.; Irani, M.; Roshanfekr Rad, L.; Heydari Haratameh, A.; Haririan, I. Removal of Cu^{2+}, Pb^{2+} and Cr^{6+} from Aqueous Solutions Using a Chitosan/Graphene Oxide Composite Nanofibrous Adsorbent. *RSC Adv.* **2015**, *5*, 16532–16539. [CrossRef]
63. Tighadouini, S.; Radi, S.; Bacquet, M.; Degoutin, S.; Zaghrioui, M.; Jodeh, S.; Warad, I. Removal Efficiency of Pb(II), Zn(II), Cd(II) and Cu(II) from Aqueous Solution and Natural Water by Ketoenol–Pyrazole Receptor Functionalized Silica Hybrid Adsorbent. *Sep. Sci. Technol.* **2017**, *52*, 608–621. [CrossRef]
64. Raji, F.; Saraeian, A.; Pakizeh, M.; Attarzadeh, F. Removal of Pb(II) from Aqueous Solution by Mesoporous Silica MCM-41 Modified by $ZnCl_2$: Kinetics, Thermodynamics, and Isotherms. *RSC Adv.* **2015**, *5*, 37066–37077. [CrossRef]
65. Huang, J.; Ye, M.; Qu, Y.; Chu, L.; Chen, R.; He, Q.; Xu, D. Pb (II) Removal from Aqueous Media by EDTA-Modified Mesoporous Silica SBA-15. *J. Colloid Interface Sci.* **2012**, *385*, 137–146. [CrossRef] [PubMed]
66. Unob, F.; Wongsiri, B.; Phaeon, N.; Puanngam, M.; Shiowatana, J. Reuse of Waste Silica as Adsorbent for Metal Removal by Iron Oxide Modification. *J. Hazard. Mater.* **2007**, *142*, 455–462. [CrossRef]
67. Nicola, R.; Costişor, O.; Ciopec, M.; Negrea, A.; Lazău, R.; Ianăşi, C.; Picioruş, E.-M.; Len, A.; Almásy, L.; Szerb, E.I.; et al. Silica-Coated Magnetic Nanocomposites for Pb^{2+} Removal from Aqueous Solution. *Appl. Sci.* **2020**, *10*, 2726. [CrossRef]
68. Qureashi, A.; Pandith, A.H.; Bashir, A.; Manzoor, T.; Malik, L.A.; Sheikh, F.A. Citrate Coated Magnetite: A Complete Magneto Dielectric, Electrochemical and DFT Study for Detection and Removal of Heavy Metal Ions. *Surf. Interfaces* **2021**, *23*, 101004. [CrossRef]
69. Melnyk, I.V.; Pogorilyi, R.P.; Zub, Y.L.; Vaclavikova, M.; Gdula, K.; Dąbrowski, A.; Seisenbaeva, G.A.; Kessler, V.G. Protection of Thiol Groups on the Surface of Magnetic Adsorbents and Their Application for Wastewater Treatment. *Sci. Rep.* **2018**, *8*, 8592. [CrossRef]
70. Zhang, X.; Cheng, T.; Chen, C.; Wang, L.; Deng, Q.; Chen, G.; Ye, C. Synthesis of a Novel Magnetic Nano-Zeolite and Its Application as an Efficient Heavy Metal Adsorbent. *Mater. Res. Express* **2020**, *7*, 085007. [CrossRef]
71. Kaliannan, D.; Palaninaicker, S.; Palanivel, V.; Mahadeo, M.A.; Ravindra, B.N.; Jae-Jin, S. A Novel Approach to Preparation of Nano-Adsorbent from Agricultural Wastes (*Saccharum Officinarum* Leaves) and Its Environmental Application. *Environ. Sci. Pollut. Res.* **2019**, *26*, 5305–5314. [CrossRef]
72. Yang, P.; Li, F.; Wang, B.; Niu, Y.; Wei, J.; Yu, Q. In Situ Synthesis of Carbon Nanotube-Steel Slag Composite for Pb(II) and Cu(II) Removal from Aqueous Solution. *Nanomaterials* **2022**, *12*, 1199. [CrossRef] [PubMed]
73. Kotsyuda, S.S.; Tomina, V.V.; Zub, Y.L.; Furtat, I.M.; Lebed, A.P.; Vaclavikova, M.; Melnyk, I.V. Bifunctional Silica Nanospheres with 3-Aminopropyl and Phenyl Groups. Synthesis Approach and Prospects of Their Applications. *Appl. Surf. Sci.* **2017**, *420*, 782–791. [CrossRef]
74. Mahmoud, M.E.; Fekry, N.A.; El-Latif, M.M.A. Nanocomposites of Nanosilica-Immobilized-Nanopolyaniline and Crosslinked Nanopolyaniline for Removal of Heavy Metals. *Chem. Eng. J.* **2016**, *304*, 679–691. [CrossRef]
75. Ramkumar, J.; Majeed, J.; Chandramouleeswaran, S. Insight to Sorption Mechanism Employing Nanocomposite: Case Study of Toxic Species Removal. *Microporous Mesoporous Mater.* **2021**, *314*, 110858. [CrossRef]
76. De, S.; Malik, S.; Ghosh, A.; Saha, R.; Saha, B. A review on natural surfactants. *RSC Adv.* **2015**, *5*, 65757. [CrossRef]
77. Putz, A.-M.; Wang, K.; Len, A.; Plocek, J.; Bezdicka, P.; Kopitsa, G.P.; Khamova, T.V.; Ianăşi, C.; Săcărescu, L.; Mitróová, Z.; et al. Mesoporous Silica Obtained with Methyltriethoxysilane as Co-Precursor in Alkaline Medium. *Appl. Surf. Sci.* **2017**, *424*, 275–281. [CrossRef]

78. Kuklin, A.I.; Soloviov, D.V.; Rogachev, A.V.; Utrobin, P.K.; Kovalev, Y.S.; Balasoiu, M.; Ivankov, O.I.; Sirotin, A.P.; Murugova, T.N.; Petukhova, T.B.; et al. New Opportunities Provided by Modernized Small-Angle Neutron Scattering Two-Detector System Instrument (YuMO). *J. Phys. Conf. Ser.* **2011**, *291*, 12013. [CrossRef]
79. Kuklin, A.I.; Rogov, A.D.; Gorshkova, Y.E.; Utrobin, P.K.; Kovalev, Y.S.; Rogachev, A.V.; Ivankov, O.I.; Kutuzov, S.A.; Soloviov, D.V.; Gordeliy, V.I. Analysis of Neutron Spectra and Fluxes Obtained with Cold and Thermal Moderators at IBR-2 Reactor: Experimental and Computer-Modeling Studies. *Phys. Part. Nucl. Lett.* **2011**, *8*, 119. [CrossRef]
80. Kuklin, A.I.; Islamov, A.K.; Gordeliy, V.I. Scientific Reviews: Two-Detector System for Small-Angle Neutron Scattering Instrument. *Neutron News* **2005**, *16*, 16–18. [CrossRef]
81. Soloviev, A.G.; Solovjeva, T.M.; Ivankov, O.I.; Soloviov, D.V.; Rogachev, A.V.; Kuklin, A.I. SAS program for two-detector system: Seamless curve from both detectors. *J. Phys. Conf. Ser.* **2017**, *848*, 012020. [CrossRef]
82. Strunz, P.; Saroun, J.; Mikula, P.; Lukas, P.; Eichhorn, F. Double-Bent-Crystal Small-Angle Neutron Scattering Setting and its Applications. *J. Appl. Cryst.* **1997**, *30*, 844–848. [CrossRef]

gels

MDPI

Article

4-Dialkylamino-2,5-dihydroimidazol-1-oxyls with Functional Groups at the Position 2 and at the Exocyclic Nitrogen: The pH-Sensitive Spin Labels

Dmitrii G. Trofimov [1], Yuri I. Glazachev [2], Artem A. Gorodetsky [1], Denis A. Komarov [1], Tatyana V. Rybalova [1] and Igor A. Kirilyuk [1,*]

[1] N. N. Vorozhtsov Novosibirsk Institute of Organic Chemistry SB RAS, 630090 Novosibirsk, Russia; ditrof@nioch.nsc.ru (D.G.T.); gorodaa@nioch.nsc.ru (A.A.G.); dkomarov@nioch.nsc.ru (D.A.K.); rybalova@nioch.nsc.ru (T.V.R.)

[2] Voevodsky Institute of Chemical Kinetics and Combustion SB RAS, 630090 Novosibirsk, Russia; glaza@kinetics.nsc.ru

* Correspondence: kirilyuk@nioch.nsc.ru

Abstract: Local acidity and electrostatic interactions are associated both with catalytic properties and the adsorption activity of various materials, and with the vital functions of biomolecules. The observation of acid–base equilibria in stable free radicals using EPR spectroscopy represents a convenient method for monitoring pH changes and the investigation of surface electrostatics, the advantages of which are especially evident in opaque and turbid samples and in porous materials such as xerogels. Imidazoline nitroxides are the most commonly used pH-sensitive spin probes and labels due to the high sensitivity of the parameters of the EPR spectra to pH changes, their small size, and their well-developed chemistry. In this work, several new derivatives of 4-(*N*,*N*-dialkylamino)-2,5-dihydrioimidazol-1-oxyl, with functional groups suitable for specific binding, were synthesized. The dependence of the parameters of their EPR spectra on pH was studied. Several showed a pK_a close to 7.4, following the pH changes in a normal physiological range, and some demonstrated a monotonous change of the hyperfine coupling constant by 0.14 mT upon pH variation by four units.

Keywords: nitroxide; spin label; spin probe; EPR; local pH; surface electrostatics; near-surface layer

Citation: Trofimov, D.G.; Glazachev, Y.I.; Gorodetsky, A.A.; Komarov, D.A.; Rybalova, T.V.; Kirilyuk, I.A. 4-Dialkylamino-2,5-dihydroimidazol-1-oxyls with Functional Groups at the Position 2 and at the Exocyclic Nitrogen: The pH-Sensitive Spin Labels. *Gels* **2022**, *8*, 11. https://doi.org/10.3390/gels8010011

Academic Editors: Francesco Caridi, Giuseppe Paladini and Andrea Fiorati

Received: 2 December 2021
Accepted: 17 December 2021
Published: 23 December 2021

1. Introduction

Interfacial phenomena and local protonation effects play an important role in biophysics, biochemistry, and in the chemistry of heterogeneous systems [1]. Catalytic and sorption properties of various materials are dependent on the local acidity and electrostatic interactions inside the pores [2]. Measurements of the local acidity and electrostatic potential of the inner pore surfaces represent a problem of great practical interest. Several methods have been developed for the characterization of the acid–base properties of different surface locations [3].

EPR spectroscopy of ionizable nitroxides is a convenient method for the investigation of the above-mentioned phenomena [2,4,5], and is fully applicable to opaque or turbid materials [5,6]. Nitroxide spin probes are small enough to penetrate directly into the pores and to be adsorbed onto the surface of the material under study. The protonation of basic centers in specially designed spin probes affects the hyperfine coupling *A*-tensor and *g*-factor matrix, as well as the rotational dynamics of the nitroxide molecule in the proximity of charged surfaces, and this is reflected in the EPR spectra [7]. An analysis of these data gives information about the acidic centers in the material and the local surface electrostatic potential. Recently, EPR studies using pH-sensitive spin probes were successfully used for the investigation of binary TiO_2-SiO_2 xerogels [8].

Imidazoline nitroxides are the most commonly used pH-sensitive spin probes and labels due to the high sensitivity of the parameters of the EPR spectra to pH changes, their small size, and their well-developed chemistry. A large number of pH-sensitive nitroxides of the imidazoline series have been prepared [1,4,9]. Some of them are highly sensitive to pH changes in physiologically important regions. The development of a convenient method for the synthesis of 4-(*N*,*N*-dialkylamino)-2,5-dihydrioimidazol-1-oxyls from 4*H*-imidazole-3-oxides [10] allowed for easy variation of the substituents in position two of the heterocycle to prepare useful spin probes. Examples illustrating the benefits of this strategy include the synthesis of nitroxides with two pK_a values showing high sensitivity in a broad range of pH [11,12], e.g., Scheme 1, label **1**, and pH-sensitive alkylating spin, labels **2a–c** [13–15], which were used to prepare the hydrophilic spin probes from glutathione [13,14], thiol-specific pH-sensitive spin, label **3**, for site-directed labeling of proteins and lipids [16,17], and siloxane-derived spin, label **4**, capable of binding to silica or alumina surfaces [7]. Despite the significant advances in this area, the broad variety of potential research objects produces a request for new pH-sensitive spin labels capable of specific attachment. Here we describe a new set of pH-sensitive imidazoline nitroxides with various functional groups in the side chain. Some of them may find an application in material science or in biophysics.

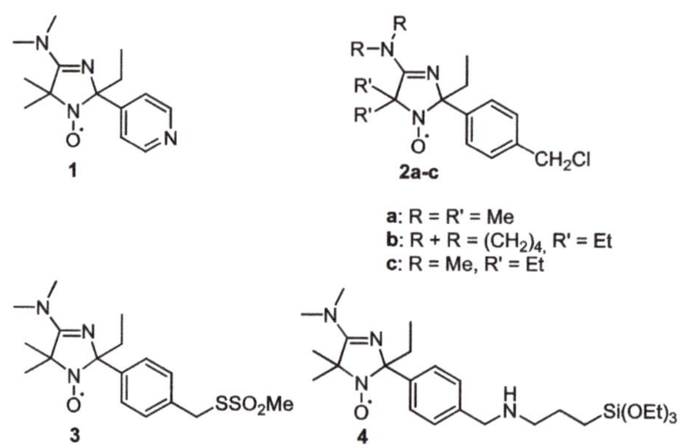

Scheme 1. Structure of the nitroxides **1–4**.

2. Results and Discussion

The high reactivity of **2** in nucleophilic substitution reactions offers easy access to new functional derivatives. Expectedly, **2b** readily reacts with sodium azide to produce **5** with a nearly quantitative yield. Nitroxide azides can be used for the spin labeling of acetylene-modified molecules via Huisgen 1,3-dipolar cycloaddition [18,19]. In analogy to the literature [20], a reaction of **5** with tetraisopropyl but-3-yne-1,1-diyldiphosphonate **6** in the presence of Cu(II) salt and ascorbic acid after subsequent re-oxidation produced **7** (Scheme 2).

The addition of nitroxides with a terminal acetylene group to azide-modified biomolecules, e.g., nucleic acids, is another way to use the Huisgen-click reaction for spin labeling. Terminal acetylenes can also be attached via Pd-catalyzed coupling [21]. A spin label with a terminal acetylene group was prepared from **8** in two steps (Scheme 3). The oxidation of benzyl alcohol **8** with the activated manganese dioxide in methanol smoothly led to the formation of the corresponding aldehyde **9**, with the nitroxyl group and the amidine moiety being unaffected. Alternatively, **8** can be oxidized to the aldehyde **9** with 1-oxo-2,2,6,6-tetramethylpiperidinium chloride **10** with similar yield. The aldehyde **9** readily reacts with Bestmann-Ohira reagent to produce **11** with a yield of 84% [22]. The structure of **11** was confirmed with X-ray analysis data (Figure S1).

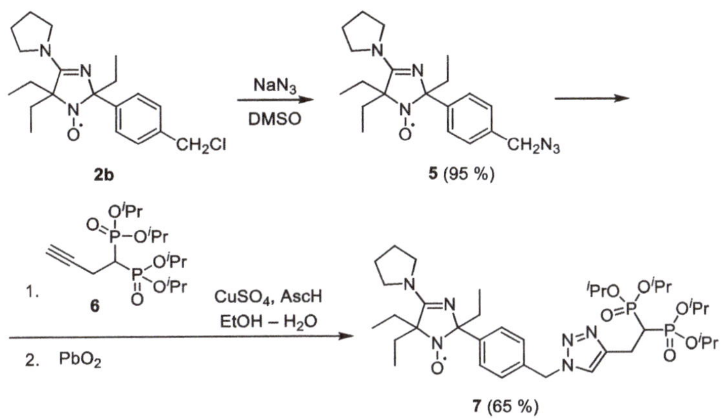

Scheme 2. Synthesis of the nitroxide **7**.

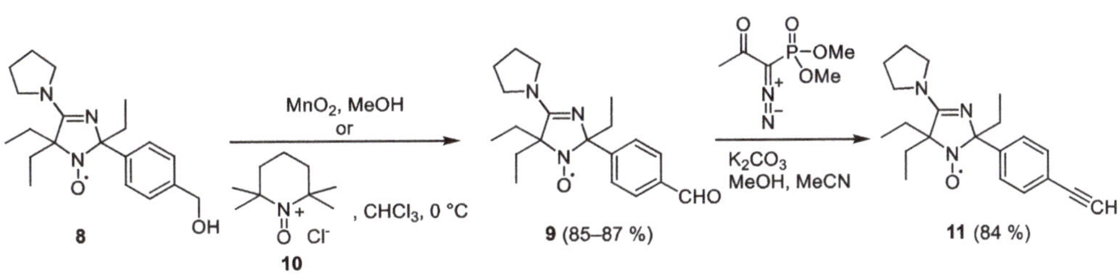

Scheme 3. Synthesis of the nitroxide **11**.

Unless the attachment of a pH-sensitive nitroxide to a primary amino group is successfully performed via alkylation with **2b** [7], the acylation reaction is more selective, allowing for the binding of a single nitroxide to the site. Carboxylic acids can be easily prepared from **8**. Here we used a convenient one-pot process, where the reaction of **8** with an oxoammonium salt **10** was followed by the Lindgren-Kraus-Pinnick procedure [23] (Scheme 4). The carboxylic acid **12** was isolated with a 95% yield. To activate carboxylic group for acylation, the nitroxide **12** was treated with $SOCl_2$ in the presence of pyridine. The chloroanhydride formed readily reacted with ethanol to give ester **14**. The reaction of **13** with *N*-hydroxysuccinimide (NHS) produced spin label **15**.

Another nitroxide with a carboxylic group on a longer spacer was prepared in one step from **8** via acylation with succinic anhydride (Scheme 5).

The EPR spectra of nitroxides **7**, **11**, **12**, **15**, and **16** are strongly pH-dependent with $\Delta a_N > 0.1$ mT, and a pK_a between 6 and 6.7 (Table 1). The titration curves demonstrate optimal sensitivity in slightly acidic media [14,15], but the sensitivity is not optimal in the normal physiological range of 7.35–7.45 [24]. Similar structures without aromatic substituents are known to show higher pK_a values [11].

To prepare 2-functionalized nitroxides with a pK_a above 7, the reaction of 4*H*-imidazol-3-oxide **17** with Grignard reagents was used (Scheme 6). The treatment of **17** with alkenylmagnesium bromides produced nitroxides with a terminal ethylene bond **18a,b**. The hydroboration of **18a,b** with 9-BBN, followed by oxidation with hydrogen peroxide, was performed using the protocol developed by Hideg for 2-allyl pyrrolidine nitroxides [25]. The reaction produced alcohols **19a,b**, which were then treated with carbonyldiimidazole (CDI) to give **20a,b**. To demonstrate the feasibility of the carbonylimidazole pH-sensitive spin labels for binding to primary amino groups, **20b** was allowed to react with *N*,*N*-diethyl-1,3-diaminopropane.

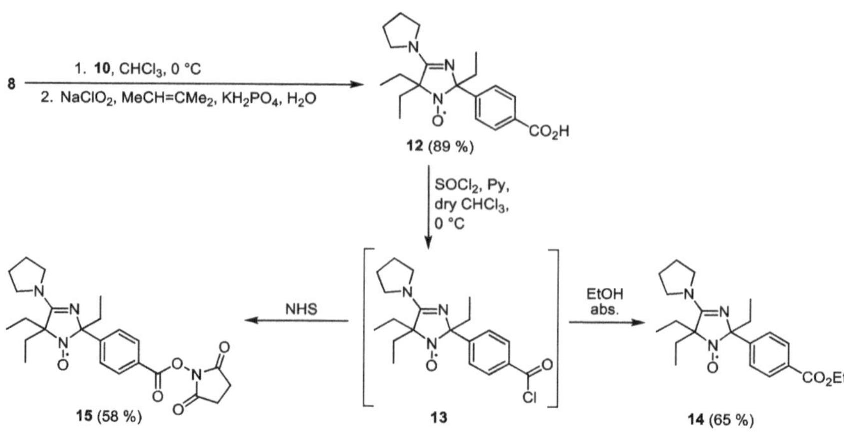

Scheme 4. Synthesis and esterification of **12**.

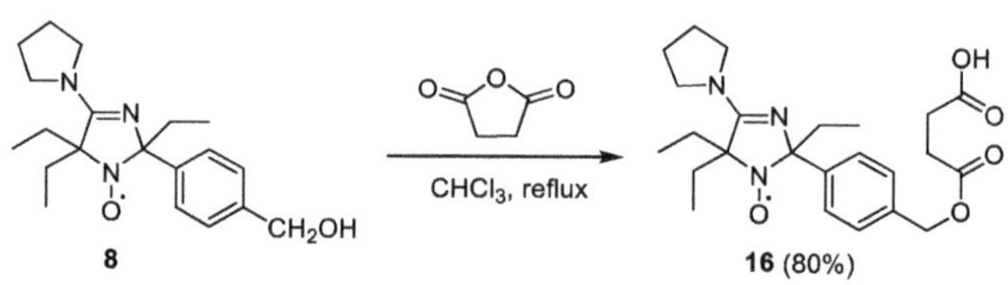

Scheme 5. A reaction of **8** with succinic anhydride.

Table 1. Titration data for the pH-sensitive niroxides: HFC constants for protonated and unprotonated forms, changes in hyperfine splitting between these forms, $\Delta a_{N,}$, and pK_a values.

Nitroxide	a_N, mT		Δa_N, mT	pK_a
	R·H+	R·		
5	1.434	1.539	0.105	6.42
7	1.429	1.531	0.102	6.24
8	1.424	1.529	0.105	6.54
11	1.436	1.539	0.103	6.21
12	1.417	1.522	0.105	6.50
15	1.422	1.531	0.109	6.49
16	1.466	1.575	0.109	6.64
18a	1.479	1.586	0.107	7.14
18b	1.468	1.565	0.097	7.20
19a	1.493	1.589	0.096	7.25
19b	1.472	1.567	0.095	7.28
20a	1.459	1.555	0.096	6.95
20b	1.448	1.545	0.097	7.24 ± 0.05
21	1.472	1.565	0.093	7.19
22	1.468	1.562	0.094	7.64
23	1.471	1.569	0.098	7.09
24	1.434	1.529	0.095	6.81 ± 0.05
25	1.482	1.587	0.105	7.50

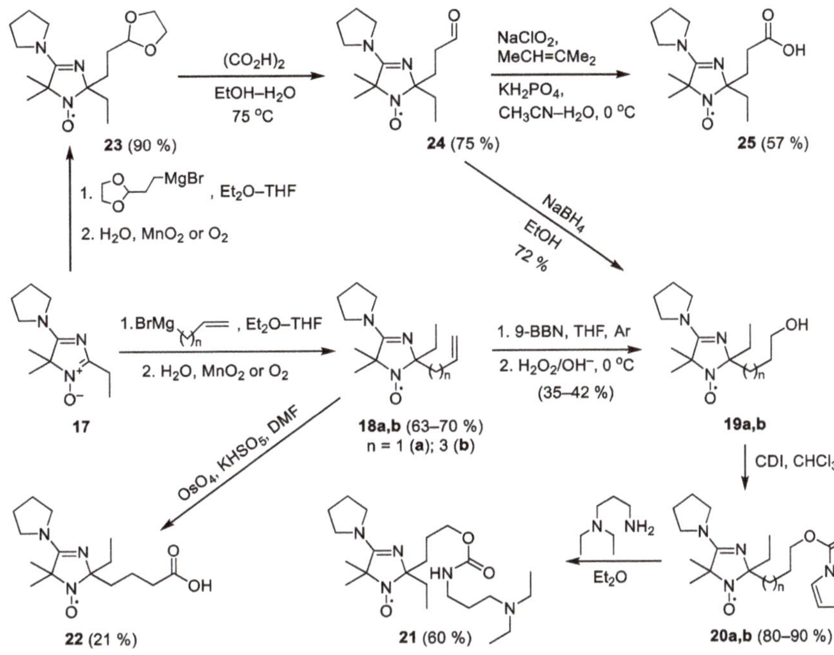

Scheme 6. Synthesis of spin labels and spin probes from **17**.

The nitroxide **21** was isolated with a 60% yield. The oxidative cleavage of the terminal double carbon–carbon bond in **18b** with osmium tetraoxide–oxone system yielded carboxylic acid **22**.

The addition of 2-(1,3-dioxolan-2-yl) ethylmagnesium bromide to **17** is another convenient way to create 2-functionalized pH-sensitive spin labels. The reaction produced nitroxide **23** with a high yield. The dioxolane protection group in **23** was readily removed under relatively mild conditions to give the corresponding aldehyde **24**, which can be either oxidized to carboxylic acid **25** using the Lindgren-Kraus-Pinnick procedure, or reduced with sodium borohydride to **19a**. The sequence **17** → **23** → **24** → **19a** gives a remarkably higher yield of the target nitroxide than the addition of allylmagnesium bromide with subsequent hydroboration. A titration of the nitroxides **20**, **22**, and **25** showed that they may be valuable spin labels and probes with high sensitivities to changes of pH within the physiological range (see Table 1, Figure 1 and Supplementary Materials).

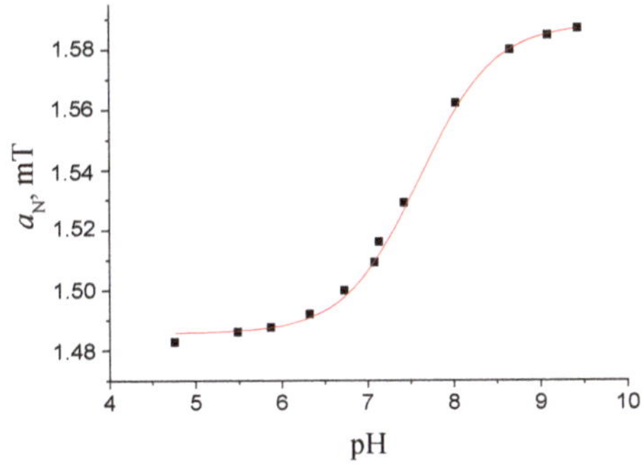

Figure 1. Titration curve of nitroxide **25**.

An investigation of the surfaces of many inorganic and organo-inorganic materials (catalysts, sorbents, etc.) requires nitroxides with a high sensitivity to acidity changes within a broad range of pH. A good example of such a spin probe is two-pK_a nitroxide **1**, which was successfully used in numerous studies [8,26–31]. A covalent attachment of similar nitroxides to the surface of a catalyst or a sorbent may provide a useful method for studies of the near-surface layer in these materials. Here we designed analogs of **1** with a functional group in a substituent at the exocyclic nitrogen atom of the amidine moiety.

N-(4-(1,3-dioxolan-2-yl)benzyl)-*N*-methylamine **26** was prepared in two steps from tereftaldicarboxaldehyde **27** (Scheme 7).

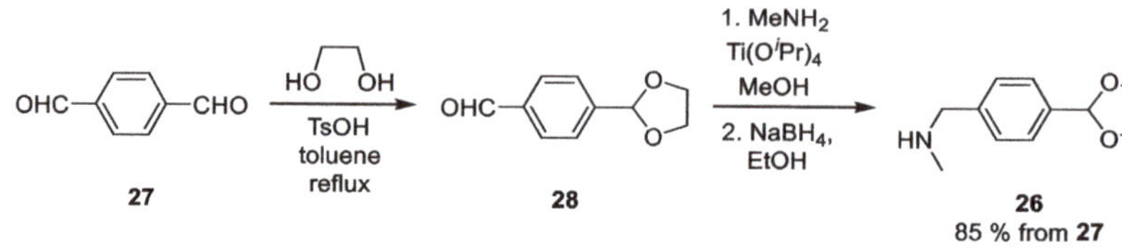

Scheme 7. Synthesis of *N*-(4-(1,3-dioxolan-2-yl)benzyl)-*N*-methylamine (**26**).

A reaction of the 5-cyano-4*H*-imidazole-3-oxide **29** with **26** resulted in cyanide substitution with the formation of **30**, and the latter was treated with an excess of ethylmagnesium bromide (Scheme 8). The nitroxide **31** was isolated after a quenching of the reaction mixture with water and oxidation. To hydrolyze the dioxolane ring, **31** was heated to reflux in 0.5 M aqueous HCl. The resulting aldehyde **32** was reduced with sodium borohydride to the corresponding alcohol **34**, or oxidized with sodium chlorite to carboxylic acid **33** as described above for **23**. Similarly to **11**, the nitroxide **33** was converted into succinimidyl ester **35** via a reaction of in situ generated chloroanhydride with NHS.

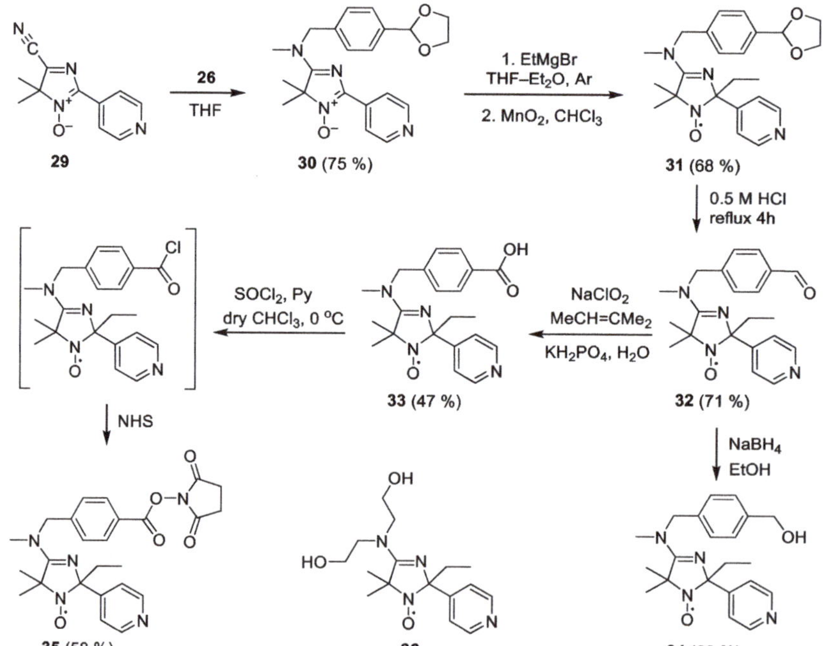

Scheme 8. Synthesis of two-pK_a nitroxides. Structure of the nitroxide **36**.

Titration of the nitroxides **32–35** showed a gradual monotonous increase of HFC on the nitroxide nitrogen atom by ca. 0.14 mT upon a pH change from 1.5 to 5.5 (see Figure 2, Table 2, and Supplementary Materials). The shape of the titration curve perfectly corresponded to a two-step acid–base equilibrium, and fitting with the Henderson-Hassellbalch function (Equation (2), see experimental part) gave two pK_a values for each nitroxide (Table 1), corresponding to the sequential protonation of the basic centers, amidine group and pyridine nitrogen.

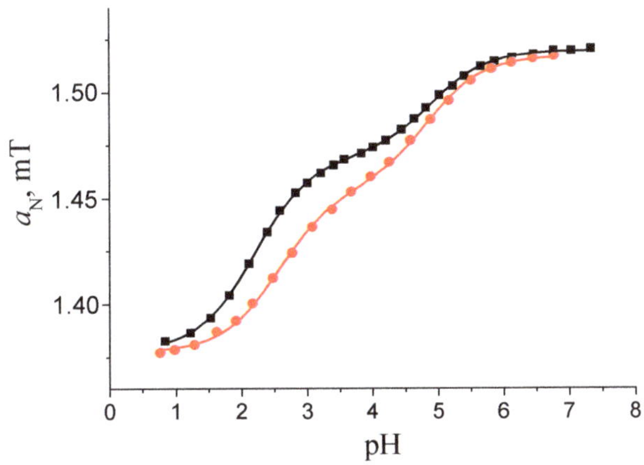

Figure 2. Titration curves of nitroxides **34** (•) and **35** (.).

Table 2. Titration data for the pH-sensitive niroxides: HFC constants for double protonated, mono-protonated, and unprotonated forms, changes in hfi splitting, Δa_N, between these forms, pK values.

Nitroxide	a_N, mT			Δa_N, mT	pK
	$R{\cdot}H_2^{2+}$	$R{\cdot}H^+$	R		
32	1.376	1.460	1.514	0.084 0.054	2.24 ± 0.02 4.73 ± 0.02
33	1.377	1.457	1.516	0.080 0.059	2.36 ± 0.02 4.86 ± 0.02
34	1.377	1.454	1.517	0.077 0.063	2.58 ± 0.02 4.85 ± 0.03
35	1.377	1.469	1.519	0.092 0.050	2.20 ± 0.01 4.89 ± 0.01

In accordance with the general concept of basicity, the pK_a value of the amidine fragment should be higher than that of the pyridine one. However, according to the simulation, the protonation of the center with a more acidic pK_a is accompanied by a change in the hyperfine constant by 0.077–0.092 mT, which is typical of the amidine group in 4-amino-2,5-dihydroimidazol-1-oxyls, while the higher pK_a (4.73–4.89) corresponds to a smaller change in the hyperfine constant (0.05–0.063 mT), which may correspond to pyridine moiety protonation. Moreover, the basic pK_a showed minor dependence on the nature of the substituent at the exocyclic nitrogen, while the acidic pK_a varies from 2.58 for **34** to 2.19 for **35**. Meanwhile, a comparison of **34** and **35** shows that an increase in the electron-withdrawing character of the substituent at the exocyclic nitrogen leads to an increase of Δa_N in the more acidic region, and a decrease of that correspondent to higher pK_a. A comparison of the titration data for **1** and **36** [11] gives similar results. Data in the literature show that pK_a values for 4-amino-2,5-dihydroimidazol-1-oxyls are strongly dependent on substituents at exocyclic nitrogen and can go below four [32,33]. Thus, it is obvious that the pyridine nitrogen and amidine group in **31–35** have similar basicity and the contribution of different monoprotonated forms is varying depending on the electronic effect of the substituents at the exocyclic nitrogen.

3. Conclusions

In this paper feasibility of the approach to synthesis of pH-sensitive spin labels of 4-amino-2,5-dihydrioimidazol-1-oxyl series was once again demonstrated. We have showed that various functional groups can be easily placed in the substituents both in position 2 and to exocyclic nitrogen to make the spin probe suitable for a specific purpose. The potential of this synthetic scheme is still far from exhaustion.

4. Materials and Methods

4.1. General Information

The nitroxides and 4*H*-imidazol-3-oxides **2b**, **8**, **17** and **29** were prepared according to the literature protocols [10,11,14]. 1,1′-Carbonyldiimidazole and Ohira-Bestmann Reagent (dimethyl (1-diazo-2-oxopropyl)phosphonate, 10% solution in acetonitrile) were purchased from TCI Europe N.V. (Zwijndrecht, Belgium); 2-(2-bromoethyl)-1,3-dioxolan and 9-BBN 0.5 M solution in THF were purchased from Acros Organics B.V.B.A. (Geel, Belgium).

The IR spectra were recorded on a Bruker Vector 22 FT-IR spectrometer (Bruker, Billerica, MA, USA) in KBr pellets (1:150 ratio) or in neat samples (for oily compounds). UV spectra were acquired on a HP Agilent 8453 spectrometer (Agilent Technologies, Santa Clara, CA, USA) in ethanol solutions (concentration ~10^{-4} M). NMR spectra ^1H and ^{13}C were recorded on a Bruker AV-300 (300.132 and 75.467 MHz), AV-400 (400.134 and 100.614 MHz). ^1H and ^{13}C chemical shifts (δ) were internally referenced to the residual solvent peak. The nitroxides were reduced to diamagnetic compounds with PhSH [34], N_2D_4 [35], Zn/CF_3COOH [36], or $Zn/ND_4Cl/D_2O$ [37] prior to recording the ^1H NMR spectra. HRMS analyses were performed with High Resolution Mass Spectrometer DFS (Thermo Electron, Brehmen, Germany). Reactions were monitored by TLC carried out using UV light 254 nm or 1% aqueous permanganate. Column chromatography was performed on silica gel 60 (70−230 mesh).

The X-ray diffraction experiment was carried out on a Bruker KAPPA APEX II (Bruker, Billerica, MA, USA) diffractometer (graphite-monochromated Mo Kα radiation). Reflection intensities were corrected for absorption by SADABS-2016 program [38]. The structure of compound **11** was solved by direct methods using the SHELXT-2014 program [39] and refined by anisotropic (isotropic for all H atoms) full-matrix least-squares method against F^2 of all reflections by SHELXL-2018 [40]. The positions of the hydrogen atoms were calculated geometrically and refined in riding model. One of the geminal ethyl groups is disordered due to thermal motion at approximate ratio 3:2. Crystallographic data for **11** have been deposited at the Cambridge Crystallographic Data Centre as supplementary publication no. CCDC 2124865. Copy of the data can be obtained, free of charge, by application to CCDC, 12 Union Road, Cambridge CB21EZ, UK (Fax: +44-122-3336033 or e-mail: deposit@ccdc.cam.ac.uk; internet: www.ccdc.cam.ac.uk (accessed on 29 November 2021)). The details are shown in Supplementary Materials.

4.2. Synthesis

2-(4-(Azidomethyl)phenyl)-2,5,5-triethyl-4-pyrrolidino-2,5-dihydro-1H-imidazol-1-oxyl (**5**)

A mixture of **2b** hydrochloride [15] (320 mg, 0.82 mmol), sodium hydrocarbonate (250 mg, 3 mmol), diethyl ether (30 mL), and water (20 mL) was vigorously stirred until powder of 2b completely dissolved. The ether solution was separated and concentrated in vacuum without heating. The residue was dissolved in DMSO (5 mL), a solution of NaN_3 (0.5 g, 7.7 mmol) was added, and the mixture was stirred at 60 °C for 10 h. The mixture was diluted with water (20 mL) and saturated solution of NaCl (50 mL) and extracted with diethyl ether. The extract was washed with saturated solution of NaCl and dried with Na_2CO_3, concentrated in vacuum, and the residue was separated using column chromatography on silica gel, eluent chloroform, to give **5**, yield 290 mg (95%), yellow crystals, m.p. 63–65 °C (hexane). Elemental analysis, found: C, 65.33; H, 8.01; N, 22.47; calcd. for $C_{20}H_{29}N_6O$: C, 65.01; H, 7.91; N, 22.75%. IR (KBr) ν_{max}: 2091 (N_3), 1597 and 1574 (C=N, C=C).

Tetraisopropyl but-3-yne-1,1-diyldiphosphonate (**6**)

(In analogy to procedure by C. Li and C. Yuan [41]) Tetraisopropyl methylenediphosphonate (5 g, 14.5 mmol) was added dropwise to a stirred suspension of NaH (0.8 g 50% content, 16.7 mmol) in dry THF (50 mL) under argon. After hydrogen evolution ceased, propargyl bromide (1.1 mL, 14.5 mmol) was added dropwise under argon to the stirred suspension. The mixture was stirred for 3 h, then the mixture was diluted with water (50 mL) and pH was adjusted to neutral with hydrochloric acid. The mixture was extracted with diethyl ether, the extract was dried with Na_2CO_3 and concentrated in vacuum. The residue was separated using column chromatography on silica gel, eluent chloroform, to give **6**, yield 1.4 g (25%), colorless liquid. Elemental analysis, found: C, 50.35; H, 8.49; P, 16.10; calcd. for $C_{16}H_{32}O_6P_2$: C, 50.26; H, 8.44; P, 16.20%; IR (neat) ν_{max} (cm^{-1}): 2122 (C≡C). ^1H NMR (400 MHz; $CDCl_3$, δ): 1.25 (24H, m, CH_3), 1.93 (1H, t, J 2.3, ≡CH), 2.43 (1H, br tt, J_{t1} 24, J_{t2} 5.9, P−CH−P), 2.64 (2H, tdd, J_t 16, J_{d1} 5.9, J_{d2} 2.3, CH_2), 4.7 (4H, m, O−CH<); $^{13}C\{^1H\}$ NMR (150 MHz; $CDCl_3$, δ): 15.87 (t, J_P 4.8, CH_2), 23.70 (dd, $J_{P\text{-}1}$ 5.8, $J_{P\text{-}2}$ 1.4, CH_3), 24.05 (t, J_P 3.5, CH_3), 37.87 (t, J_P 135.7, P_2CH), 69.66 (s, ≡CH), 71.36 (dd, $J_{P\text{-}1}$ 5.7, $J_{P\text{-}2}$ 6.5, OCH), 81.60 (t, J_P 9.7, –C≡).

2-(4-((4-(2,2-Bis(diisopropoxyphosphoryl)ethyl)-1H-1,2,3-triazol-1-yl)methyl)phenyl)-2,5,5-triethyl-4-pyrrolidino-2,5-dihydro-1H-imidazol-1-oxyl (**7**)

Ascorbic acid (140 mg, 0.79 mmol) was added to a mixture of **5** (241 mg, 0.65 mmol), 6 (259 mg, 0.67 mmol), EtOH (1.5 mL), H_2O (1.5 mL), and saturated solution of $CuSO_4$ in water (0.15 mL). The mixture was stirred for 2 h, then PbO_2 (1 g, 4.17 mmol) was added, the mixture was stirred for 1 h, then the precipitate was filtered off and washed with ethanol. The combined solutions were evaporated in vacuum and the residue was separated using column chromatography on silica gel, eluent chloroform, to give **7**, yield 320 mg (65%), yellow oil. Elemental analysis, found: C, 57.23; H, 8.30; N, 10.98; P, 8.35; calcd. for $C_{36}H_{61}N_6O_7P_2$: C, 57.51; H, 8.18; N, 11.18; P, 8.24%; IR (neat) ν_{max} (cm^{-1}): 1595, 1576 (C=N, C=C). ^1H NMR (300 MHz; CD_3OD–$CDCl_3$, reduced with Zn in ND_4Cl/D_2O, δ): 0.76 (3H, t, J 7, CH_3), 0.90 (3H, t, J 7, CH_3), 1.04 (3H, t, J 7, CH_3), 1.26 (24H, m, $C(CH_3)_2$), 1.45 (2H, m, CH_2Me), 1.83 (1H, m, CH_2Me), 2.12 (7H, br m, CH_2Me and C-CH_2CH_2-C), 2.86 (1H (partly exchanged), tt, J_P 23, J_H 6, P–CH–P), 3.26 (2H, br t, J_P 16.5, P_2C-CH_2-), 3.7 (4H, m, CH_2-N-CH_2), 4.69 (4H, septet, J 6, O−CH), 5.51 (2H, s, Ar−CH_2), 7.32 (2H, d, J 8, CH Ar), 7.39 (1H, s, OH), 7.52 (2H, d, J 8, CH Ar), 7.69 (1H, s, N−CH=); ^1H NMR (300 MHz; CD_3OD–$CDCl_3$, reduced with Zn/CF_3COOH in CD_3OD, 65 °C, δ): 0.85 (3H, t, J 7, CH_3), 0.89 (3H, t, J 7, CH_3), 1.04 (3H, t, J 7, CH_3), 1.19 (6H, d, J 6, $C(CH_3)_2$), 1.27 (6H, d, J 6, $C(CH_3)_2$), 1.33 (12H, d, J 6, $C(CH_3)_2$), 1.49 (2H, m, CH_2Me), 2.02 (8H, br m, CH_2Me and C−CH_2CH_2−C), 2.99 (1H (partly exchanged), tt, J_P 24, J_H 6, P−CH-P), 3.26 (2H, m, P_2C-CH_2-), 3.7 (4H, m, CH_2-N-CH_2), 4.75 (4H, septet, J 6, O−CH), 5.56 (2H, s, Ar−CH_2), 7.37 (2H, d, J 8, CH Ar) and 7.60 (2H, d, J 8, CH Ar), 7.75 (1H, s, N−CH=); ^{31}P NMR (121.497 MHz; CD_3OD-$CDCl_3$, reduced with Zn in ND_4Cl/D_2O, δ): 20.44, 20.47.

2,5,5-Triethyl-2-(4-formylphenyl)-4-pyrrolidino-2,5-dihydro-1H-imidazol-1-oxyl (**9**)

Method A

Activated manganese dioxide (4 g, 46 mmol) was added to a stirred solution of **8** (0.4 g, 1.16 mmol) in methanol (50 mL). The mixture was stirred for 4 h, manganese oxides were filtered off through celite 281, the solvent was distilled off in vacuum and the residue was separated using column chromatography on silica gel, eluent chloroform, to give **9**, yield 340 mg (85%), yellow crystals, m.p. 84–85 °C dec. (chroloform-hexane). Elemental analysis, found: C, 70.20; H, 8.15; N, 12.31; calcd. for $C_{20}H_{28}N_3O_2$: C, 70.14; H, 8.24; N, 12.27%. IR (KBr) ν_{max} (cm^{-1}): 1697 (C=O); 1593, 1570 (C=N, C=C); UV (EtOH) λ_{max} (log ε): 229 (4.22), 253 (4.26).

Method B

A powder of 2,2,6,6-tetramethylpiperidinium chloride (0.4 g, 2.03 mmol) was added to a solution of **8** (0.5 g, 1.5 mmol) in chloroform (10 mL) and the solution was stirred for 2 h at room temperature. The solvent was distilled off in vacuum, the residue was separated using column chromatography on silica gel, eluent chloroform, to give **9**, yield 440 mg (85%).

2,5,5-Triethyl-2-(4-ethynylphenyl)-4-pyrrolidino-2,5-dihydro-1H-imidazol-1-oxyl (**11**)

A solution of dimethyl (1-diazo-2-oxopropyl)phosphonate in acetonitrile (10%, 0.7 mL, 0.31 mmol) was added to a mixture of **8** (100 mg, 0.29 mmol), freshly annealed K_2CO_3 (84 mg, 0.61 mmol) and anhydrous methanol (5 mL). The mixture was stirred overnight, methanol was distilled off in vacuum, the residue was triturated with ethyl acetate, the precipitate was filtered off and washed with ethyl acetate, the combined solution was concentrated in vacuum and separated by column chromatography on silica gel, eluent diethyl ether–hexane 1:1 to give **11**, yield 82 mg (84%), orange crystals, m.p. 156–158 °C (hexane-ethyl acetate). Elemental analysis, found: C, 74.81; H, 7.97; N, 12.50; calcd. for $C_{21}H_{28}N_3O$: C, 74.52; H, 8.34; N, 12.41%. IR (KBr) ν_{max} (cm^{-1}): 3151 (≡C–H), 2094 (C≡C); 1587, 1554 (C=N, C=C).

2-(4-Carboxyphenyl)-2,5,5-triethyl-4-pyrrolidino-2,5-dihydro-1H-imidazol-1-oxyl (**12**)

A solution of **8** (0.5 g, 1.5 mmol) in $CHCl_3$ (10 mL) was cooled to 0 °C and 2,2,6,6-tetramethyloxopiperidinium chloride (**10**) (0.4 g, 2.0 mmol) was added in one portion. The mixture was stirred for 2 h at 0 °C. Then 2-methylbut-2-ene (1.8 mL, 17.4 mmol) was added to reaction mixture followed by addition of a solution of $NaClO_2$ (0.9 g, 9.8 mmol) and KH_2PO_4 (1.3 g, 9.8 mmol) in H_2O (44 mL). The mixture was stirred for 2 h, the organic layer was separated, washed with saturated aqueous solution of Na_2CO_3 (3 × 20 mL) and concentrated in vacuum. The residue was separated by column chromatography on silica gel using $CHCl_3$–EtOH mixture (100:16) as an eluent to give light-yellow crystals of **12**, yield 463 mg (89%), m.p. 205–207 °C (AcOEt—*i*-PrOH 10:1). Elemental analysis, found: C, 66.85; H, 7.87; N, 11.71; calcd. for $C_{20}H_{28}N_3O_3$: C, 67.01; H, 7.87; N, 11.72%. IR (KBr) ν_{max} (cm^{-1}): 2974 (C–H), 1693 (C=O), 1591 (C=N), 1571 (C=C). UV (EtOH) λ_{max} (log ε): 232 (4.41). ^1H NMR (400 MHz; CD_3OD–$CDCl_3$, reduced with PhSH, δ): 0.78 (3H, t, *J* 7.3, CH_3), 0.85 (3H, t, *J* 7.3, CH_3), 0.95 (3H, t, *J* 7.3, CH_3), 1.00–1.12 (2H, m, CH_2, Et), 1.36, 1.75 (2H, AB, CH_2, Et), 1.85–1.99 (2H, m, CH_2, Et), 2.00 (4H, m, 4CH_2, Pyrr), 3.50–3.55 (4H, m, CH_2−N−CH_2, Pyrr), 7.66 (2H, d, *J* 8, CH Ar), 8.01 (2H, d, *J* 8, CH Ar).

2-(4-(Ethoxycarbonyl)phenyl)-2,5,5-triethyl-4-pyrrolidino-2,5-dihydro-1H-imidazol-1-oxyl (**14**)

Pyridine (340 μL, 4.2 mmol) was added to a suspension of **12** (0.5 g, 1.4 mmol) acid in dry $CHCl_3$ (10 mL). The resulting solution was stirred at 0 °C, and $SOCl_2$ (130 μL, 1.8 mmol) was added dropwise. The stirring continued for 3 h, then ethanol (1 mL, 17 mmol) was added in one portion. The mixture was stirred for 2 h, the solvent was removed in vacuum, and the residue was separated using column chromatography on silica gel, eluent $CHCl_3$–EtOH 200:1, to give **14**, yield 352 mg (65%), yellow crystals, m.p. 85–90 °C (hexane). Elemental analysis, found: C, 68.60; H, 8.10; N, 10.80; calcd. for $C_{23}H_{32}N_3O_3$: C, 68.27; H, 8.35; N, 10.87). IR (KBr) ν_{max} (cm^{-1}): 2970 (C–H), 1718 (C=O), 1593 (C=N), 1571 (C=C). UV (EtOH) λ_{max} (log ε): 231 (4.45). ^1H NMR (300 MHz; $CDCl_3$-CD_3OD, reduced with Zn/CF_3COOH in CD_3OD, 65 °C, δ): 0.62 (3H, t, *J* 7.2, CH_3), 0.68 (3H, t, *J* 7.4, CH_3), 0.81 (3H, t, *J* 7.5, CH_3), 0.88–1.12 (2H, m, CH_2, Et), 1.17 (3H, t, *J* 7.1, CH_3CH_2O), 1.2, 1.60–1.84 (4H, m, CH_2, Et_2), 1.89 (4H, m, C−CH_2CH_2−C, Pyrr), 3.44 (4H, m, CH_2–N–CH_2, Pyrr), 4.15 (2H, q, *J* 7.1, CH_2O), 7.44 (2H, d, *J* 8, CH Ar), 7.80 (2H, d, *J* 8, CH Ar).

2-(4-((2,5-Dioxopyrrolidinooxy)carbonyl)phenyl)-2,5,5-triethyl-4-pyrrolidino-2,5-dihydro-1H-imidazol-1-oxyl (**15**)

Pyridine (100 μL, 1.2 mmol) was added to a suspension of **12** (0.138 g, 0.39 mmol) in dry $CHCl_3$ (5 mL), the resulting solution was stirred at 0 °C, and $SOCl_2$ (30 μL, 0.4 mmol) was added dropwise. The reaction mixture was stirred for 3 h, then *N*-hydroxysuccinimide (44 mg, 0.39 mmol) was added in one portion. The mixture was stirred for 1 h, the solvent was removed in vacuum, and residue was separated using column chromatography on silica gel, eluent $CHCl_3$–EtOH (100:1), to give **15**, yield 102 mg (58%), yellow crystals, m.p.

107–108 °C (Et_2O—hexane 1:2). Elemental analysis, found: C, 63.09; H, 6.86; N, 11.92; calcd. for $C_{24}H_{31}N_4O_5$: C, 63.28; H, 6.86; N, 12.30%. IR (KBr) ν_{max} (cm^{-1}): 2974 (C–H), 1774, 1743 (C=O), 1591 (C=N), 1571 (C=C). UV (EtOH) λ_{max} (log ε): 235 (4.41).

2-(4-((3-Carboxypropanoyloxy)methyl)phenyl)-2,5,5-triethyl-4-pyrrolidino-2,5-dihydro-1H-imidazol-1-oxyl (**16**)

Succinic anhydride (0.15 g, 1.5 mmol) was added to a solution of **8** (0.2 g, 0.6 mmol) in $CHCl_3$ (10 mL) and the reaction mixture was heated to reflux for 2 h. The resulting solution was washed with H_2O (10 mL), dried with Na_2SO_4, and the solvent was removed in vacuum. The solid residue was triturated with ether, the crystalline precipitate of **16** was filtered off and washed with diethyl ether, yield 213 mg (80%), yellow crystals, m.p. 168–169 °C dec. (Et_2O). Elemental analysis, found: C, 64.37; H, 7.52; N, 9.48; calcd. for $C_{24}H_{34}N_3O_5$: C, 64.24; H, 7.71; N, 9.45%. IR (KBr) ν_{max} (cm^{-1}): 2969 (C-H), 1731 (C=O ester), 1587 (C=N), 1569 (C=O carboxy). UV (EtOH) λ_{max} (log ε): 220 (4.03). ^1H NMR (300 MHz; $CDCl_3$-CD_3OD, reduced with Zn/CF_3COOH in CD_3OD, 65 °C, δ): 0.66 (6H, m, CH_3), 0.81 (3H, m, CH_3), 1.00, 1.24 (2H, m, CH_2, Et), 1.56–1.81 (4H, m, 2 × CH_2, Et_2), 1.89 (4H, m, C−CH_2CH_2−C, Pyrr), 2.44 (4H, m, $CH_2CH_2CO_2H$), 3.44 (4H, m, CH_2–N–CH_2, Pyrr), 4.9 (2H, m, CH_2O), 7.14 (2H, d, *J* 8, CH Ar), 7.31 (2H, d, *J* 8, CH Ar).

2-Allyl-2-ethyl-5,5-dimethyl-4-(pyrrolidino)-2,5-dihydroimidazol-1-oxyl (**18a**)

A solution of allylmagnesium bromide prepared from allyl bromide (1.69 mL, 20 mmol) and Mg (0.5 g, 20.5 mmol) in diethyl ether (15 mL) under argon was added dropwise to a stirred solution of nitrone **17** (0.83 g, 4.0 mmol) in THF (15 mL). The reaction mixture was stirred for 1 h, then water (30 mL) was added dropwise under vigorous stirring. Then manganese dioxide (5 g, 57 mmol) was added and the reaction mixture was stirred for 1 h. The manganese oxides were filtered off and the precipitate was washed with *tert*-butylmethyl ether. The organic layer was separated, the water solution was saturated with NaCl and extracted with *tert*-butylmethyl ether. The combined organic extracts were concentrated in vacuum and the residue was separated using column chromatography on Al_2O_3, eluent *tert*-butylmethyl ether–hexane (1:1) to give **18a**, yield 630 mg (63%), yellow crystals, m.p. 55–57 °C (hexane). Elemental analysis, found: C, 67.22; H, 10.23; N, 16.88; calcd. for $C_{14}H_{24}N_3O$: C, 67.16; H, 9.66; N, 16.78%. IR (KBr) ν_{max} (cm^{-1}): 2975 (C–H), 1645 (C=C), 1590 (C=N). UV (EtOH) λ_{max} (log ε): 225 (4.17). ^1H NMR (400 MHz; CD_3OD, reduced with N_2D_4, δ): 0.91 (3H, t, *J* 7, CH_3), 1.47 (6H, d, *J* 2.2, CH_3), 1.60–1.83 (4H, m, CH_2, Et), 2.01 (4H, m, CH_2–CH_2 (pyrr)), 2.39–2.63 (2H, m, CH_2–CH=CH_2), 3.56 (4H, s, CH_2–N–CH_2), 5.08 (2H, m, CH_2=CH), 5.97 (1H, tdd, J_t 7, J_{d1} 10.7, J_{d2} 17.2, CH_2=CH).

2-Ethyl-5,5-dimethyl-2-(pent-4-enyl)-4-(pyrrolidino)-2,5-dihydroimidazol-1-oxyl (**18b**)

A solution of pent-4-enylmagnesium bromide was prepared from 5-bromopentene (1.6 g, 12 mmol) and Mg (335 mg, 14 mmol) in THF (20 mL) under argon. This solution was added dropwise to a stirred solution of **17** (1 g, 4.8 mmol) in THF (20 mL). The reaction mixture was stirred overnight, then water (4 mL) was added dropwise under vigorous stirring. The reaction mixture was vigorously stirred in air for 1 h, then organic layer was separated, and the aqueous layer was extracted with Et_2O–EtOH (100:1). The combined organic extracts were dried with Na_2SO_4, solvents were distilled off in vacuum, and the residue was separated by column chromatography on Al_2O_3 using hexane–$CHCl_3$ mixture (2:1) as an eluent to give **18b**. Yield 931 mg (70%), yellow oil. Elemental analysis, found: C, 68.93; H, 9.80; N, 15.00; calcd. for $C_{16}H_{28}N_3O$: C, 69.02; H, 10.14; N, 15.09%. IR (KBr) ν_{max} (cm^{-1}): 2973 (C–H), 1639 (C=C), 1594 (C=N). UV (EtOH) λ_{max} (log ε): 225 (4.19). ^1H NMR (300 MHz; $CDCl_3$-CD_3OD, reduced with Zn/CF_3COOH in CD_3OD, 65 °C, δ): 0.65 (3H, m, CH_3, Et), 1.16 (2H, m, CH_2, Et), 1.23 (2H, s, CH_3), 1.32 (4H, s, CH_3), 1.36–1.64 (4H, m, CH_2–CH_2–Allyl), 1.77 (2H, m, CH_2–C=), 1.85 (4H, m, CH_2–CH_2–CH_2–CH_2), 3.23, 3.43 (4H, m, CH_2–N–CH_2), 4.60–4.76 (2H, m, CH_2=), 5.48 (1H, tdd, =CH–, J_t 7, J_{d1} 10.3, J_{d2} 17,1).

2-Ethyl-2-(3-hydroxypropyl)-5,5-dimethyl-4-(pyrrolidino)-2,5-dihydroimidazol-1-oxyl (**19a**)

Method A

A solution of 9-BBN in THF (0.5 M, 8 mL, 4.1 mmol) was added dropwise to a stirred solution of **18a** (400 mg, 1.6 mmol) in THF (10 mL) under argon. The reaction mixture was vigorously stirred for 4 h, then cooled to 0 °C and cold (0 °C) aqueous NaOH (20%, 10 mL) and cold (0 °C) H_2O_2 (30%, 3 mL) were added dropwise successively. The mixture was allowed to warm to room temperature upon stirring (ca. 2 h), organic layer was separated, dried with Na_2CO_3, and the solvent was distilled off in vacuum. The residue was dissolved in $CHCl_3$ (25 mL), anhydrous Na_2CO_3 (1 g) was added, and mixture was allowed to stand overnight in air. The solution was concentrated in vacuum and separated by column chromatography on silica gel using $CHCl_3$–EtOH mixture (100:4) as an eluent to give **19a**. Yield 150 mg (35%), yellow oil. Elemental analysis, found: C, 62.53; H, 9.49; N, 15.45; calcd. for $C_{14}H_{26}N_3O_2$: C, 62.65; H, 9.76; N, 15.66%. IR (KBr) ν_{max} (cm^{-1}): 3386 (br., OH), 1592 (C=N). UV (EtOH) λ_{max} (log ε): 225 (4.07). ^{1}H NMR (400 MHz; CD_3OD, reduced with N_2D_4, δ): 0.94 (3H, t, *J* 7.2, CH_3, Et), 1.43 (6H, s, CH_3), 1.53–1.92 (6H, m, CH_2), 1.98 (4H, m, CH_2-CH_2-CH_2-CH_2), 3.52 (4H, s, CH_2-N-CH_2), 3.57 (2H, br. s, CH_2O). *2-Ethyl-2-(5-hydroxypentyl)-5,5-dimethyl-4-(pyrrolidino)-2,5-dihydroimidazol-1-oxyl* (**19b**) was prepared similarly from **18b**. Yield 42%, yellow crystals, m.p. 68–73 °C (Et_2O). Elemental analysis found: C, 65.17; H, 10.56; N, 14.08; calcd. for $C_{16}H_{30}N_3O_2$: C, 64.83; H, 10.20; N, 14.18%. IR (KBr) ν_{max} (cm^{-1}): 3261 (br., OH), 1593 (C=N). UV (EtOH) λ_{max} (log ε): 225 (4.1). ^{1}H NMR (300 MHz; $CDCl_3$–CD_3OD, reduced with Zn/CF_3COOH in CD_3OD, 65 °C, δ): 0.62 (3H, m, CH_3, Et), 1.06 (4H, br. m, CH_2–CH_2–$(CH_2)_2OH$), 1.21–1.28 (8H, m, 2 × CH_3, CH_2, Et), 1.40 (2H, m, CH_2–CH_2OH), 1.57 (2H, m, >C(Et)–CH_2), 1.79 (4H, br. m, C–CH_2CH_2–C, Pyrr), 3.20, 3.44 (4H, m, CH_2–N–CH_2, Pyrr), 3.26 (2H, t, *J* 6.5, CH_2O).

Method B

Sodium borohydride (60 mg, 1.6 mmol) was added portionwise to a stirred solution of **24** (400 mg, 1.5 mmol) in EtOH (10 mL) at 0 °C. The reaction was controlled with TLC, Silufol UV-254, eluent $CHCl_3$–EtOH (25:1). Inorganic residue was filtered off, the solution was distilled off in vacuum, and the residue separated by column chromatography as described above to give **19a**. Yield 309 mg (72%).

2-(3-(1H-Imidazole-1-carbonyloxy)propyl)-2-ethyl-5,5-dimethyl-4-(pyrrolidino)-2,5-dihydro-1H-imidazol-1-oxyl (**20a**)

Carbonyldiimidazole (80 mg, 0.49 mmol) was added to a solution of alcohol **19a** (114 mg, 0.43 mmol) in dry $CHCl_3$ (5 mL) and the mixture was allowed to stand for 24 h. The solution was washed with brine, dried with Na_2SO_4, and concentrated in vacuum. The residue was separated by column chromatography on silica gel using $CHCl_3$–EtOH mixture (100:2) as an eluent, producing **20a** as yellow oil. Yield 139 mg (90%). Elemental analysis, found: C, 59.69; H, 7.72; N, 19.45; calcd. for $C_{18}H_{28}N_5O_3$: C, 59.65; H, 7.79; N, 19.32%. IR (KBr) ν_{max} (cm^{-1}): 1760 (C=O), 1592 (C=N). UV (EtOH) λ_{max} (log ε): 223 (4.17). *2-(5-(1H-Imidazole-1-carbonyloxy)pentyl)-2-ethyl-5,5-dimethyl-4-(pyrrolidino)-2,5-dihydro-1H-imidazol-1-oxyl* (**20b**) was prepared similarly, yield 80%, yellow oil. Elemental analysis, found: C, 61.30; H, 8.26; N, 17.70; cacld. for $C_{20}H_{32}N_5O_3$: C, 61.51; H, 8.26; N, 17.93%. IR (KBr) ν_{max} (cm^{-1}): 1762 (C=O), 1593 (C=N). UV (EtOH) λ_{max} (log ε): 226 (3.99).

2-(3-(3-(Diethylamino)propylcarbamoyloxy)propyl)-2-ethyl-5,5-dimethyl-4-(pyrrolidino)-2,5-dihydro-1H-imidazol-1-oxyl (**21**)

N,N-Diethyl-1,3-diaminopropane (50 mg, 0.38 mmol) was added to a solution of **20a** (126 mg, 0.35 mmol) in dry Et_2O (5 mL), and mixture was allowed to stay for 24 h. The solution was concentrated in vacuum, and the residue was separated by column chromatography on Al_2O_3 using $CHCl_3$ as an eluent to give **21** (Figure 3). Yield 82 mg (60%), yellow oil. Elemental analysis, found: C, 62.21; H, 10.01; N, 16.51; calcd. for $C_{22}H_{42}N_5O_3$: C, 62.23; H, 9.97; N, 16.49%. IR (KBr) ν_{max} (cm^{-1}): 1718 (C=O), 1593 (C=N). UV (EtOH) λ_{max} (log ε): 225 (4.19). ^{1}H NMR (300 MHz; $CDCl_3$-CD_3OD, reduced with Zn/CF_3COOH in CD_3OD, 65 °C, δ): 0.75 (3H, t, *J* 7.2, CH_3), 1.11 (6H, t, *J* 7.3, 2 × CH_3) 1.33 (6H, br., CH_3), 1.40 (6H, br., CH_3), 1.43–1.79 (8H, m, CH_3CH_2C, $^1CH_2$, $^2CH_2$, $^5CH_2$,), 1.90 (4H, br. m, $^{10}CH_2$, $^{11}CH_2$), 2.87–3.03 (8H, m, $^4CH_2$, $^6CH_2$, $^7CH_2$, $^8CH_2$), 3.36, 3.52 (4H, m, $^9CH_2$, $^{12}CH_2$), 3.84 (2H, m, CH_2O).

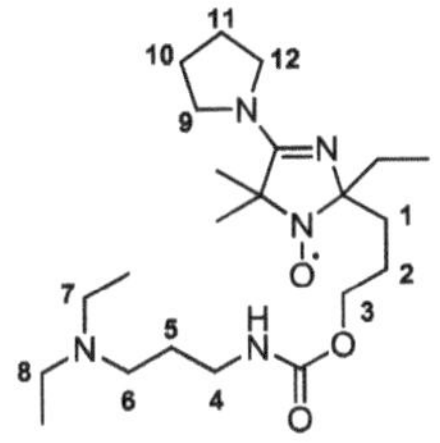

Figure 3. The structure of nitroxide 21.

2-(3-Carboxypropyl)-2-ethyl-5,5-dimethyl-4-(pyrrolidino)-2,5-dihydro-1H-imidazol-1-oxyl (22)

Osmium tetroxide (30 mg, 0.4 mmol) and oxone (1.77 g, 5.8 mmol) were added successively to a solution of **18b** (400 mg, 1.4 mmol) in DMF (20 mL) and the mixture was stirred for 3 h. A powder of Na_2SO_3 (10 g, 63 mmol) was added in one portion. Inorganic precipitate was filtered off and washed with EtOH, the combined solution was evaporated to dryness in vacuum and the residue was separated by column chromatography on silica gel using EtOH as an eluent to give **22**, yield 90 mg (21%), yellow oil. M^+ (cacld./found) 296.1969/296.1972. IR (neat) ν_{max} (cm^{-1}): 2977 (C-H), 1664 (C=O), 1592 (C=N). λ_{max} (EtOH)/nm: 225 (lgε 4.19).

2-(2-(1,3-Dioxolan-2-yl)ethyl)-2-ethyl-5,5-dimethyl-4-(pyrrolidin-1-yl)-2,5-dihydro-1H-imidazol-1-oxyl (23)

A solution of 2-(1,3-dioxolan-2-yl)ethylmagnesium bromide was prepared from 2-(2-bromoethyl)-1,3-dioxolan (4.3 g, 24 mmol) and Mg (670 mg, 28 mmol) in 20 mL THF under a stream of argon. This solution was added dropwise to a stirred solution of nitrone **17** (850 mg, 4 mmol) in 20 mL Et_2O and 6 mL THF. The reaction mixture was stirred overnight, then water (5 mL) was added dropwise under vigorous stirring. The reaction mixture was allowed to air for 1 h, then organic layer was separated, inorganic residue was quenched with Et_2O–EtOH (100:1). An isolated organic layer was dried over Na_2SO_4, solvents were removed in vacuum. The residue was separated using column chromatography on Al_2O_3 using $CHCl_3$ as an eluent to give **23**, yield 1.13 g (90%), yellow oil. Elemental analysis, found: C, 62.08; H, 9.21; N, 13.43; calcd for $C_{16}H_{28}N_3O_3$: C, 61.91; H, 9.09; N, 13.54%. IR (KBr) ν_{max} (cm^{-1}): 2972 (C-H), 1593 (C=N), 1143 (C–O). λ_{max} (EtOH)/nm: 225 (lgε 3.90).

2-Ethyl-5,5-dimethyl-2-(3-oxopropyl)-4-(pyrrolidin-1-yl)-2,5-dihydro-1H-imidazol-1-oxyl (24)

A solution of oxalic acid (180 mg, 2 mmol) in water (6 mL) was added to a solution of nitroxide 23 (250 mg, 0.8 mmol) in EtOH (4 mL). The reaction mixture was stirred for 3 h under reflux, then ethanol was removed in vacuum, saturated aqueous $KHCO_3$ (10 mL) was added to a residue. The product was extracted with $CHCl_3$–*i*-PrOH mixture (50:1) (3 × 15 mL). An isolated organic layer was dried over Na_2SO_4, the solvents were removed in vacuum, the residue was separated using column chromatography on silica gel using $CHCl_3$-EtOH mixture (50:1) as an eluent to give **24**, yield 161 mg (75%), yellow oil. Elemental analysis, found: %: C, 63.08; H, 9.18; N, 15.63; calcd. for $C_{14}H_{24}N_3O_2$: C, 63.13; H, 9.08; N, 15.78. IR (neat) ν_{max} (cm^{-1}): 2972 (C-H), 1720 (C=O), 1593 (C=N). λ_{max} (EtOH)/nm: 225 (lgε 4.16).

2-(2-Carboxyethyl)-2-ethyl-5,5-dimethyl-4-(pyrrolidin-1-yl)-2,5-dihydro-1H-imidazol-1-oxyl (25)

Trimethylethylene (1 mL, 9.0 mmol) was added to a cooled (0 °C) solution of aldehyde **24** (200 mg, 0.8 mmol) in 10 mL CH_3CN followed by addition of a solution of $NaClO_2$ (480 mg, 5.3 mmol) and KH_2PO_4 (710 mg, 5.3 mmol) in H_2O (20 mL). Progress of the reaction was monitored by TLC (silica gel, $CHCl_3$–EtOH (50:1), developing with 1% aq $KMnO_4$). CH_3CN was removed in vacuum, the product was extracted from water by $CHCl_3$–*i*-PrOH mixture (100:1) (5 × 15 mL). An isolated organic layer was dried over Na_2SO_4, the solvents were removed in vacuum, the residue was separated using column chromatography on silica gel using $CHCl_3$–EtOH mixture (5:2) as an eluent to give **25**, yield 121 mg (57%), yellow oil, M^+ (calcd./found) 282.1812/282.1811. IR (neat) ν_{max} (cm^{-1}): 2973 (C–H), 1729 (C=O), 1591 (C=N). λ_{max} (EtOH)/nm: 223 (lgε 4.04). ^{1}H NMR (300 MHz; $CDCl_3$-CD_3OD, reduced with Zn/CF_3COOH in CD_3OD, 65 °C, δ) 0.75 (3H, m, CH_3, Et),

0.98–1.14 (2H, m, CH_2, Et), 1.34, 1.40 (6H, m, 2 × CH_3), 1.47–1.61 (2H, m, $CH_2CH_2CO_2H$), 1.91 (4H, m, CH_2–CH_2–CH_2–CH_2, Pyrr), 2.14–2.27 (2H, m, $CH_2CH_2CO_2H$), 3.36, 3.55 (4H, m, CH_2–N–CH_2, Pyrr).

1-(4-(1,3-Dioxolan-2-yl)phenyl)-N-methylmethanamine (**26**)

p-Toluene sulfonic acid monohydrate (0.5 g, 74.6 mmol) was added to a solution of terephthalic aldehyde **27** (10 g, 74.6 mmol) in 175 mL $PhCH_3$. Water was distilled off with Dean-Stark tube. The reaction mixture was then quenched with aqueous $NaHCO_3$, dried over Na_2CO_3, the solvent was remove in vacuum, and residue was dissolved in methanol saturated with methylamine (20 mL). The resulting solution was added to the previously maintained under vigorous stirring for 10 min in a solution of Ti(O*i*-Pr)$_4$ (14 mL, 47 mmol) in methanol saturated with methylamine (30 mL). The mixture was stirred for 5 h, then $NaBH_4$ (1.34 g, 33.6 mmol) was added portionwise, and mixture was stirred for 2 h. Water (7 mL) was added dropwise, solvents were removed in vacuum, brine was added to a residue, and the product was extracted by ether. Organic layer was dried over NaOH. Residue was separated using column chromatography on silica gel using Et_2O–EtOH mixture (10:1) as an eluent, yield 12.24 g (85%), colorless oil. Elemental analysis, found: C, 67.84; H, 7.91; N, 6.94; calcd for $C_{11}H_{15}NO_2$: C, 68.37; H, 7.82; N, 7.25%. ^1H NMR (300 MHz; $CDCl_3$, δ): 2.38 (3H, s, CH_3), 3.70 (2H, s, N–CH_2), 3.95–4.09 (4H, m, –O–CH_2–CH_2–O–), 5.75 (1H, s, O–CH–O), 7.29, 7.38 (4H, AA′BB′, C_6H_4)), ^{13}C NMR (75 MHz; $CDCl_3$, δ): 35.77 (N–CH_3), 55.56 (N–CH_2), 65.06 (O–CH_2–CH_2–O), 103.44 (O–CH–O), 126.31 (CH–C–CH_2NHCH_3), 127.89 (CH–C–CH), 136.32 (C–CH_2NHCH_3), 141.10 (C–CH). IR (neat) ν_{max} (cm^{-1}): 3325 (N–H), 1082 (O–C–O). λ_{max} (EtOH)/nm: 210 (logε 3.94), 260 (logε 2.36).

5-((4-(1,3-Dioxolan-2-yl)benzyl)(methyl)amino)-4,4-dimethyl-2-(pyridin-4-yl)-4H-imidazole 3-oxide (**30**)

1-(4-(1,3-Dioxolan-2-yl)phenyl)-*N*-methylmethanamine **26** (6.72 g, 34.8 mmol) was added to a solution of 5-cyano-4,4-dimethyl-2-(pyridin-4-yl)-4*H*-imidazole 3-oxide **29** (2.98 g, 13.9 mmol) in THF (25 mL) and the mixture was allowed to stand at r.t. for 24 h. The solvent was removed in vacuum, residue was triturated with ether and crystallization from CH_3CN to give **30**, yield 3.97 g (75%), dirty-yellow crystals, m.p. 160 °C (dec.). Elemental analysis, found: C, 65.81; H, 6.25; N, 14.41; calcd for $C_{21}H_{24}N_4O_3$: C, 66.30; H, 6.36; N, 14.73%. ^1H NMR (400 MHz; $CDCl_3$, δ) 1.66 (6H, s, 2 × CH_3), 3.06 (3H, s, N–CH_3), 3.86–4.16 (4H, m, O–CH_2–CH_2–O), 4.77 (2H, br. s, N–CH_2–Ar), 5.75 (1H, s, O–CH–O), 7.26, 7.46 (4H, AA′BB′, C_6H_4)), 8.46, 8.70 (4H, AA′BB′, Py). ^{13}C NMR (75 MHz; $CDCl_3$, δ) 21.50 (2 × Me), 35.22 (N–CH_3), 53.11 (N–CH_2), 65.00 (O–CH_2–CH_2–O), 75.73 (Me_2C), 102.82 (O–CH–O), 121.00 (3,5–Py), 126.77 (br., CH (C_6H_4)), 133.86 (Py, *i*), 136.36 (C–CH_2NCH_3), 137.47 (C–CH), 144.67 (C=N→O), 149.82 (2,6-Py), 172.33 (C=N). IR (KBr) ν_{max} (cm^{-1}): 1597 (C=N), 1082 (O–C–O). λ_{max} (EtOH)/nm: 263 (logε 4.30), 389 (logε 3.78).

4-((4-(1,3-Dioxolan-2-yl)benzyl)(methyl)amino)-2-ethyl-5,5-dimethyl-2-(pyridin-4-yl)-2,5-dihydro-1H-imidazol-1-oxyl (**31**)

A solution of ethylmagnesium bromide was prepared from ethyl bromide (2.73 g, 25 mmol) and Mg (630 mg, 26 mmol) in 35 mL Et_2O under a stream of argon. This solution was added dropwise to a stirred solution of nitrone **27** (1 g, 2.6 mmol) in 15 mL THF. The reaction mixture was allowed to stand for 1 h. Then water (3 mL) was added dropwise under vigorous stirring followed by MnO_2 (3 g, 34.5 mmol) addition. Progress of the reaction was monitored by TLC (silica gel, $CHCl_3$–EtOH (100:3), developing with 1% aq. $KMnO_4$). The mixture was stirred vigorously for 2 h, the oxidant was filtered off and the residue was washed by $CHCl_3$ and MeOH. The solvent from filtrate was removed in vacuum and the residue was separated by column chromatography on silica gel using $CHCl_3$-EtOH (100:3) as an eluent. The product **31** was isolated as a hydrochloride. Yield 797 mg (68%), yellow oil. Elemental analysis, found: C, 62.18; H, 6.83; N, 12.48; Cl, 6.70; calcd. for $C_{23}H_{30}ClN_4O_3$: C, 61.94; H, 6.78; N, 12.56; Cl, 6.95%. IR (neat) ν_{max} (cm^{-1}): 1593 (C=N), 1082 (O–C–O). λ_{max} (EtOH)/nm: 216 (logε 4.34).

2-Ethyl-4-((4-formylbenzyl)(methyl)amino)-5,5-dimethyl-2-(pyridin-4-yl)-2,5-dihydro-1H-imidazol-1-oxyl (**32**)

A solution of nitroxide **30** (1.8 g, 4.4 mmol) in 15 mL 0.5 M aq. HCl was refluxed for 4 h, then Na_2CO_3 added to the end of gas evolution. A product was extracted by mixture of 20 mL $CHCl_3$ + 1 mL *i*-PrOH three times, organic layer was dried over Na_2CO_3, the solvents were removed in vacuum, and the nitroxide **29** was isolated from the residue by column chromatography on silica gelusing $CHCl_3$ as an eluent. Yield 1.14 g (71%), yellow oil. Elemental analysis, found: C, 68.73; H, 6.88; N, 14.92; calcd. for $C_{21}H_{25}N_4O_2$: C, 69.02; H, 6.90; N, 15.33%. IR (KBr) ν_{max} (cm^{-1}): 1701 (C=O), 1593 (C=N). λ_{max} (EtOH)/nm: 252 (logε 4.29).

4-((4-Carboxybenzyl)(methyl)amino)-2-ethyl-5,5-dimethyl-2-(pyridin-4-yl)-2,5-dihydro-1H-imidazol-1-oxyl (**33**)

Trimethylethylene (1.33 g, 19.2 mmol) was added to a cooled (0 °C) solution of aldehyde **29** (583 mg, 1.6 mmol) in 20 mL $CHCl_3$ followed by addition of a solution of $NaClO_2$ (1.02 g, 11.2 mmol) and KH_2PO_4 (1.5 g, 11.2 mmol) in H_2O (50 mL). Progress of the reaction was monitored by TLC (silica gel, $CHCl_3$–EtOH (50:1), developing with 1% aq. $KMnO_4$). The organic layer was separated, the product was extracted from water by $CHCl_3$—*i*-PrOH mixture (20:1) (2 × 20 mL). A combined organic extracts were washed with brine, dried over Na_2SO_4, the solvents were removed in vacuum, the residue was separated using column chromatography on silica gel using AcOEt–EtOH mixture (10:1) as an eluent. Yield 285 mg (47%), yellow crystals, compound **33** was isolated as a crystal solvate 3 (**33**) × 2 EtOH (ether–EtOH 100:2), m.p. 204 °C (dec.). Elemental analysis, found: C, 65.12; H, 6.42; N, 13.12; calcd. for $C_{67}H_{87}N_{12}O_{11}$: C, 65.08; H, 7.09; N, 13.59%. IR (KBr) ν_{max} (cm^{-1}): 2474 (O–H), 1708 (C=O), 1597 (C=N), λ_{max} (EtOH)/nm: 242 (logε 4.20).

2-Ethyl-4-((4-(hydroxymethyl)benzyl)(methyl)amino)-5,5-dimethyl-2-(pyridin-4-yl)-2,5-dihydro-1H-imidazol-1-oxyl (**34**)

$NaBH_4$ (54 mg, 1.4 mmol) was added portionwise to a cooled (0 °C) solution of aldehyde **32** (511 mg, 1.4 mmol) in EtOH (20 mL). The reaction mixture was stirred until the reaction was complete (TLC, Silufol UV-254, eluent AcOEt). The solvent was removed in vacuum, the residue was separated using column chromatography on silica gel using AcOEt as an eluent. Yield 308 mg (60%), yellow crystals, compound **34** was isolated as a crystal solvate 2 (**34**) × 3 H_2O (ether), m.p. 147–148 °C. Elemental analysis, found: C, 66.31; H, 7.12; N, 14.55; calcd. for $C_{63}H_{85}N_{12}O_8$: C, 66.47; H, 7.53; N, 14.76%. IR (KBr) ν_{max} (cm^{-1}): 3178 (O-H), 1595 (C=N). λ_{max} (EtOH)/nm: 220 (logε 4.30). ^{1}H NMR (400 MHz; CD_3OD–$CDCl_3$, reduced with Zn/CF_3COOH in CD_3OD, 65 °C, δ): 1.03 (3H, t, *J* 7.2, CH_3 Et_2), 1.29 (3H, br s, CH_3), 1.79 (2H, q, *J* 7.2, CH_2), 1.90 (3H, s, CH_3), 3.20 (3H, br s, NCH_3), 4.65 (2H, s CH_2OH), 4.97 (2H, br s, N–CH_2), 7.26 (2H, m, Ar), 7.46 (2H, m, Ar), 7.94 (2H, d, *J* 6.5, Py), 8.73 (2H, d, *J* 6.5, Py)

4-((4-(((2,5-Dioxopyrrolidin-1-yl)oxy)carbonyl)benzyl)(methyl)amino)-2-ethyl-5,5-dimethyl-2-(pyridin-4-yl)-2,5-dihydro-1H-imidazol-1-oxyl (**35**)

Pyridine (240 µL, 3 mmol) was added to a cooled (0 °C) suspension of acid **33** (228 mg, 0.6 mmol) in 10 mL of dry $CHCl_3$ followed by addition of $SOCl_2$ (90 µL, 1.2 mmol). The reaction mixture was vigorously stirred for 3 h, then *N*-hydroxysuccinimide (138 mg, 1.2 mmol) was added and the mixture was allowed to stand for 24 h. The solvents were then removed in vacuum, residue was separated using column chromatography on silica gel using $CHCl_3$–EtOH mixture (100:2) as an eluent to give **35**, yield 123 mg (40%), yellow crystals, compound **35** was isolated as a hydrochloride (hexane), m.p. 58 °C (dec.). Elemental analysis, found: C, 58.42; H, 5.47; N, 13.25; Cl, 6.56; calcd for $C_{25}H_{29}ClN_5O_5$: C, 58.31; H, 5.68; N, 13.60; Cl, 6.88%. IR (KBr) ν_{max} (cm^{-1}): 2976 (C-H), 1770 (O=C-N-C=O), 1741 (C=O), 1593 (C=N). λ_{max} (EtOH)/nm: 239 (logε 4.25)

4.3. EPR Experiments

EPR experiments were performed on X-band EPR (9.8 GHz) spectrometer Bruker ER-200D. Titrations of the radicals (~0.2 mM) were performed in a buffer mixture of acetate-phosphate-borate (0.5 mM of each) in a pH range of 2–10 starting from the acidic value. Small aliquots of NaOH solution were used for titration to a higher pH. The observed hfi constants were measured as a distance between low field and central lines of nitroxide triplet spectra.

Single proton exchange:

$$R\cdot + H^+ \rightleftarrows R\cdot H^+;\ pK_a$$

was fitted with single pK_a titration curve function:

$$a_N\ (pH) = \frac{a_N(R\cdot) + a_N(R\cdot H^+) \times 10^{pK_a - pH}}{1 + 10^{pK_a - pH}}$$

Double proton exchange implies the serial protonation,

$$R\cdot + H^+ \rightleftarrows R\cdot H^+;\ pK_{a1}$$

$$R\cdot H^+ + H^+ \rightleftarrows\ R\cdot H_2^{2+};\ pK_{a2}$$

which results in the double pK_a titration curve function:

$$a_N\ (pH) = \frac{a_N(R\cdot) + a_N(R\cdot H^+) \times 10^{pK_{a1} - pH} + a_N(R\cdot {H_2}^{2+}) \times 10^{pK_{a1} - pH} \times 10^{pK_{a2} - pH}}{1 + 10^{pK_{a1} - pH} + 10^{pK_{a1} - pH} \times 10^{pK_{a2} - pH}}$$

EPR settings. Microwave power, 5 mW; modulation amplitude, 0.08–0.12 mT. SD for pK_a, 0.05; for hfc, 0.005 mT.

Supplementary Materials: The following are available online at https://www.mdpi.com/article/10.3390/gels8010011/s1, Figure S1: "The molecular structure of 2,5,5-triethyl-2-(4-ethynylphenyl)-4-pyrrolidino-2,5-dihydro-1*H*-imidazol-1-oxyl (**11**)", IR and NMR spectra of synthesized compounds, titration curves of pH-sensitive nitroxides.

Author Contributions: Conceptualization, I.A.K.; methodology, D.G.T.; validation, I.A.K., D.G.T., and Y.I.G.; formal analysis, Y.I.G.; investigation, D.G.T., A.A.G., D.A.K. and T.V.R.; writing—original draft preparation, D.G.T.; writing—review and editing, I.A.K. All authors have read and agreed to the published version of the manuscript.

Funding: This research was funded by the Russian Scientific Foundation, grant number 21-14-00219.

Institutional Review Board Statement: Not applicable.

Informed Consent Statement: Not applicable.

Data Availability Statement: Data are shown within the article or in supplementary materials. Crystallographic data for **11** have been deposited at the Cambridge Crystallographic Data Centre as supplementary publication no. CCDC 2124865. Copy of the data can be obtained, free of charge, by application to CCDC, 12 Union Road, Cambridge CB21EZ, UK (fax: +44-122-3336033 or e-mail: deposit@ccdc.cam.ac.uk; internet: www.ccdc.cam.ac.uk (accessed on 29 November 2021)).

Acknowledgments: The authors would like to acknowledge the Multi-Access Chemical Research Center SB RAS for spectral and analytical measurements.

Conflicts of Interest: The authors declare no conflict of interest.

References

1. Voinov, M.A.; Smirnov, A.I. Spin labels and spin probes for measurements of local pH and electrostatics by EPR. *Electron Paramagn. Reson.* **2010**, *22*, 71–106.
2. Kovaleva, E.; Molochnikov, L. pH-Sensitive Nitroxide Radicals for Studying Inorganic and Organo-Inorganic Materials and Systems. In *Nitroxides: Applications in Chemistry, Biomedicine and Material Science*; Likhtenshtein, G.I., Yamauchi, J., Nakatsuji, S., Smirnov, A., Tamura, R., Eds.; Wiley—VCH Verlag, GmbH & Co. KGaA: Weinheim, Germany, 2008; Chapter 7; pp. 211–246. [CrossRef]
3. Védrine, J.C. Acid–base characterization of heterogeneous catalysts: An up-to-date overview. *Res. Chem. Intermed.* **2015**, *41*, 9387–9423. [CrossRef]

4. Khramtsov, V.V.; Volodarsky, L.B. Use of imidazoline nitroxides in studies of chemical reactions: ESR measurements of the concentration and reactivity of protons, thiols and nitric oxide. In *Biological Magnetic Resonance*; Berliner, L.J., Ed.; Plenum Press: New York, NY, USA, 1998; Volume 14, pp. 109–178. [CrossRef]
5. Smirnova, T.I.; Voinov, M.A.; Smirnov, A. Spin Probes and Spin Labels. In *Encyclopedia of Analytical Chemistry: Applications, Theory and Instrumentation*; John Wiley & Sons, Ltd.: Chichester, UK, 2009; p. 31. [CrossRef]
6. Khramtsov, V.; Marsh, D.; Weiner, L.; Reznikov, V. The application of pH-sensitive spin labels to studies of surface potential and polarity of phospholipid membranes and proteins. *Biochim. Biophys. Acta* **1992**, *1104*, 317–324. [CrossRef]
7. Kovaleva, E.G.; Molochnikov, L.S.; Tambasova, D.; Marek, A.; Chestnut, M.; Osipova, V.A.; Antonov, D.O.; Kirilyuk, I.A.; Smirnov, A. Electrostatic Properties of Inner Nanopore Surfaces of Anodic Aluminum Oxide Membranes upon High Temperature Annealing Revealed by EPR of pH-sensitive Spin Probes and Labels. *J. Membr. Sci.* **2020**, *604*, 118084. [CrossRef]
8. Antonov, D.O.; Tambasova, D.P.; Shishmakov, A.B.; Kirilyuk, I.A.; Kovaleva, E.G. Acidic and Electrosurface Properties of Binary TiO_2–SiO_2 Xerogels Using EPR of pH-Sensitive Nitroxides. *Gels* **2021**, *7*, 119. [CrossRef]
9. Khramtsov, V.V.; Vainer, L.M. Photon Transfer Reactions in Free Radicals. Spin pH Probes. *Russ. Chem. Rev.* **1988**, *57*, 824–838. [CrossRef]
10. Kirilyuk, I.A.; Shevelev, T.G.; Morozov, D.A.; Khromovskih, E.L.; Skuridin, N.G.; Khramtsov, V.V.; Grigor'ev, I.A. Grignard Reagent Addition to 5-Alkylamino-4H-Imidazole 3-Oxides: Synthesis of New pH-Sensitive Spin Probes. *Synthesis* **2003**, *6*, 871–878. [CrossRef]
11. Kirilyuk, I.A.; Bobko, A.A.; Khramtsov, V.V.; Grigor'ev, I.A. Nitroxides with two pK values—Useful spin probes for pH monitoring within a broad range. *Org. Biomol. Chem.* **2005**, *3*, 1269–1274. [CrossRef]
12. Potapenko, D.I.; Foster, M.A.; Lurie, D.J.; Kirilyuk, I.A.; Hutchison, J.M.S.; Grigor'ev, I.A.; Bagryanskaya, E.G.; Khramtsov, V.V. Real-time monitoring of drug-induced changes in the stomach acidity of living rats using improved pH-sensitive nitroxides and low-field EPR techniques. *J. Magn. Reson.* **2006**, *182*, 1–11. [CrossRef]
13. Woldman, Y.Y.; Semenov, S.V.; Bobko, A.A.; Kirilyuk, I.A.; Polienko, J.F.; Voinov, M.A.; Bagryanskaya, E.G.; Khramtsov, V.V. Design of liposome-based pH sensitive nanoSPIN probes: Nano-sized particles with incorporated nitroxides. *Analyst* **2009**, *134*, 904–910. [CrossRef]
14. Bobko, A.A.; Eubank, T.D.; Voorhees, J.L.; Efimova, O.V.; Kirilyuk, I.A.; Petryakov, S.; Trofimov, D.G.; Marsh, C.B.; Zweier, J.L.; Grigor'ev, I.A.; et al. In Vivo Monitoring of pH, Redox Status, and Glutathione Using L-Band EPR for Assessment of Therapeutic Effectiveness in Solid Tumors. *Magn. Reson. Med.* **2012**, *67*, 1827–1836. [CrossRef]
15. Kovaleva, E.G.; Molochnikov, L.S.; Stepanova, D.P.; Pestov, A.V.; Trofimov, D.G.; Kirilyuk, I.A.; Smirnov, A.I. Interfacial electrostatic properties of hydrated mesoporous and nanostructured alumina powders by spin labeling EPR. *Cell BioChem. Biophys.* **2017**, *75*, 159–170. [CrossRef] [PubMed]
16. Smirnov, A.; Voinov, M.; Kirilyuk, I. Spin-labeled pH-sensitive phospholipids for interfacial pKa determination: Synthesis and characterization in aqueous and micellar solutions. *J. Phys. Chem. B.* **2009**, *113*, 3453–3460.
17. Voinov, M.A.; Scheid, C.T.; Kirilyuk, I.A.; Trofimov, D.G.; Smirnov, A.I. IKMTSL-PTE, a Phospholipid-Based EPR Probe for Surface Electrostatic Potential of Biological Interfaces at Neutral pH: Effects of Temperature and Effective Dielectric Constant of the Solvent. *J. Phys. Chem. B* **2017**, *121*, 2443–2453. [CrossRef]
18. Kálai, T.; Hubbell, W.L.; Hideg, K. Click Reactions with Nitroxides. *Synthesis* **2009**, *8*, 1336–1340. [CrossRef]
19. Jakobsen, U.; Shelke, S.A.; Vogel, S.; Sigurdsson, S.T. Site-Directed Spin-Labeling of Nucleic Acids by Click Chemistry: Detection of Abasic Sites in Duplex DNA by EPR Spectroscopy. *J. Am. Chem. Soc.* **2010**, *132*, 10424–10428. [CrossRef]
20. Skarpos, H.; Osipov, S.N.; Vorob'eva, D.V.; Odinets, I.L.; Lork, E.; Röschenthaler, G.-V. Synthesis of functionalized bisphosphonates via click chemistry. *Org. Biomol. Chem.* **2007**, *5*, 2361–2367. [CrossRef]
21. Shelke, S.A.; Sigurdsson, S.T. Site-Directed Nitroxide Spin Labeling of Biopolymers. In *Structural Information from Spin-Labels and Intrinsic Paramagnetic Centres in the Biosciences*; Springer: Berlin/Heidelberg, Germany, 2011; pp. 121–162. [CrossRef]
22. Ur, G.; Kálai, T.; Balog, M.; Bognár, B.; Gulyás-Fekete, G.; Hideg, K. Synthesis of New Pyrroline Nitroxides with Ethynyl Functional Group. *Synthetic Commun.* **2015**, *45*, 2122–2129. [CrossRef]
23. Bal, B.S.; Childers, W.E., Jr.; Pinnick, H.W. Oxidation of α,β-unsaturated aldehydes. *Tetrahedron* **1981**, *37*, 2091. [CrossRef]
24. Hopkins, E.; Sanvictores, T.; Sharma, S. Physiology, Acid Base Balance. In *StatPearls*; StatPearls Publishing: Treasure Island, FL, USA, 2021. Available online: https://www.ncbi.nlm.nih.gov/books/NBK507807/ (accessed on 14 September 2021).
25. Hideg, K.; Lex, L. Synthesis of new 2-mono- and 2,5-di-functionalized pyrrolidin-1-oxyl spin labels. *J. Chem. Soc. Perkin Trans.* **1987**, *1*, 1117–1121. [CrossRef]
26. Kovaleva, E.G.; Molochnikov, L.S.; Golovkina, E.L.; Hartmann, M.; Kirilyuk, I.A.; Grigoriev, I.A. Electrical potential near hydrated surface of ordered mesoporous molecular sieves assessed by EPR of molecular pH-probes. *Micropor. Mesopor. Mater.* **2015**, *203*, 1–7. [CrossRef]
27. Kovaleva, E.G.; Molochnikov, L.S.; Parshina, E.V.; Shishmakov, A.B.; Mikushina, Y.u.V.; Kirilyuk, I.A.; Grigor'ev, I.A. Effect of the surface charge on the complexing and catalytic properties of Cu^{2+}-containing composite materials based on zirconia and powdered cellulose. *Russ. J. Phys. Chem. B* **2014**, *8*, 317–325. [CrossRef]
28. Kovaleva, E.G.; Molochnikov, L.S.; Golovkina, E.L.; Hartmann, M.; Kirilyuk, I.A.; Grigor'ev, I.A. Dynamics of pH-sensitive nitroxide radicals in water adsorbed in ordered mesoporous molecular sieves by EPR Spectroscopy. *Micropor. Mesopor. Mat.* **2013**, *179*, 258–264. [CrossRef]

29. Mekhaev, A.V.; Pestov, A.V.; Molochnikov, L.S.; Kovaleva, E.G.; Pervova, M.G.; Yaltuk, Y.u.G.; Grigor'ev, I.A.; Kirilyuk, I.A. Structure and Characteristics of Chitosan Cobalt-Containing Hybrid Systems, the Catalysts of Olefine Oxidation. *Russ. J. Phys. Chem. A* **2011**, *85*, 1155–1161. [CrossRef]
30. Mekhaev, A.V.; Pestov, A.V.; Molochnikov, L.S.; Kovaleva, E.G.; Yatluk, Y.u.G.; Grigor'ev, I.A.; Kirilyuk, I.A. Investigation of the Structure of Chitosan Hybrid Systems by pH-Sensitive Nitroxyl Radical. *Russ. J. Phys. Chem. A* **2011**, *85*, 987–992. [CrossRef]
31. Parshina, E.V.; Molochnikov, L.S.; Kovaleva, E.G.; Shishmakov, A.B.; Mikushina, Y.V.; Kirilyuk, I.A.; Grigor'ev, I.A. Medium acidity and catalytic properties of composite materials based on silica and titania and powder cellulose in the presence of Cu2+ ions. *Russ. J. Phys. Chem. A* **2011**, *85*, 452–456. [CrossRef]
32. Polienko, J.F.; Schanding, T.; Gatilov, Y.V.; Grigor'ev, I.A.; Voinov, M.A. Studies toward the synthesis of 4-(2-R-ethyl)amino-2,2,5,5-tetramethyl-3-imidazoline 1-oxyls. Nucleophilic substitution of bromide in the N-alkyl chain of the 1,2,4-oxadiazol-2-one precursor. *J. Org. Chem.* **2008**, *73*, 502–510. [CrossRef]
33. Voinov, M.A.; Polienko, J.F.; Schanding, T.; Bobko, A.A.; Khramtsov, V.V.; Gatilov, Y.V.; Rybalova, T.V.; Smirnov, A.I.; Grigor'ev, I.A. Synthesis, structure, and X-band (9.5 GHz) EPR characterization of the new series of pH-sensitive spin probes: N,N-disubstituted 4-amino-2,2,5,5-tetramethyl-3-imidazoline 1-oxyls. *J. Org. Chem.* **2005**, *70*, 9702–9711. [CrossRef]
34. Kirilyuk, I.A.; Polienko, Y.F.; Krumkacheva, O.A.; Strizhakov, R.K.; Gatilov, Y.V.; Grigor'ev, I.A.; Bagryanskaya, E.G. Synthesis of 2,5-bis(spirocyclohexane)-substituted Nitroxides of Pyrroline and Pyrrolidine series, including Thiol-specific spin label: An analogue of MTSSL with long relaxation time. *J. Org. Chem.* **2012**, *77*, 8016–8027. [CrossRef]
35. Polienko, Y.F.; Vinogradova, V.I.; Sagdullaev, S.S.; Abdullaev, N.D.; Gatilov, Y.V.; Grigor'ev, I.A. First spin-labeled cytisine derivatives. *Chem. Nat. Compd.* **2013**, *49*, 311–315. [CrossRef]
36. Dobrynin, S.A.; Usatov, M.S.; Zhurko, I.F.; Morozov, D.A.; Polienko, Y.F.; Glazachev, Y.I.; Parkhomenko, D.A.; Tyumentsev, M.A.; Gatilov, Y.V.; Chernyak, E.I.; et al. A Simple Method of Synthesis of 3-Carboxy-2,2,5,5-Tetraethylpyrrolidine-1-oxyl and Preparation of Reduction-Resistant Spin Labels and Probes of Pyrrolidine Series. *Molecules* **2021**, *26*, 5761. [CrossRef] [PubMed]
37. Antimonova, A.N.; Petrenko, N.I.; Shults, E.E.; Polienko, Y.F.; Shakirov, M.M.; Irtegova, I.G.; Pokrovskii, M.A.; Sherman, K.M.; Grigor'ev, I.A.; Pokrovskii, A.G. Synthetic transformations of higher triterpenoids. XXX: Synthesis and cytotoxic activity of betulonic acid amides with fragments of nitroxyl radicals. *Russ. J. Bioorg. Chem.* **2013**, *39*, 181–185. [CrossRef] [PubMed]
38. Krause, L.; Herbst-Irmer, R.; Sheldrick, G.M.; Stalke, D. Comparison of silver and molybdenum microfocus X-ray sources for single-crystal structure determination. *J. Appl. Cryst.* **2015**, *48*, 3–10. [CrossRef]
39. Sheldrick, G.M. A short history of SHELX. *Acta Crystallogr. Sect. A* **2008**, *A64*, 112–122. [CrossRef] [PubMed]
40. Sheldrick, G.M. Crystal structure refinement with SHELXL. *Acta Crystallogr. Sect. C* **2015**, *C71*, 3–8. [CrossRef]
41. Li, C.; Yuan, C. Studies on organophosphorus compounds 81. A novel synthetic approach to substituted cyclopentane-1,1-diylbisphosphonates via Pd(0) catalyzed enyne cyclization. *Heteroat. Chem.* **1993**, *4*, 517–520. [CrossRef]

gels

Review

Insights into the Role of Biopolymer-Based Xerogels in Biomedical Applications

H. P. S. Abdul Khalil [1,2,*], Esam Bashir Yahya [1,2], Husnul Azan Tajarudin [1], Venugopal Balakrishnan [3] and Halimatuddahliana Nasution [4]

1 School of Industrial Technology, Universiti Sains Malaysia, Penang 11800, Malaysia; essam912013@gmail.com (E.B.Y.); azan@usm.my (H.A.T.)
2 Cluster of Green Biopolymer, Coatings and Packaging, School of Industrial Technology, Universiti Sains Malaysia, Penang 11800, Malaysia
3 Institute for Research in Molecular Medicine, Universiti Sains Malaysia, Penang 11800, Malaysia; venugopal@usm.my
4 Department of Chemical Engineering, Faculty of Engineering, Universitas Sumatera Utara, Medan 20155, Indonesia; halimatuddahliana@usu.ac.id
* Correspondence: akhalilhps@gmail.com

Abstract: Xerogels are advanced, functional, porous materials consisting of ambient, dried, cross-linked polymeric networks. They possess characteristics such as high porosity, great surface area, and an affordable preparation route; they can be prepared from several organic and inorganic precursors for numerous applications. Owing to their desired properties, these materials were found to be suitable for several medical and biomedical applications; the high drug-loading capacity of xerogels and their ability to maintain sustained drug release make them highly desirable for drug delivery applications. As biopolymers and chemical-free materials, they have been also utilized in tissue engineering and regenerative medicine due to their high biocompatibility, non-immunogenicity, and non-cytotoxicity. Biopolymers have the ability to interact, cross-link, and/or trap several active agents, such as antibiotic or natural antimicrobial substances, which is useful in wound dressing and healing applications, and they can also be used to trap antibodies, enzymes, and cells for biosensing and monitoring applications. This review presents, for the first time, an introduction to biopolymeric xerogels, their fabrication approach, and their properties. We present the biological properties that make these materials suitable for many biomedical applications and discuss the most recent works regarding their applications, including drug delivery, wound healing and dressing, tissue scaffolding, and biosensing.

Keywords: xerogel; biopolymers; dried gels; porous materials; biomedical applications

Citation: Abdul Khalil, H. P. S.; Yahya, E.B.; Tajarudin, H.A.; Balakrishnan, V.; Nasution, H. Insights into the Role of Biopolymer-Based Xerogels in Biomedical Applications. *Gels* **2022**, *8*, 334. https://doi.org/10.3390/gels8060334

Academic Editor: Qiang Chen

Received: 24 March 2022
Accepted: 25 May 2022
Published: 29 May 2022

1. Introduction

In the past few years, we have witnessed the development of various novel functional materials from different precursors. Xerogels and aerogels are two examples of porous, structured materials that result from the different drying techniques of wet gels [1]. The attractive and unique properties of such porous materials arise from the extraordinary flexibility and resilience of the sol–gel developing process, which is combined with either ambient drying (xerogel) [2] or supercritical drying (aerogel) [3]. These materials have been prepared from several precursors, including silica [4], carbon [5], synthetic [6], and biopolymers [7]. Biopolymeric xerogels possess different physical, chemical, mechanical, and biological properties, depending on several factors, including precursor material/s, solvent medium, and drying conditions [7]. These factors also influence the shrinking of the biopolymeric gels, leading to an increased density and reduced porosity [8].

The structure, shape, and morphology of xerogels can be controlled in both the synthesizing and drying phases, but their porosity remains less than that of aerogels of the same materials [9]. Recently, xerogels have been widely synthesized from biopolymeric materials such as cellulose, chitosan, alginate, and pectin [10–15]. Such precursor materials are known for their biocompatibility and non-toxicity, which make them suitable for many biomedical applications, such as drug delivery, wound healing and dressing, tissue scaffolding, and biosensing applications [16]. Xerogels of the same material differ from aerogels in terms of their shrinking ratio, density, porosity, and specific surface area [8,17]. Although aerogels are higher in porosity and specific surface area, the synthesis of the xerogels under ambient pressure drying, which does not require energy-consuming instruments, has made the xerogels desirable materials, especially in the cases of operational risks and economic issues [18]. The intermediate porosity of xerogels is highly preferable in sustained drug release; the high porosity of aerogels may lead to fast drug release, which is not desirable in some cases, such as cancer drug therapy [19,20]. Several research papers have been recently published regarding the fabrication and characterization of biopolymeric xerogels for different medical and biomedical applications [21–23], which are increasing by the day. However, to the best of our knowledge, a limited number of review papers on the applications of biopolymeric xerogels in biomedical applications have been published [24,25]. Salimian et al. [26] generally reviewed aerogel/polymer nanocomposites, and our previous reviews also regarded the applications of aerogels, but not xerogels, for biomedical applications [27]. In this review, we present, for the first time, an overview of the xerogels, their fabrication approach, and their properties, including the biological properties that make these materials suitable for many biomedical applications. We also highlight the most recent works regarding the biomedical applications of biopolymeric xerogels, including utilizing them for drug delivery, wound healing and dressing, tissue engineering applications, and the development of smart biosensors.

2. Xerogel Functional Material

A xerogel is defined as a porous, structural material that can be obtained via the evaporative drying of any precursor's wet gel. Although the porosity and surface area of xerogels are lower than the aerogels, they are characterized by their easy and unexpansive fabrication, better mechanical stability, and higher density compared with aerogels [28].

2.1. Fabrication Techniques

The fabrication of xerogels generally consists of forming the polymeric hydrogel and drying that hydrogel in a way that retains (at least in part) its porous texture after the drying [29]. The process varies from one polymer to another, and drying conditions also differ based on the used solvent and precursor material/s. Pectin xerogel has been prepared from its alcogel. The authors used meld temperature (60 °C) for the drying purpose under vacuum conditions for 4 days until the complete drying of the alcogel [12]. The authors reported that in order to prevent a major collapse during the drying process, ionic gelation is a necessary step. A massive shrinkage of around 90 vol% commonly occurs after evaporative drying due to structural collapse, leading to an increase in the density of the material and a reduction in its porosity. The attractive properties of the biopolymeric porous hydrogels arise from their extraordinary flexibility during the sol–gel phase, which is mostly combined with various drying techniques, leading to the formation of the desired xerogel. Cellulose xerogel has been fabricated using a facile approach consisting of three steps: the partial ionic liquid dissolution of cellulose suspension, non-solvent rinsing, and drying [30]. In a different study, cellulose nanofiber xerogels were fabricated by Toivonen et al. [31] through a solvent exchange process (with octane), the vacuum filtration of their solvent dispersion, and finally, ambient drying. The authors reported a mesoporous xerogel with good porosity and surface area. Melone et al. [32] suggested a new, economically affordable synthesization protocol for the design of novel xerogels based on the cross-linking of TEMPO-oxidized cellulose nanofibers (TOUS-CNFs) and branched polyethyleneimine. The xerogel exhibited high adsorption capability for different organic pollutants, indicating its potential for water decontamination. In a different work, the authors were able to prepare different xerogels with attractive properties by cross-linking TEMPO-oxidized and ultra-sonicated cellulose nanofibers [33]. The drying step is the most important in most cases of biopolymeric xerogel fabrication. It directly affects most of the physical and morphological properties of the material. Xerogels and aerogels are the two closest relatives of polymeric substances, with slight differences in terms of fabrication approaches and properties. Unlike aerogels, xerogels cannot be formed from pure nanocellulose or any non-gel forming polymers [27,34]. Such biopolymers require cross-linking in order to form gels in them, then drying these gels to obtain xerogels [35]. Chitosan-silica xerogel was prepared by sol-gel and emulsification-crosslinking [36]. The addition of 20 wt% of SiO_2 was found to be enough to make the xerogels exhibit a regular spherical shape with sufficient dispersity and a uniform microstructure for drug delivery applications. However, compared with the pure chitosan xerogel-based microspheres, this hybrid showed significantly improved in vitro bioactivity in addition to good drug loading capacity and sustained release. Figure 1 presents an illustration of biopolymeric xerogel fabrication and the difference between biopolymeric xerogels and biopolymeric aerogels.

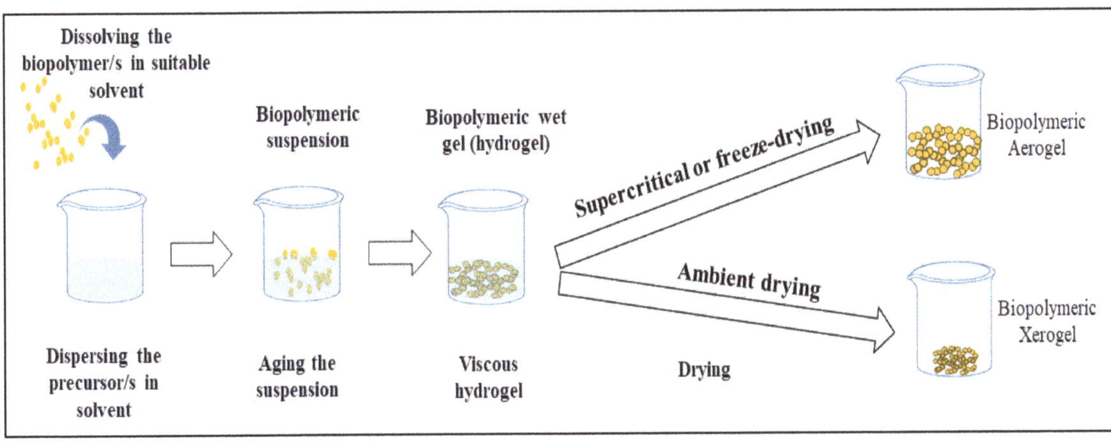

Figure 1. Schematic drawing of biopolymeric xerogel fabrication process and the difference between xerogels and aerogels.

2.2. Properties and Advantages of Xerogels

A xerogel is a solid, porous material resulting from the slow drying of hydrogels at room temperature, with unconstrained shrinkage depending on the type of precursor/s. Xerogels differ from aerogels in many aspects, including their shrinkage ratio, porosity, specific surface area, and bulk density [37]. Xerogels generally possess higher shrinkage than aerogels, and thus, they have lower porosity, lower surface area, and greater bulk density. Groult et al. [12] compared the properties of pectin xerogels and aerogels and found that in order to prevent a major collapse during the drying process, ionic gelation is a necessary step. The xerogels had bulk density and porosity of 1.057 g/cm^3 and 29.5%, respectively, compared with the pectin aerogels, which had 0.083 g/cm^3 and 94.4% for the bulk density and porosity, respectively. The xerogels exhibited a higher loading efficiency of 94% compared with the aerogels' loading efficiency, which was recorded to be 62%. The mechanical properties of xerogels vary depending on the type of precursor materials; in most cases, xerogels possess better mechanical properties than aerogels due to their lower porosity and higher bulk density [38]. Similarly, Ganesan et al. [39] prepared cellulose-based xerogels and aerogels and compared their characteristics, as presented in Figure 2. The authors found that the aerogels possessed significantly higher porosity, ranging between 92.7 and 96.4%, while the xerogels only possessed a porosity of 70.2 to 80.3%. The properties of biopolymeric xerogels are highly influenced by two main factors:

the precursor material/s and the liquid–vapour interface, in addition to the solvent medium, which affects the drying process [13]. Thus, changing these factors will lead to xerogels with different physical and morphological properties.

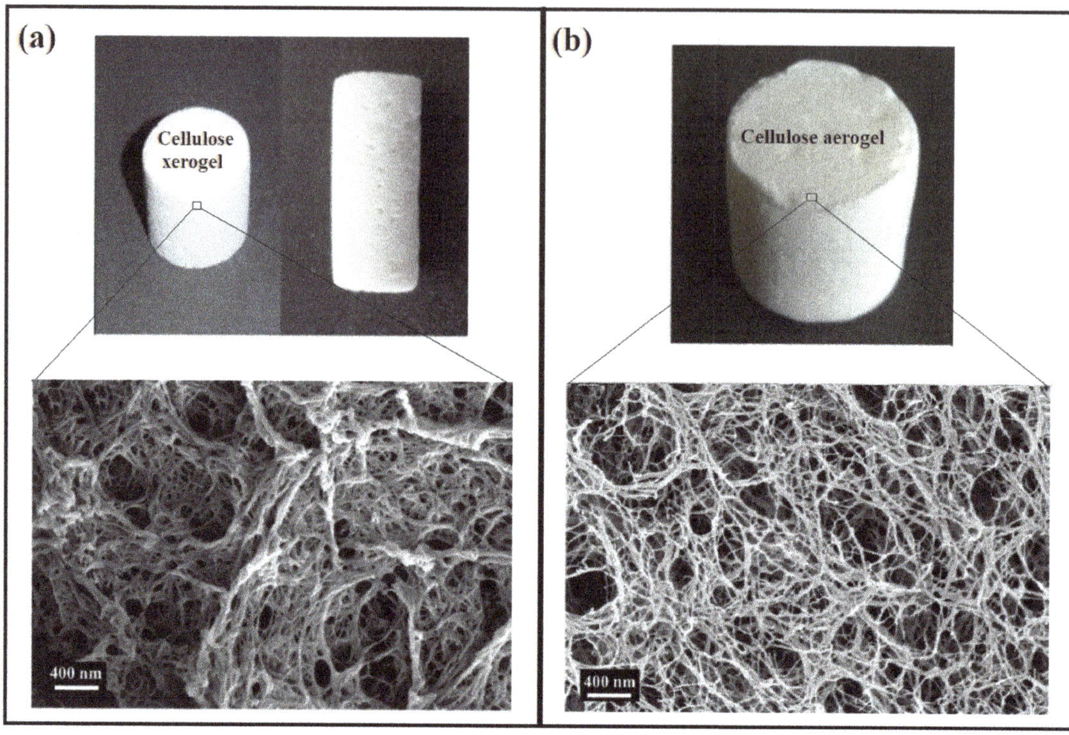

Figure 2. Comparison between cellulose xerogel and aerogel in term of shrinking and porosity; (**a**) present the xerogel sample of cellulose-based xerogel and the SEM image, and (**b**) present its aerogel and the SEM image. Adapted with permission from Ganesan et al. [39]. Copyright 2016 Elsevier.

Solvents such as ethanol have similar surface tension values to isopropanol, and research has reported that using these two solvents to prepare xerogel in the same condition could yield xerogels with different physical properties due to the change in vapour pressures [40]. Pramanik et al. [9] used nanocellulose in different mass ratios to improve the mechanical strength of polyvinyl alcohol xerogels. The authors reported that increasing the nanocellulose content led to a significant enhancement in the thermal properties of the xerogel. However, a xerogel rupture occurred in the case of a higher quantity of nanocellulose (18%) due to the formation of weak cellulose-rich regions. The addition of this much nanocellulose in the polymeric matrix increased the brittleness of the xerogels, which is the main cause of xerogel fracture. Silk fibroin-based xerogels possess great water absorption capacity, and Cheng et al. [23] reported that their xerogels were able to absorb up to 90 times its own mass of water within a minute in addition to its great hemostatic properties, making such material suitable for absorbing other body exudates. Several attempts have been made to produce aerogel-like xerogels under ambient conditions to minimize the shrinkage. However, the resulting xerogels in most of the cases inevitably took the form of thin films with relatively low porosity [7]. Prakash et al. [41] developed a unique approach to exchange the hydrogel's solvent for a solvent with a lower polarity than water, such as pentane or hexane, to reduce the capillary force and thus produce xerogels with higher porosity. Other materials, such as organosilicons, have been introduced to the xerogels to enhance the optical transparency of the xerogels and make them exhibit rubbery compression [42]. Cellulose nanofiber xerogels were fabricated through a solvent exchange process with mesoporous and in a film-like shape [31]. The xerogel possessed 60% porosity and 200 m^2/g specific surface area, which is considered close to the properties of aerogels. The characteristics of biopolymeric xerogels are highly influenced by the preparation conditions, as they directly affect the shrinkage of the hydrogels.

2.3. Suitability of Biopolymeric Xerogels in Biomedical Applications

Toxicity evaluation is very important when it comes to any medical applications, and the material will directly attach to the human or animal cells. Although many of the natural materials did not show significant toxicity to living cells, the preparation conditions may alter the chemistry of these materials and alter their biological effects [43,44]. Biopolymeric xerogels are dried forms of the biopolymer/s precursor; they have the chemical and biological characteristics of that biopolymer/s [45]. Several xerogels have been prepared without any need for further chemical addition or modification, but in other cases, natural compounds are added to extend the applications such as adding essential oils as an antibacterial agent. Biopolymers are known for being biocompatible and non-cytotoxic; they have been evaluated in several forms including the raw biopolymers [46], films [47], membranes [48], composites [49], hydrogels [50], aerogels [51], and even xerogels [14]. Although the number of cytotoxicity evaluations regarding biopolymeric xerogels is limited compared with aerogels, despite the drying process, aerogels and xerogels are prepared with the same principle, and thus, both of them are highly biocompatible, non-cytotoxic, and allow the attachment and migration of cells [52]. Refer to Table 1 for a summary of the cytotoxicity and biocompatibility evaluations of biopolymeric xerogels.

Table 1. Illustration of biocompatibility and cytotoxicity studies of biopolymer-based xerogels.

Type of Xerogel	Experiment	Type of Cells	Conclusion	Ref
Chitosan-gelatin xerogel	Hemocompatibility, cytotoxicity assays	Mouse embryonic fibroblast cells	Good platelet activation, good biocompatibility, and thrombin generation activities.	[14]
Collagen-silica xerogel	Cell culture experiments	Human monocytes	The xerogel promoted the differentiation of monocytes into osteoclast-like cells.	[53]
Carbon xerogel	Cytotoxicity test	Fibroblast cell	The xerogel was biocompatible; the presence of carbon fibers increases the cell's proliferation.	[54]
Chitosan coated mesoporous silica xerogels	Cytotoxicity assays	Mouse myoblast cells line	No obvious cytotoxicity was reported for the xerogel even after 7 days of the exposure.	[55]
Silk Fibroin Protein Xerogel	Hemostasis experiments	In-vitro and in-vivo rabbit ear	Good hemostatic properties were observed both in vitro and in vivo for the xerogel.	[23]
Chitosan–poly(vinyl alcohol) xerogel	Cytotoxicity and migration rate	Mouse embryonic fibroblast	The xerogel exhibited significant cell proliferation & migration rates and high biocompatibility.	[56]
Alginate-hydroxyapatite aerogel	Cytotoxicity, viability, and migration	Mesenchymal stem cells	Highly biocompatible, allowed attachment and migration.	[57]
Collagen–silica xerogel	Cell proliferation assay	Preosteoblast cells	Good biocompatibility and high level of osteoblast differentiation	[58]

3. Biopolymeric Xerogels in Biomedical Applications

Biopolymeric xerogels are porous networks of many unique and desirable properties that have been widely studied for different biomedical applications including controlled and sustained drug delivery, wound dressing and healing applications, tissue engineering scaffolds, and other applications [23]. Owing to the biocompatibility, non-cytotoxicity, and non-immunogenicity of the biopolymers, biopolymeric xerogels are considered to be a safer option than inorganic and synthetic materials in medical applications [1,59].

3.1. Drug Delivery

Xerogels have been extensively studied for their potential use in drug delivery since their discovery. Owing to their porous texture, their ability to control pore structure, and their large surface area, they attracted the attention of scientists in many pharmaceutical applications. Such desirable characteristics are favoured by drug loading and allow for better control of the drug release behavior [60]. Zhou et al. [16] used a poly (ε-caprolactone)-chitosan-silica xerogel for tetracycline hydrochloride delivery by green fabrication route. The presence of silica in the xerogel significantly enhanced the thermal stability and endowed good in vitro bioactivity and drug release behavior for the xerogel. The ability to modify the surfaces of biopolymers within the xerogel facilitates the drug incorporation in higher loading capacity and more sustained release. In a recent study, an alginate-based xerogel was modified using g-poly (methacrylic acid; AGM2S) for insulin delivery toward wound care [61]. The authors reported significant improvement in the physical stability, good swelling, and low degradation of the modified xerogel. More than 70% of loaded insulin was released from the xerogel in two days, which modulated the healing response [61]. In a different study, a novel xerogel was prepared from silica and poly(ethylene glycol) by the facile sol–gel route and showed sustained release of an enrofloxacin antibiotic drug [62]. The unique properties and facile fabrication of xerogels permit the slow release of drugs, making them a better option for sustained drug delivery applications. Different precursors consisting of naturally available diatomaceous earth microparticles have been used for the first time in xerogel fabrication [20]. Such unique xerogels were modified to enhance their drug loading capacity by using a facile sol–gel method resulting in a pH-sensitive micro drug carrier, which was evaluated for diclofenac sodium drug delivery. The authors reported a significant increase in drug loading capacity and sustained drug release fitting the zero-order model. Križman et al. [63] fabricated silk fibroin-based xerogels and evaluated their potential for long-acting hormone estradiol delivery (Figure 3). Ethanol was used in the preparation process and acted as a dissolving agent for the drug in addition to an accelerator for the gelation process. The authors were able to achieve a sustained drug release of up to 129 days from the xerogel delivery system, suggesting the great potential of such biopolymeric xerogel in the prolonged release of hydrophobic drugs.

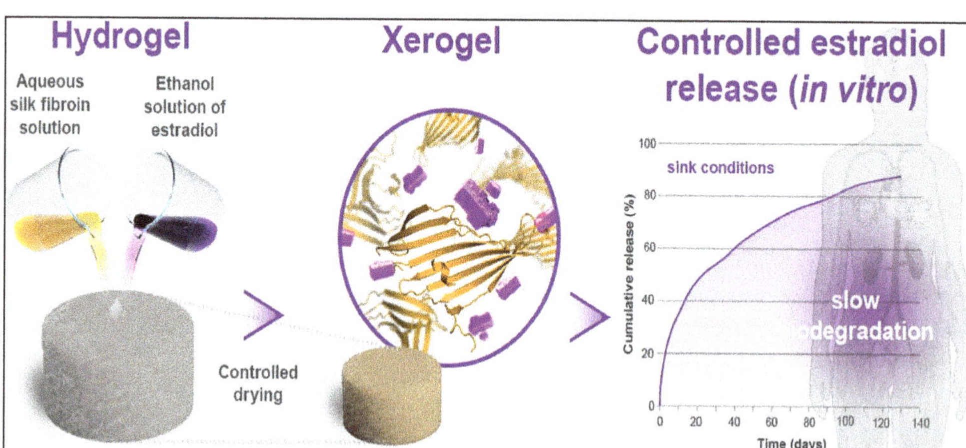

Figure 3. Illustration of silk fibroin based xerogels in controlled release of estradiol drug. Adapted with permission from Križman et al. [63]. Copyright 2022 Elsevier.

3.2. Antibacterial and Wound Healing Applications

The process of wound healing is a complex and dynamic process consisting of several stages that lasts days or even weeks depending on multiple factors, such as the type of wound, its depth, microbial colonization, and the patient's immune system, to enable the injured skin to restore itself [64]. Hydrogels' antibacterial materials [65–67] have been widely used in wound healing applications, but they have the drawback of requiring gauze or other adjuvants to be applied to a bleeding wound. Furthermore, the overly moist environment caused by hydrogel is not conducive to promoting wound healing and the scabbing effect, especially at the early stages of wound formation [68]. Deep wounds may favor the growth of anaerobic bacteria, leading to severe inflammation and suppuration [69]. Xerogels have been used to overcome these drawbacks, which can be customized to be super-hydrophobic and/or super-adhesive functional materials [70]. In a recent investigation, Huang et al. [71] fabricated a novel xerogel with good mechanical properties, using silver nanoparticles as an antibacterial agent. The hybrid xerogel was able to rapidly capture bacteria and kill 99.9% of *E. coli* and 99.85% of *S. aureus* through the electrostatic interactions of the disulfide groups. Although silver nanoparticles have been linked with minor adverse health effects, the authors reported the good biocompatibility and non-toxicity results of the xerogel [72]. Natural antibacterial agents, such as plant essential oils and extracts, could be also loaded into the xerogel and used for wound healing. Plant polysaccharide-based xerogels are characterized by their high biocompatibility, large biodegradability, and high water absorption capacity [34]. Owing to the excessive distribution of surface functional groups, they have the potential to cross-link with natural antibacterial agents. Chitin and chitosan are the most used animal-based biopolymers in terms of wound healing application due to their special properties, including bactericidal and antifungal characteristics, high permeability to oxygen, and healing activities by stimulating fibroblast proliferation [51]. Deon et al. [73] used a silica/titania magnetic xerogel to immobilize chitosan-stabilized gold nanoparticles as an antibacterial system. Owing to the synergistic effect of chitosan and gold nanoparticles, the surface reactivity of titania, and the porous and magnetic response of silica, the xerogel system possessed strong antibacterial activity, even at an extremely low gold content. Using two or more biopolymers in xerogel fabrication was found to enhance the properties of the material and limit the shrinkage after drying; a porous xerogel was fabricated using chitosan in combination with sodium polyacrylate, polyethylene glycol wound treatment, and hemorrhage control [74]. Chitosan was used as antimicrobial agent that always cross-linked with different organic or inorganic materials, such as gelatin and tannic acid, which played a hemostatic role [75]. Gelatin is a biopolymer that is extensively used in wound and skin care applications due to its ability to activate platelet aggregation, and it can also act as an absorbable hemostatic agent [75]. Patil et al. [14] used the two biopolymers to prepare a highly porous xerogel for an efficient, multimodal topical hemostat (Figure 4). The authors ionically cross-linked gelatin and chitosan with sodium tripolyphosphate, and they were able to achieve in vitro >16-fold improved blood clotting compared to the available commercial materials. The xerogel content displayed good platelet activation and promoted the generation of thrombin, which is very important in wound healing applications. The same authors conducted an in vivo study of their xerogel on a lethal femoral artery injury and reported 2.5 min hemostasis, which is significantly faster than the commercial Gauze (4.6 min) and Celox (3.3 min), in addition to easy removal from the wound. Although xerogels have not been used for commercialization purposes yet, in the coming years, we will witness the utilization of these materials in wound healing applications, as they have great potential as topical hemostatic agents and can be used to save precious lives.

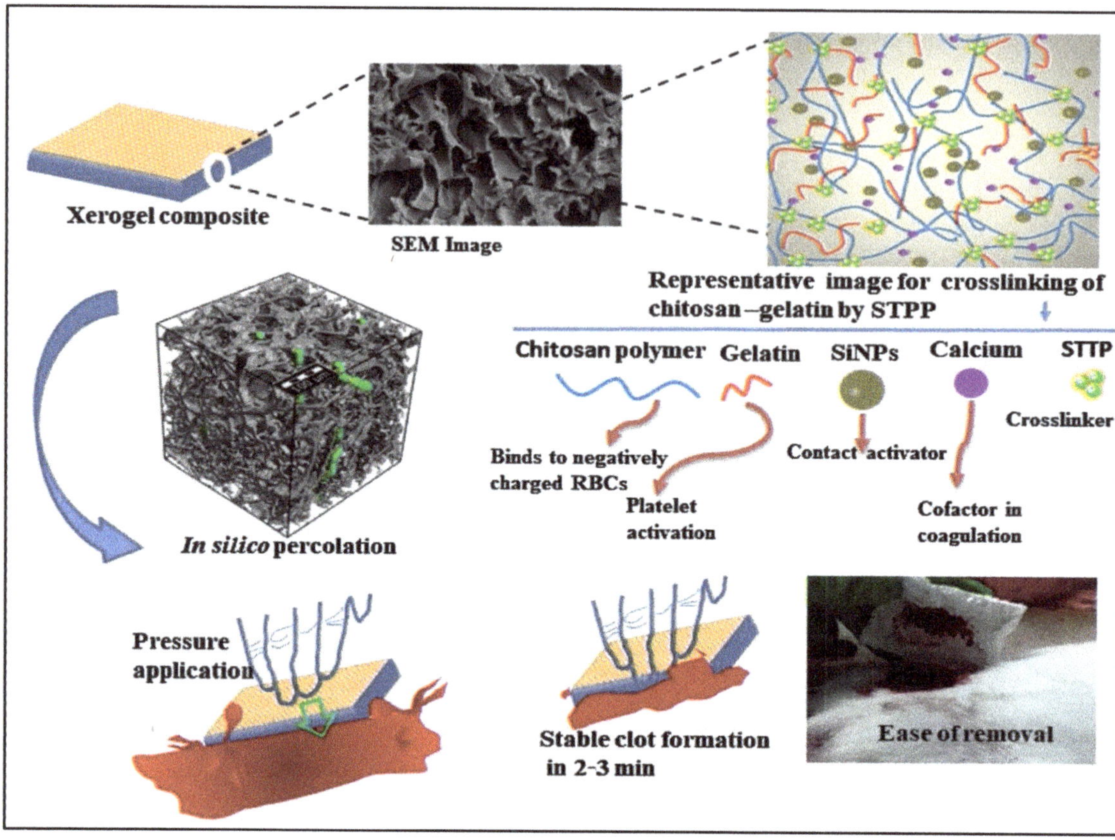

Figure 4. Chitosan-gelatin xerogel composite loaded with silica nanoparticles and calcium for rapid halting blood loss, showing the interaction between the biopolymers and its wound healing properties. Adapted with permission from Patil et al. [14]. Copyright 2022 Elsevier.

3.3. Tissue Engineering

Porous biopolymeric xerogels have been also used in tissue engineering scaffolds, as the easy adjustment of pore size and structure, in addition to their high biocompatibility, makes them a highly favorable form of the materials in such an application. A porous chitosan/berberine hydrochloride composite xerogel was prepared for tissue regeneration and hemostatic applications [76]. This biopolymeric xerogel exhibited good antibacterial activity, hemostatic properties, and fast degradability after immersion in phosphate-buffered saline. The authors reported good biocompatibility and strong hemostatic potential, as it was only composed of natural materials, which implies that it is a promising material for skin regeneration and hemostatic applications. The unique properties of some biopolymers, such as the antimicrobial activity of chitosan and promoting cell growth in collagen and silk fibrin, made their xerogels highly favorable in tissue engineering and regenerative medicine [77]. Wu et al. [78] fabricated a novel bioactive hybrid xerogel based on silk fibroin as precursor material, silica to enhance the mechanical properties, and CaO–P_2O_5 to enhance the xerogel's properties for bone regeneration applications. The authors reported excellent porosity and pore structures for their xerogel and adding the silica significantly enhanced the mechanical properties. The xerogel exhibited profound bioactivity once immersed in a simulated fluid due to the hydroxyapatite layers on its surfaces. The xerogel was biocompatible, although it showed little toxicity to MC3T3-E1 cells, which was due to the effect of silica on the cells. In a similar study, Lee et al. [58] fabricated a hybrid xerogel from calcium, silica, and collagen for bone regeneration applications. The authors used calcium to promote the bone cells' proliferation and silica to enhance the mechanical properties of collagen. Owing to the homogenous mixing and the incorporation of silica in the collagen matrix, the xerogel did not form any by-products, and it showed excellent bioactive characteristics. The hybrid xerogel expressed a better osteoblastic phenotype than the xerogels of pure collagen and pure silica. Elshishiny & Mamdouh [56] reported the fabrication of novel tri-layered, asymmetric, porous xerogel scaffolds for skin regeneration applications. The xerogel scaffold consisted of two layers: an upper layer of electrospun chitosan–poly(vinyl alcohol) and a lower layer of their regular xerogel. The authors fixed the two layers together by using a third material, fibrin glue, as a middle layer. This novel fabrication showed promising scaffold-swelling capability in addition to a high absorption capacity in regard to wound exudates. The porosity of the xerogel provided an optimum environment for the fibroblast cells' migration and proliferation. In a recent study, Rößler et al. [79] used the three-dimensional (3D) plotting of a silica and collagen hybrid xerogel scaffold in another biopolymeric matrix consisting of alginate (Figure 5). The authors used viscoelastic alginate as a matrix to enhance the biocompatibility and binding properties of the xerogel scaffold, and they reported that alginate concentration is the golden key to controlling the shape regularity of the xerogel granules.

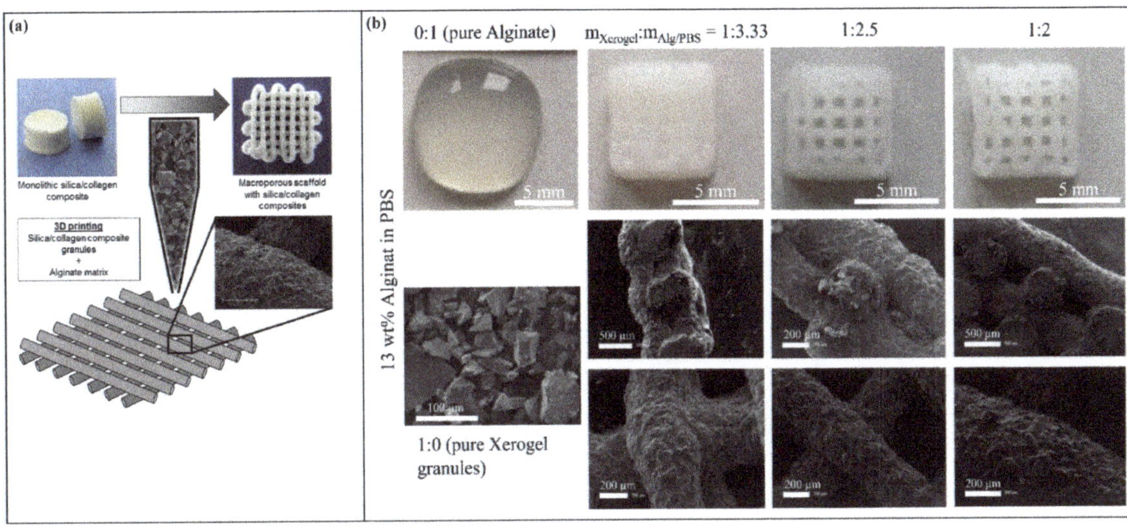

Figure 5. Illustration of 3D plotting approach of silica–collagen hybrid xerogel granules in an alginate matrix; (**a**) the fabrication approach, and (**b**) photographs and SEM images of a different ratio of xerogels. Adapted from Rößler et al. [79].

3.4. Biosensing

Biopolymeric xerogels have also been utilized in the sensing applications of many medically important parameters such as glucose level, uric acid, cholesterol, etc. Xerogels possess desired biosensor properties, such as porous structure and high surface area, making them highly advanced detection tools [80]. Khattab et al. [13] developed an easy-to-use, smart, microporous cellulose xerogel-based colorimetric sensor by immobilizing bromocresol purple chromophore into a cross-linked carboxymethyl cellulose xerogel matrix. Proton shifting from the hydroxyl group in the bromocresol purple dye to ammonia nitrogen enabled the identification of ammonia gas. Unlike dense material and metal-based xerogels, biopolymeric xerogels are distinguished by their non-toxicity, lighter weight, and larger surface area; thus, they are suitable for the identification of different parameters both in liquids and gaseous analytes [81]. The home-based detection and quantification of common analytes such as glucose monitoring, in addition to environmental routine monitoring, is exceedingly challenging and requires high measurement accuracy. Xerogel-based biosensors attracted attention for this purpose, as they are inexpensive, robust, and reusable materials able to meet all the requirements of biosensors [27]. The fabrication of xerogel-based biosensors began with the immobilization of active agents that are able to detect the desired parameters. Numerous active compounds, such as antibodies, active receptors, enzymes, cells, regulatory proteins, etc. have been used for this reason [82]. Figure 6 presents an illustration and several examples of active agents and their role in xerogel-based biosensors.

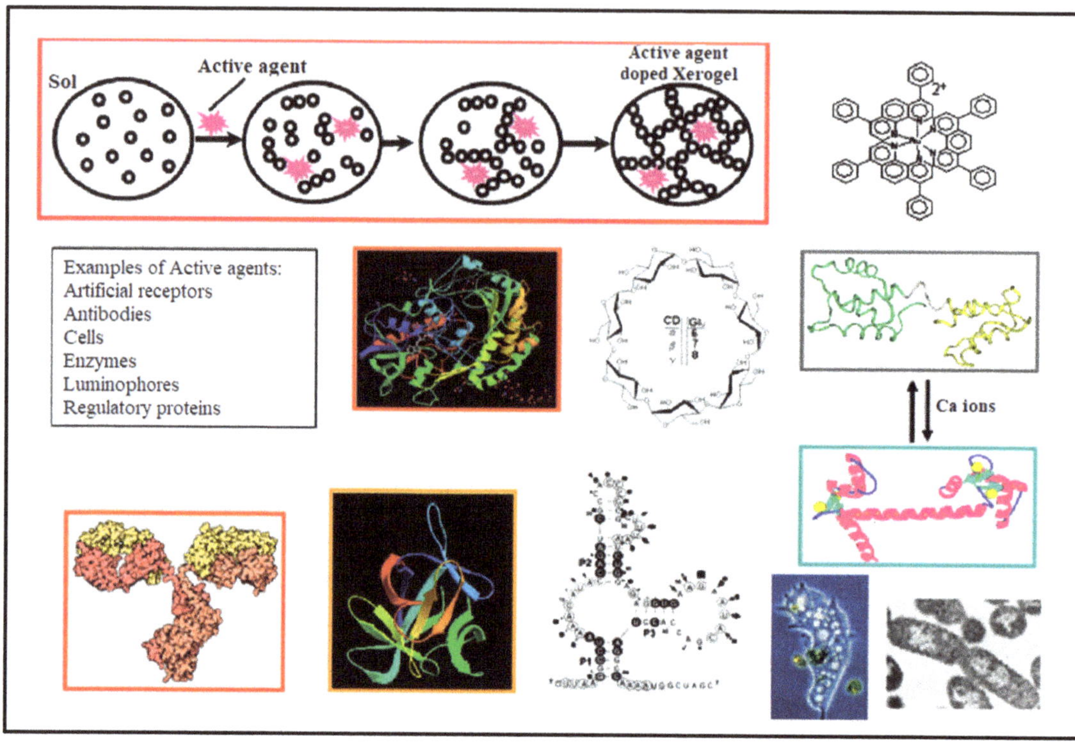

Figure 6. The role of active agents in the fabrication of xerogels based biosensors, presenting different examples of active agents. Reprinted with permission from Holthoff et al. [82]. Copyright 2007 American Chemical Society.

Three main approaches have been reported for the immobilization of active agents in the xerogels, including entrapment, physisorption, and covalent attachment [83]. The physisorption approach is the simplest, but it has the drawback of the random orientation of active agents on the xerogel, which could lead to them being unable to access the target molecule, thus lowering the accuracy of the xerogels [84,85]. To solve this issue, covalent attachment, which generally forms more stable interfaces, was developed. However, this approach also severs from the partial orientation of some kinds of active agents in addition to being a more expensive and time-consuming approach [82]. Freeman et al. [86] prepared the first generation of novel amperometric glucose biosensors, but they used several synthetic materials instead of biopolymers, which has the drawback of toxicity. To solve this issue, Alharthi et al. [87] recently used a nanocellulose acetate-based xerogel for the colorimetric detection of urea. The authors reported that their sponge-like, microporous xerogel was highly sensitive to urea because it used a urease enzyme as a catalytic agent and triarylmethane as a spectroscopic chromophore. The porous xerogel allowed for the in-situ integration of the triarylmethane probe, which enhanced the detection process and increased the accuracy of detection. Similarly, Abdelrahman et al. [88] developed a highly sensitive, reversible, and cost-effective biopolymeric xerogel for ammonia vapor detection. The microporous cellulose xerogel exhibited naked-eye colorimetric responsiveness immediately upon exposure to ammonia vapour. The application of biopolymeric xerogel in biosensing has not yet been extensively studied; limited works have been established, but we believe that these materials have great potential in regards to this application.

4. Challenges and Future Prospective

Biopolymeric porous materials such as xerogels and aerogels are still in their initial experimental stage in many biomedical applications. A limited number of materials entered the clinical trials, and most of them are still in the developmental and laboratory experimental phases. Although there are a significant number of studies that have proven that biopolymeric xerogels are highly suitable for biomedical applications, clinical and long-term evaluations of these materials are highly necessary before they can be commercialized. Smart and controlled delivery has been achieved in some experiments, especially in the case of cancer [89], but long-term evaluation and different cell evaluation have yet to be explored. Numerous challenges remain for other pathologies, as most of the current research focuses only on delivering specific drugs, particularly anticancer, anti-diabetic, and antimicrobial agents, and many of them involve in vitro studies or only short-term in vivo studies without considering the effect of these materials on other human bodies' biological parameters. Toxicity and biocompatibility experiments are, in most cases, carried out by using one type of cell in simulation conditions [90], and the real conditions inside our bodies might be different, so such materials may not be as biocompatible as they seem. The full effects of these biopolymeric materials on the human body have not yet been determined. The future generation of therapeutic biopolymeric materials with antibiotics, antibodies, hormones, peptides, genes, etc. should minimize undesirable side effects, not increase them. The future of biopolymeric xerogels requires serious collaboration among worldwide researchers, different industries, and regulatory agencies to maintain and ensure the safety and effectiveness of these therapeutic platforms to evaluate the potential and the possibilities of xerogel production in adequate quantities and of adequate quality to meet the expected demands of society.

Author Contributions: Conceptualization, A.K.H.P.S. and E.B.Y.; software, H.N. and H.A.T.; validation, E.B.Y. and V.B.; formal analysis, A.K.H.P.S. and E.B.Y.; resources, A.K.H.P.S.; writing—original draft preparation, E.B.Y.; visualization, H.N. and V.B.; supervision, A.K.H.P.S.; project administration, A.K.H.P.S. and E.B.Y.; funding acquisition, A.K.H.P.S. All authors have read and agreed to the published version of the manuscript.

Funding: This research was funded by Ministry of Higher Education Malaysia for Fundamental Research Grant Scheme-Malaysia Research Star Award (FRGS-MRSA) with Project Code: FRGS/1/2019/TK05/USM/01/6.

Data Availability Statement: Not applicable.

Acknowledgments: Acknowledgement to "Ministry of Higher Education Malaysia for Fundamental Research Grant Scheme Malaysia Research Star Award (FRGS-MRSA) with Project Code: FRGS/1/2019/TK05/USM/01/6".

Conflicts of Interest: The authors declare no conflict of interest.

References

1. Yahya, E.B.; Amirul, A.; Abdul Khalil, H.P.S.; Olaiya, N.G.; Iqbal, M.O.; Jummaat, F.; AK, A.S.; Adnan, A. Insights into the role of biopolymer aerogel scaffolds in tissue engineering and regenerative medicine. *Polymers* **2021**, *13*, 1612. [CrossRef] [PubMed]
2. Niu, Z.; He, X.; Huang, T.; Tang, B.; Cheng, X.; Zhang, Y.; Shao, Z. A facile preparation of transparent methyltriethoxysilane based silica xerogel monoliths at ambient pressure drying. *Microporous Mesoporous Mater.* **2019**, *286*, 98–104. [CrossRef]

3. Darpentigny, C.; Nonglaton, G.; Bras, J.; Jean, B. Highly absorbent cellulose nanofibrils aerogels prepared by supercritical drying. *Carbohydr. Polym.* **2020**, *229*, 115560. [CrossRef] [PubMed]
4. Elma, M.; Setyawan, H. Synthesis of Silica Xerogels Obtained in Organic Catalyst via Sol Gel Route. *IOP Conf. Ser. Earth Environ. Sci.* **2018**, *175*, 012008. [CrossRef]
5. Ptaszkowska-Koniarz, M.; Goscianska, J.; Pietrzak, R. Removal of rhodamine B from water by modified carbon xerogels. *Colloids Surf. A Physicochem. Eng. Asp.* **2018**, *543*, 109–117. [CrossRef]
6. Adamova, L.; Safronov, A.; Terziyan, T.; Shabadrov, P.; Klyukina, A. Thermodynamics of Swelling of Polyacrylamide and Poly (methacrylic acid) Lyophilized Xerogels in Water. *Polym. Sci. Ser. A* **2018**, *60*, 190–197. [CrossRef]
7. Yamasaki, S.; Sakuma, W.; Yasui, H.; Daicho, K.; Saito, T.; Fujisawa, S.; Isogai, A.; Kanamori, K. Nanocellulose Xerogels with high porosities and large specific surface areas. *Front. Chem.* **2019**, *7*, 316. [CrossRef]
8. Kaya, G.G.; Deveci, H. Synergistic effects of silica aerogels/xerogels on properties of polymer composites: A review. *J. Ind. Eng. Chem.* **2020**, *89*, 13–27. [CrossRef]
9. Pramanik, R.; Ganivada, B.; Ram, F.; Shanmuganathan, K.; Arockiarajan, A. Influence of nanocellulose on mechanics and morphology of polyvinyl alcohol xerogels. *J. Mech. Behav. Biomed. Mater.* **2019**, *90*, 275–283. [CrossRef]
10. Awadallah-F, A.; Al-Muhtaseb, S.A. Influence of Chitosan Addition on Resorcinol–Formaldehyde Xerogel Structure. *Appl. Sci.* **2019**, *9*, 4582. [CrossRef]
11. Rbihi, S.; Laallam, L.; Sajieddine, M.; Jouaiti, A. Characterization and thermal conductivity of cellulose based composite xerogels. *Heliyon* **2019**, *5*, e01704. [CrossRef] [PubMed]
12. Groult, S.; Buwalda, S.; Budtova, T. Pectin hydrogels, aerogels, cryogels and xerogels: Influence of drying on structural and release properties. *Eur. Polym. J.* **2021**, *149*, 110386. [CrossRef]
13. Khattab, T.A.; Dacrory, S.; Abou-Yousef, H.; Kamel, S. Development of microporous cellulose-based smart xerogel reversible sensor via freeze drying for naked-eye detection of ammonia gas. *Carbohydr. Polym.* **2019**, *210*, 196–203. [CrossRef] [PubMed]
14. Patil, G.; Torris, A.; Suresha, P.; Jadhav, S.; Badiger, M.V.; Ghormade, V. Design and synthesis of a new topical agent for halting blood loss rapidly: A multimodal chitosan-gelatin xerogel composite loaded with silica nanoparticles and calcium. *Colloids Surf. B Biointerfaces* **2021**, *198*, 111454. [CrossRef]
15. Bilanovic, D.; Starosvetsky, J.; Armon, R.H. Preparation of biodegradable xanthan–glycerol hydrogel, foam, film, aerogel and xerogel at room temperature. *Carbohydr. Polym.* **2016**, *148*, 243–250. [CrossRef]
16. Zhou, H.-J.; Teng, S.-H.; Zhou, Y.-B.; Qian, H.-S. Green Strategy to Develop Novel Drug-Containing Poly (ε-Caprolactone)-Chitosan-Silica Xerogel Hybrid Fibers for Biomedical Applications. *J. Nanomater.* **2020**, *2020*, 6659287. [CrossRef]
17. Kaya, G.G.; Yilmaz, E.; Deveci, H. Synthesis of sustainable silica xerogels/aerogels using inexpensive steel slag and bean pod ash: A comparison study. *Adv. Powder Technol.* **2020**, *31*, 926–936. [CrossRef]
18. Zhu, J.; Xie, J.; Lü, X.; Jiang, D. Synthesis and characterization of superhydrophobic silica and silica/titania aerogels by sol–gel method at ambient pressure. *Colloids Surf. A Physicochem. Eng. Asp.* **2009**, *342*, 97–101. [CrossRef]
19. Zhang, P.; Lu, T.; Xia, X.; Wu, L.; Shao, L.; Zhou, J.; Li, J. How biomimetic amino modified mesoporous silica xerogel regulates loading and in vitro sustained delivery of levorotary ofloxacin. *Mater. Sci. Eng. C* **2020**, *107*, 110266. [CrossRef]
20. Uthappa, U.; Sriram, G.; Brahmkhatri, V.; Kigga, M.; Jung, H.-Y.; Altalhi, T.; Neelgund, G.M.; Kurkuri, M.D. Xerogel modified diatomaceous earth microparticles for controlled drug release studies. *New J. Chem.* **2018**, *42*, 11964–11971. [CrossRef]
21. Antonov, D.O.; Tambasova, D.P.; Shishmakov, A.B.; Kirilyuk, I.A.; Kovaleva, E.G. Acidic and Electrosurface Properties of Binary TiO2-SiO2 Xerogels Using EPR of pH-Sensitive Nitroxides. *Gels* **2021**, *7*, 119. [CrossRef] [PubMed]
22. El-Naggar, M.E.; Othman, S.I.; Allam, A.A.; Morsy, O.M. Synthesis, drying process and medical application of polysaccharide-based aerogels. *Int. J. Biol. Macromol.* **2020**, *145*, 1115–1128. [CrossRef] [PubMed]
23. Cheng, K.; Tao, X.; Qi, Z.; Yin, Z.; Kundu, S.C.; Lu, S. Highly Absorbent Silk Fibroin Protein Xerogel. *ACS Biomater. Sci. Eng.* **2021**, *7*, 3594–3607. [CrossRef] [PubMed]
24. Mahmood, A.; Patel, D.; Hickson, B.; DesRochers, J.; Hu, X. Recent Progress in Biopolymer-Based Hydrogel Materials for Biomedical Applications. *Int. J. Mol. Sci.* **2022**, *23*, 1415. [CrossRef]
25. Soorbaghi, F.P.; Isanejad, M.; Salatin, S.; Ghorbani, M.; Jafari, S.; Derakhshankhah, H. Bioaerogels: Synthesis approaches, cellular uptake, and the biomedical applications. *Biomed. Pharmacother.* **2019**, *111*, 964–975. [CrossRef]
26. Salimian, S.; Zadhoush, A.; Talebi, Z. Interpenetrating organic–inorganic network: A short review on aerogel as a nanoporous filler in epoxy nanocomposite. *Mater. Des. Process. Commun.* **2019**, *1*, e107. [CrossRef]
27. Abdul Khalil, H.P.S.; Adnan, A.; Yahya, E.B.; Olaiya, N.; Safrida, S.; Hossain, M.; Balakrishnan, V.; Gopakumar, D.A.; Abdullah, C.; Oyekanmi, A. A review on plant cellulose nanofibre-based aerogels for biomedical applications. *Polymers* **2020**, *12*, 1759. [CrossRef]
28. Gutierrez Cano, V.; Menelaou, M.; Kastyl, J.; Cihlář, J.; Tkachenko, S.; González, J.A.; Kalmár, J.; Fabian, I.; Lázár, I.; Čelko, L. Rare-Earth Zirconate Ln2Zr2O7 (Ln: La, Nd, Gd, and Dy) Powders, Xerogels, and Aerogels: Preparation, Structure, and Properties. *Inorg. Chem.* **2019**, *58*, 14467–14477.
29. Paladini, G.; Venuti, V.; Crupi, V.; Majolino, D.; Fiorati, A.; Punta, C. FTIR-ATR analysis of the H-bond network of water in branched polyethyleneimine/TEMPO-oxidized cellulose nano-fiber xerogels. *Cellulose* **2020**, *27*, 8605–8618. [CrossRef]
30. Aiello, A.; Cosby, T.; McFarland, J.; Durkin, D.P.; Trulove, P.C. Mesoporous xerogel cellulose composites from biorenewable natural cotton fibers. *Carbohydr. Polym.* **2022**, *282*, 119040. [CrossRef]

31. Toivonen, M.S.; Kaskela, A.; Rojas, O.J.; Kauppinen, E.I.; Ikkala, O. Ambient-dried cellulose nanofibril aerogel membranes with high tensile strength and their use for aerosol collection and templates for transparent, flexible devices. *Adv. Funct. Mater.* **2015**, *25*, 6618–6626. [CrossRef]
32. Melone, L.; Rossi, B.; Pastori, N.; Panzeri, W.; Mele, A.; Punta, C. TEMPO-oxidized cellulose cross-linked with branched polyethyleneimine: Nanostructured adsorbent sponges for water remediation. *ChemPlusChem* **2015**, *80*, 1408–1415. [CrossRef] [PubMed]
33. Paladini, G.; Venuti, V.; Almásy, L.; Melone, L.; Crupi, V.; Majolino, D.; Pastori, N.; Fiorati, A.; Punta, C. Cross-linked cellulose nano-sponges: A small angle neutron scattering (SANS) study. *Cellulose* **2019**, *26*, 9005–9019. [CrossRef]
34. Yahya, E.B.; Jummaat, F.; Amirul, A.; Adnan, A.; Olaiya, N.; Abdullah, C.; Rizal, S.; Mohamad Haafiz, M.; Khalil, H. A review on revolutionary natural biopolymer-based aerogels for antibacterial delivery. *Antibiotics* **2020**, *9*, 648. [CrossRef]
35. Rizal, S.; Yahya, E.B.; Abdul Khalil, H.P.S.; Abdullah, C.; Marwan, M.; Ikramullah, I.; Muksin, U. Preparation and Characterization of Nanocellulose/Chitosan Aerogel Scaffolds Using Chemical-Free Approach. *Gels* **2021**, *7*, 246. [CrossRef]
36. Niu, N.; Teng, S.-H.; Zhou, H.-J.; Qian, H.-S. Synthesis, characterization, and in vitro drug delivery of chitosan-silica hybrid microspheres for bone tissue engineering. *J. Nanomater.* **2019**, *2019*, 7425787. [CrossRef]
37. Chong, S.; Riley, B.J.; Peterson, J.A.; Olszta, M.J.; Nelson, Z.J. Gaseous iodine sorbents: A comparison between Ag-loaded aerogel and xerogel scaffolds. *ACS Appl. Mater. Interfaces* **2020**, *12*, 26127–26136. [CrossRef]
38. Mariana, M.; Abdul Khalil, H.P.S.; Yahya, E.B.; Olaiya, N.; Alfatah, T.; Suriani, A.; Mohamed, A. Recent trends and future prospects of nanostructured aerogels in water treatment applications. *J. Water Process Eng.* **2022**, *45*, 102481. [CrossRef]
39. Ganesan, K.; Dennstedt, A.; Barowski, A.; Ratke, L. Design of aerogels, cryogels and xerogels of cellulose with hierarchical porous structures. *Mater. Des.* **2016**, *92*, 345–355. [CrossRef]
40. Guzel Kaya, G.; Deveci, H. Effect of aging solvents on physicochemical and thermal properties of silica xerogels derived from steel slag. *ChemistrySelect* **2020**, *5*, 1586–1591. [CrossRef]
41. Prakash, S.S.; Brinker, C.J.; Hurd, A.J.; Rao, S.M. Silica aerogel films prepared at ambient pressure by using surface derivatization to induce reversible drying shrinkage. *Nature* **1995**, *374*, 439–443. [CrossRef]
42. Kanamori, K.; Aizawa, M.; Nakanishi, K.; Hanada, T. New transparent methylsilsesquioxane aerogels and xerogels with improved mechanical properties. *Adv. Mater.* **2007**, *19*, 1589–1593. [CrossRef]
43. Tan, E.; Li, B.L.; Ariga, K.; Lim, C.-T.; Garaj, S.; Leong, D.T. Toxicity of two-dimensional layered materials and their heterostructures. *Bioconjug. Chem.* **2019**, *30*, 2287–2299. [CrossRef] [PubMed]
44. Iqbal, M.O.; Yahya, E.B.; Andleeb, S.; Ahmed, M.M.; Javaid, M.U.; Shakeel, W.; Iqbal, I. In vivo assessment of reversing Cisplatin-Induced nephrotoxicity using Jatropha mollissima crude extract and its potential cytotoxicity. *Saudi J. Biol. Sci.* **2021**, *28*, 7373–7378. [CrossRef]
45. Iskandar, M.; Yahya, E.B.; Abdul Khalil, H.P.S.; Rahman, A.; Ismail, M. Recent Progress in Modification Strategies of Nanocellulose-Based Aerogels for Oil Absorption Application. *Polymers* **2022**, *14*, 849. [CrossRef]
46. Bilal, M.; Iqbal, H.M. Naturally-derived biopolymers: Potential platforms for enzyme immobilization. *Int. J. Biol. Macromol.* **2019**, *130*, 462–482. [CrossRef]
47. Oyekanmi, A.; Abdul Khalil, H.P.S.; Rahman, A.; Mistar, E.; Olaiya, N.; Alfatah, T.; Yahya, E.B.; Mariana, M.; Hazwan, C.; Abdullah, C. Extracted supercritical CO2 cinnamon oil functional properties enhancement in cellulose nanofibre reinforced *Euchema cottoni* biopolymer films. *J. Mater. Res. Technol.* **2021**, *15*, 4293–4308. [CrossRef]
48. Kenawy, E.; Omer, A.; Tamer, T.; Elmeligy, M.; Eldin, M.M. Fabrication of biodegradable gelatin/chitosan/cinnamaldehyde crosslinked membranes for antibacterial wound dressing applications. *Int. J. Biol. Macromol.* **2019**, *139*, 440–448. [CrossRef]
49. Rizal, S.; Saharudin, N.; Olaiya, N.; ABDUL Khalil, H.P.S.; Haafiz, M.; Ikramullah, I.; Muksin, U.; Olaiya, F.G.; Abdullah, C.; Yahya, E.B. Functional Properties and Molecular Degradation of Schizostachyum Brachycladum Bamboo Cellulose Nanofibre in PLA-Chitosan Bionanocomposites. *Molecules* **2021**, *26*, 2008. [CrossRef]
50. Appuhamillage, G.A.; Berry, D.R.; Benjamin, C.E.; Luzuriaga, M.A.; Reagan, J.C.; Gassensmith, J.J.; Smaldone, R.A. A biopolymer-based 3D printable hydrogel for toxic metal adsorption from water. *Polym. Int.* **2019**, *68*, 964–971. [CrossRef]
51. Yahya, E.B.; Alzalouk, M.M.; Alfallous, K.A.; Abogmaza, A.F. Antibacterial cellulose-based aerogels for wound healing application: A review. *Biomed. Res. Ther.* **2020**, *7*, 4032–4040. [CrossRef]
52. García-González, C.A.; López-Iglesias, C.; Concheiro, A.; Alvarez-Lorenzo, C. Biomedical applications of polysaccharide and protein based aerogels. *Biobased Aerogels* **2018**, *16*, 295–323.
53. Heinemann, S.; Heinemann, C.; Bernhardt, R.; Reinstorf, A.; Nies, B.; Meyer, M.; Worch, H.; Hanke, T. Bioactive silica–collagen composite xerogels modified by calcium phosphate phases with adjustable mechanical properties for bone replacement. *Acta Biomater.* **2009**, *5*, 1979–1990. [CrossRef] [PubMed]
54. Bailon-Garcia, E.; Maldonado-Hodar, F.J.; Carrasco-Marin, F.; Perez-Cadenas, A.F.; Bosi, S.; Prato, M. The use of functionalized carbon xerogels in cells growth. *Mater. Sci. Eng. C* **2019**, *100*, 598–607. [CrossRef]
55. Dai, C.; Liu, C.; Wei, J.; Hong, H.; Zhao, Q. Molecular imprinted macroporous chitosan coated mesoporous silica xerogels for hemorrhage control. *Biomaterials* **2010**, *31*, 7620–7630. [CrossRef]
56. Elshishiny, F.; Mamdouh, W. Fabrication of Nanofibrous/Xerogel Layer-by-Layer Biocomposite Scaffolds for Skin Tissue Regeneration: In Vitro Study. *ACS Omega* **2020**, *5*, 2133–2147. [CrossRef]

57. Iglesias-Mejuto, A.; García-González, C.A. 3D-printed alginate-hydroxyapatite aerogel scaffolds for bone tissue engineering. *Mater. Sci. Eng. C* **2021**, *131*, 112525. [CrossRef]
58. Lee, E.J.; Jun, S.H.; Kim, H.E.; Koh, Y.H. Collagen–silica xerogel nanohybrid membrane for guided bone regeneration. *J. Biomed. Mater. Res. A* **2012**, *100*, 841–847. [CrossRef]
59. Jummaat, F.; Yahya, E.B.; Abdul Khalil, H.P.S.; Adnan, A.S.; Alqadhi, A.M.; Abdullah, C.K.; A.K., A.S.; Olaiya, N.; Abdat, M. The role of biopolymer-based materials in obstetrics and gynecology applications: A review. *Polymers* **2021**, *13*, 633. [CrossRef]
60. Zhao, Q.; Lin, Y.; Han, N.; Li, X.; Geng, H.; Wang, X.; Cui, Y.; Wang, S. Mesoporous carbon nanomaterials in drug delivery and biomedical application. *Drug Deliv.* **2017**, *24*, 94–107. [CrossRef]
61. Rajalekshmy, G.; Rekha, M. Synthesis and evaluation of an alginate-methacrylate xerogel for insulin delivery towards wound healing applications. *Ther. Deliv.* **2021**, *12*, 215–234. [CrossRef] [PubMed]
62. Rafati, A.; Ebadi, A.; Bavafa, S.; Nowroozi, A. Kinetic study, structural analysis and computational investigation of novel xerogel based on drug-PEG/SiO2 for controlled release of enrofloxacin. *J. Mol. Liq.* **2018**, *266*, 733–742. [CrossRef]
63. Križman, K.; Novak, S.; Kristl, J.; Majdič, G.; Drnovšek, N. Long-acting silk fibroin xerogel delivery systems for controlled release of estradiol. *J. Drug Deliv. Sci. Technol.* **2021**, *65*, 102701. [CrossRef]
64. Moeini, A.; Pedram, P.; Makvandi, P.; Malinconico, M.; d'Ayala, G.G. Wound healing and antimicrobial effect of active secondary metabolites in chitosan-based wound dressings: A review. *Carbohydr. Polym.* **2020**, *233*, 115839. [CrossRef]
65. Li, S.; Dong, S.; Xu, W.; Tu, S.; Yan, L.; Zhao, C.; Ding, J.; Chen, X. Antibacterial hydrogels. *Adv. Sci.* **2018**, *5*, 1700527. [CrossRef]
66. Dai, T.; Wang, C.; Wang, Y.; Xu, W.; Hu, J.; Cheng, Y. A nanocomposite hydrogel with potent and broad-spectrum antibacterial activity. *ACS Appl. Mater. Interfaces* **2018**, *10*, 15163–15173. [CrossRef]
67. Yahya, E.; Abdulsamad, M.A. In-vitro Antibacterial Activity of Carbopol-Essential Oils hydrogels. *J. Appl. Sci. Process Eng.* **2020**, *7*, 564–571. [CrossRef]
68. Rajalekshmy, G.; Rekha, M. Strontium ion cross-linked alginate-g-poly (PEGMA) xerogels for wound healing applications: In vitro studies. *Carbohydr. Polym.* **2021**, *251*, 117119.
69. Correa-Gallegos, D.; Jiang, D.; Christ, S.; Ramesh, P.; Ye, H.; Wannemacher, J.; Kalgudde Gopal, S.; Yu, Q.; Aichler, M.; Walch, A. Patch repair of deep wounds by mobilized fascia. *Nature* **2019**, *576*, 287–292. [CrossRef]
70. Zhang, Q.; Shi, C.-Y.; Qu, D.-H.; Long, Y.-T.; Feringa, B.L.; Tian, H. Exploring a naturally tailored small molecule for stretchable, self-healing, and adhesive supramolecular polymers. *Sci. Adv.* **2018**, *4*, eaat8192. [CrossRef]
71. Huang, B.; Liu, X.; Li, Z.; Zheng, Y.; Yeung, K.W.K.; Cui, Z.; Liang, Y.; Zhu, S.; Wu, S. Rapid bacteria capturing and killing by AgNPs/N-CD@ ZnO hybrids strengthened photo-responsive xerogel for rapid healing of bacteria-infected wounds. *Chem. Eng. J.* **2021**, *414*, 128805. [CrossRef]
72. Banu, A.; Gousuddin, M.; Yahya, E.B. Green synthesized monodispersed silver nanoparticles' characterization and their efficacy against cancer cells. *Biomed. Res. Ther.* **2021**, *8*, 4476–4482. [CrossRef]
73. Deon, M.; Morawski, F.; Passaia, C.; Dalmás, M.; Laranja, D.; Malheiros, P.; Nicolodi, S.; Arenas, L.; Costa, T.; de Menezes, E. Chitosan-stabilized gold nanoparticles supported on silica/titania magnetic xerogel applied as antibacterial system. *J. Sol-Gel Sci. Technol.* **2019**, *89*, 333–342. [CrossRef]
74. Qian, Z.; Wang, H.; Tuo, X.; Guo, H.; Xu, P.; Liu, D.; Wei, Y.; Liu, H.; Fan, Y.; Guo, X. A porous sodium polyacrylate-grafted chitosan xerogel for severe hemorrhage control synthesized from one-pot reaction. *J. Mater. Chem. B* **2017**, *5*, 4845–4851. [CrossRef] [PubMed]
75. Lan, G.; Lu, B.; Wang, T.; Wang, L.; Chen, J.; Yu, K.; Liu, J.; Dai, F.; Wu, D. Chitosan/gelatin composite sponge is an absorbable surgical hemostatic agent. *Colloids Surf. B Biointerfaces* **2015**, *136*, 1026–1034. [CrossRef] [PubMed]
76. Wang, F.; Chen, Y.; SU, X. Preparation of porous biodegradable chitosan/berberine hydrochloride composite xerogel and its antibacterial and hemostatic properties. *Chin. J. Tissue Eng. Res.* **2017**, *53*, 899–905.
77. Mirtaghavi, A.; Luo, J.; Muthuraj, R. Recent Advances in Porous 3D Cellulose Aerogels for Tissue Engineering Applications: A Review. *Journal of Composites Science* **2020**, *4*, 152. [CrossRef]
78. Wu, X.; Yan, F.; Liu, W.; Zhan, H.; Yang, W. Synthesis and characterization of silk fibroin-bioactive glass hybrid xerogels. *Biomater. Biomech. Bioeng.* **2014**, *1*, 63–71. [CrossRef]
79. Rößler, S.; Brückner, A.; Kruppke, I.; Wiesmann, H.-P.; Hanke, T.; Kruppke, B. 3D Plotting of Silica/Collagen Xerogel Granules in an Alginate Matrix for Tissue-Engineered Bone Implants. *Materials* **2021**, *14*, 830. [CrossRef]
80. Khattab, T.A.; Fouda, M.M.; Rehan, M.; Okla, M.K.; Alamri, S.A.; Alaraidh, I.A.; Al-Ghamdi, A.A.; Soufan, W.H.; Abdelsalam, E.M.; Allam, A.A. Novel halochromic cellulose nanowhiskers from rice straw: Visual detection of urea. *Carbohydr. Polym.* **2020**, *231*, 115740. [CrossRef]
81. Pottathara, Y.B.; Bobnar, V.; Finšgar, M.; Grohens, Y.; Thomas, S.; Kokol, V. Cellulose nanofibrils-reduced graphene oxide xerogels and cryogels for dielectric and electrochemical storage applications. *Polymer* **2018**, *147*, 260–270. [CrossRef]
82. Holthoff, E.L.; Bright, F.V. Molecularly imprinted xerogels as platforms for sensing. *Acc. Chem. Res.* **2007**, *40*, 756–767. [CrossRef] [PubMed]
83. Echeverría, J.C.; Faustini, M.; Garrido, J.J. Effects of the porous texture and surface chemistry of silica xerogels on the sensitivity of fiber-optic sensors toward VOCs. *Sens. Actuators B Chem.* **2016**, *222*, 1166–1174. [CrossRef]

84. Tang, J.; Zhang, F.; Liang, X.; Dai, G.; Qu, F. Abundant defects of zirconium-organic xerogels: High anhydrous proton conductivities over a wide temperature range and formic acid impedance sensing. *J. Colloid Interface Sci.* **2022**, *607*, 181–191. [CrossRef] [PubMed]
85. Wang, C.-T.; Wu, C.-L. Electrical sensing properties of silica aerogel thin films to humidity. *Thin Solid Film.* **2006**, *496*, 658–664. [CrossRef]
86. Freeman, M.H.; Hall, J.R.; Leopold, M.C. Monolayer-protected nanoparticle doped xerogels as functional components of amperometric glucose biosensors. *Anal. Chem.* **2013**, *85*, 4057–4065. [CrossRef] [PubMed]
87. Alharthi, S.; El-Naggar, M.E.; Abu-Saied, M.; Khattab, T.A.; Saleh, D.I. Preparation of biosensor based on triarylmethane loaded cellulose acetate xerogel for the detection of urea. *Mater. Chem. Phys.* **2022**, *276*, 125377. [CrossRef]
88. Abdelrahman, M.S.; Khattab, T.A.; Aldalbahi, A.; Hatshan, M.R.; El-Naggar, M.E. Facile development of microporous cellulose acetate xerogel immobilized with hydrazone probe for real time vapochromic detection of toxic ammonia. *J. Environ. Chem. Eng.* **2020**, *8*, 104573. [CrossRef]
89. Zakerzadeh, E.; Alizadeh, E.; Samadi Kafil, H.; Mohammad Hassanzadeh, A.; Salehi, R.; Mahkam, M. Novel antibacterial polymeric nanocomposite for smart co-delivery of anticancer drugs. *Artif. Cells Nanomed. Biotechnol.* **2017**, *45*, 1509–1520. [CrossRef]
90. Riss, T.; Niles, A.; Moravec, R.; Karassina, N.; Vidugiriene, J. Cytotoxicity assays: In vitro methods to measure dead cells. In *Assay Guidance Manual*; Eli Lilly & Company and the National Center for Advancing Translational: Bethesda, MD, USA, 2019.

MDPI
St. Alban-Anlage 66
4052 Basel
Switzerland
Tel. +41 61 683 77 34
Fax +41 61 302 89 18
www.mdpi.com

Gels Editorial Office
E-mail: gels@mdpi.com
www.mdpi.com/journal/gels

www.ingramcontent.com/pod-product-compliance
Lightning Source LLC
Chambersburg PA
CBHW061520210926
51850CB00008B/1983